D1559765

Unless Recalled Earlier

DATE DUE

DEMCO, INC. 38-2931

Progress in Mathematics

Volume 189

Series Editors

H. Bass
J. Oesterlé
A. Weinstein

Yves André
Francesco Baldassarri

De Rham Cohomology of Differential Modules on Algebraic Varieties

RECEIVED
APR 24 2002
MINNESOTA STATE UNIVERSITY LIBRARY
MANKATO, MN 56002-8419

Birkhäuser Verlag
Basel · Boston · Berlin

Authors:

Yves André
Institut de Mathématiques
Université Pierre et Marie Curie
Tour 46, 5e étage, Boite 247
4, place Jussieu
F-75252 Paris Cedex 05

Francesco Baldassarri
Dipartimento di Matematica Pura e Applicata
Università degli Studi di Padova
Via Belzoni 7
Italy

QA
612.3
.A53
2001

2000 Mathematics Subject Classification 14F40, 13N05

A CIP catalogue record for this book is available from the Library of Congress, Washington D.C., USA

Deutsche Bibliothek Cataloging-in-Publication Data

André, Yves:
De Rham cohomology of differential modules on algebraic varieties / Yves André ; Francesco Baldassarri. – Basel ; Boston ; Berlin : Birkhäuser, 2000
(Progress in mathematics ; Vol. 189)
ISBN 3-7643-6348-7

ISBN 3-7643-6348-7 Birkhäuser Verlag, Basel – Boston – Berlin

This work is subject to copyright. All rights are reserved, whether the whole or part of the material is concerned, specifically the rights of translation, reprinting, re-use of illustrations, broadcasting, reproduction on microfilms or in other ways, and storage in data banks. For any kind of use whatsoever, permission from the copyright owner must be obtained.

© 2001 Birkhäuser Verlag, P.O. Box 133, CH-4010 Basel, Switzerland
Memeber of the BertelsmannSpringer Publishing Group
Printed on acid-free paper produced of chlorine-free pulp. TCF ∞
Printed in Germany
ISBN 3-7643-6348-7

9 8 7 6 5 4 3 2 1

Table of Contents

1 Regularity in several variables **1**
§1 Geometric models of divisorially valued function fields 2
§2 Logarithmic differential operators 11
§3 Connections regular along a divisor 15
§4 Extensions with logarithmic poles 22
§5 Regular connections: the global case 29
§6 Exponents 32
Appendix A: A letter of Ph. Robba (Nov. 2, 1984) 40
Appendix B: Models and log schemes 42

2 Irregularity in several variables **49**
§1 Spectral norms 51
§2 The generalized Poincaré-Katz rank of irregularity 57
§3 Some consequences of the Turrittin-Levelt-Hukuhara theorem 69
§4 Newton polygons 73
§5 Stratification of the singular locus by Newton polygons 79
§6 Formal decomposition of an integrable connection at a singular divisor 85
§7 Cyclic vectors, indicial polynomials and tubular neighborhoods 97

3 Direct images (the Gauss-Manin connection) **103**
§1 Elementary fibrations 105
§2 Review of connections and De Rham cohomology 110
§3 Dévissage 117
§4 Generic finiteness of direct images 123
§5 Generic base change for direct images 128
§6 Coherence of the cokernel of a regular connection 132
§7 Regularity and exponents of the cokernel of a regular connection 138
§8 Proof of the main theorems: finiteness, regularity, monodromy, base change (in the regular case) 140
Appendix C: Berthelot's comparison theorem on $\mathcal{O}_X$- *vs.* $\mathcal{D}_X$-linear duals 142
Appendix D: Introduction to Dwork's algebraic dual theory 146

4 Complex and p-adic comparison theorems **171**
§1 Review of analytic connections and De Rham cohomology 172
§2 Abstract comparison criteria 175
§3 Comparison theorem for algebraic *vs.* complex-analytic cohomology 177

§4 Comparison theorem for algebraic *vs.* rigid-analytic cohomology (regular coefficients) 182
§5 Rigid-analytic comparison theorem in relative dimension one 184
§6 Comparison theorem for algebraic *vs.* rigid-analytic cohomology (irregular coefficients) 192
§7 The relative non-archimedean Turrittin theorem 202
Appendix E: Riemann's "existence theorem" in higher dimension, an elementary approach 206
References **209**

General Introduction

This is a study of algebraic differential modules in several variables, and of some of their relations with analytic differential modules. Let us explain its source.

The idea of computing the cohomology of a manifold, in particular its Betti numbers, by means of differential forms goes back to E. Cartan and G. De Rham. In the case of a smooth complex algebraic variety X, there are three variants:

i) using the De Rham complex of algebraic differential forms on X,
ii) using the De Rham complex of holomorphic differential forms on the analytic manifold X^{an} underlying X,
iii) using the De Rham complex of C^∞ complex differential forms on the differentiable manifold X^{dif} underlying X^{an}.

These variants turn out to be equivalent. Namely, one has canonical isomorphisms of hypercohomology:

$$H_{DR}(X) \cong H_{DR}(X^{an}) \cong H_{DR}(X^{dif}).$$

While the second isomorphism is a simple sheaf-theoretic consequence of the Poincaré lemma, which identifies both vector spaces with the complex cohomology $H(X^{top}, \mathbf{C})$ of the topological space underlying X, the first isomorphism is a deeper result of A. Grothendieck, which shows in particular that the Betti numbers can be computed algebraically.

This result has been generalized by P. Deligne to the case of nonconstant coefficients: for any algebraic vector bundle $\mathcal{M}$ on X endowed with an integrable *regular* connection, one has canonical isomorphisms

$$H_{DR}(X, \mathcal{M}) \cong H_{DR}(X^{an}, \mathcal{M}^{an}).$$

The notion of regular connection is a higher dimensional generalization of the classical notion of fuchsian differential equations (only regular singularities).

These results were to have a significant influence on the later development of the theory of $\mathcal{D}$-modules. The crucial point in their proof is a comparison between a meromorphic De Rham complex and a De Rham complex with essential singularities, carried out by using Hironaka's resolution of singularities. This also turns out to be the main point in the proof of the non-archimedean counterparts of these comparison theorems, established respectively by R. Kiehl and by the second author (under the extra assumption that the exponents of $\mathcal{M}$ are algebraic numbers).

This point became much better understood when Z. Mebkhout showed, in the complex case, that the gap between meromorphic and essentially singular De Rham complexes is measured by a certain perverse sheaf ("positivity of the irregularity"). Thanks to this deep theorem, it is no longer necessary to invoke resolution of singularities, as a *deus ex machina*, to investigate any serious question of ramification arising in the theory of $\mathcal{D}$-modules, for example in the proof of the Grothendieck-Deligne theorem.

However, the situation is not so nice in the non-archimedean setting, where no analogue of the positivity theorem is known. This motivated us to pose the following problem, which was the starting point of this work:

($*$) *to give an "elementary" proof of the comparison theorems, which is "formal" enough to apply both in the complex-analytic and rigid-analytic situations.*

Of course, such a proof should not only avoid Hironaka's theorem, but also any monodromy argument.

Our solution (IV.3,4) is inspired by M. Artin's proof of the comparison theorem between algebraic and complex-analytic étale cohomology. It relies on three new tools:

i) an algebraic construction of Deligne's canonical extensions (I.4),
ii) a dévissage, inspired by Artin's technique of elementary fibrations, which permits the reduction of many problems on direct images to the relative one-dimensional case (III.3),
iii) an abstract model of "comparison theorem" in the setting of differential algebra (IV.2).

Settling this problem also led us to revisit some fundamental results on direct images of regular differential modules by a smooth morphism. For instance, we found *elementary and algebraic proofs of the so-called finiteness, base change, regularity and monodromy theorems* (III.8). As a by-product of these techniques, we present an *elementary proof of the generalized Riemann existence theorem for coverings* (IV.Appendix E) which uses neither resolution, nor any extension theorem for analytic coherent sheaves.

The first chapter (I) provides a self-contained exposition of the algebraic theory of regularity in several variables, and (III.Appendix D) proposes a new definition of relative algebraic De Rham cohomology with compact supports. This is based on a strong comparison result (III.Appendix C), due to Berthelot, between two different notions of the dual of a differential module. Then (III.Appendix D) shows that our relative algebraic De Rham cohomology with compact supports coincides with Dwork's algebraic dual theory, which is a powerful tool in the study of direct images in the case of elementary fibrations.

What happens if one drops the assumption that the connection on $\mathcal{M}$ is regular? In the complex setting, Deligne pointed out that the natural mapping

$$H_{DR}(X, \mathcal{M}) \to H_{DR}(X^{an}, \mathcal{M}^{an})$$

may be neither injective nor surjective. In strong contrast, the second author *conjectured* ten years ago that, in the *non-archimedean* setting, the isomorphism

$$H_{DR}(X, \mathcal{M}) \cong H_{DR}(X^{an}, \mathcal{M}^{an})$$

holds even in the case of an *irregular* connection (at least if X and $\mathcal{M}$ together with its connection are defined over some number field). Different approaches (Baldassarri, B. Chiarellotto) have established the conjecture for curves, but failed in the higher dimensional case.

The final result of this book is a complete *proof of this general non-archimedean comparison theorem for De Rham cohomology* (IV.6.1). As a corollary, we obtain that *the functor of p-adic analytification of connections* (defined over some number field) *is fully faithful* (IV.6.8).

The strategy of the proof is the same as in the case of regular connections, but our dévissages à la Artin are much more delicate in the general case. Indeed, we found that a detailed study of irregularity in several variables was a necessary preliminary step. This study is carried out in chapter (II), where we introduce the *stratification* of *Z by Newton polygons*, prove a *semicontinuity theorem* for Newton polygons (II.5) and obtain a *formal decomposition* of an integrable connection at a singular divisor Z (II.6).

Note on the style. Chapters I, II, III are purely algebraic; I, II are non-cohomological. The partition into four chapters is based on logical grounds; however, we advise the reader to read I, III and the first half of IV *before* II and the second half of IV, which are much more technical (although most of the new results of this work are contained in the latter parts).

We have consistently tried to be as elementary and down-to-earth as possible, even at the cost of elegance or conciseness. For instance, we work mainly with vector bundles with integrable connection and with smooth morphisms, so that neither the general theory of holonomic modules nor the language of derived categories is needed (except in appendices C and D, independent of the rest of the text).

Acknowledgements. This work was partially sponsored by the European Network *Arithmetic Algebraic Geometry* (contract FMRX-CT96-0006 (DG 12 BDCN)). The first (resp. second) author thanks the University of Padova (resp. Paris 6) for hospitality and support during the preparation of this book.

Yves André
Institut de Mathématiques
bureau 7A43
175 rue du Chevaleret
F-75013 Paris - France
e-mail: andre@math.jussieu.fr

Francesco Baldassarri
Dipartimento di Matematica
Università di Padova
Via Belzoni 7
I-35131 Padova - Italy
e-mail: baldassa@math.unipd.it

1 Regularity in several variables

Introduction

The central topic of this chapter is the notion of regularity in several variables. For an algebraic integrable connection ∇ on the complement of a divisor Z in an algebraic variety X, the notion of regularity along Z may be defined, or characterized, in at least four different algebraic ways:

a) in terms of the iterated action of any *single* vector field D generically transversal to Z: the order of the poles occurring in the action of D^n is at most $n + constant$,
b) by the fact that the logarithmic differential operators of increasing order act with poles of bounded order at the generic point of Z,
c) via the classical notion of regularity in one variable, applied to the restriction of ∇ to sufficiently many smooth curves in X intersecting Z transversally,
d) by the existence of an extension with logarithmic poles along Z.

Although the equivalence of these points of view is "well-known", the proofs given in the literature are always transcendental, and usually rely on Hironaka's resolution of singularities (and are sometimes incomplete, *cf.* (5.9) below). J. Bernstein in his notes [Bn] requested an algebraic proof for these equivalences.

The aim of this chapter is to provide a purely algebraic, elementary, systematic treatment of these questions.

Let us describe in more detail the content of each section. Throughout this chapter the letter K will denote an algebraically closed field of characteristic 0.

In Section 1, we discuss *divisorially valued* function fields over K and their *geometric models*, namely smooth K-models carrying a marked smooth divisor. The local structure of morphisms of models is analyzed with some care. We study the existence of étale tubular neighborhoods rectifying a given transversal vector field. In Appendix B, where we profited from suggestions by L. Illusie and A. Ogus, these results are recast in the now standard framework of log geometry.

In Section 2, we examine the functoriality of logarithmic differential operators.

In Section 3, we concentrate on differential modules over a function field $\mathcal{F}$ over K, that are *regular* along a divisorial valuation v of $\mathcal{F}$. For any model (X, Z) of $(\mathcal{F}, v)$, we prove the equivalence of $a)$ - $d)$.

In Section 4, we fix a section $\tau : K/\mathbf{Z} \longrightarrow K$ of the canonical projection, with $\tau(0) = 0$. We consider a divisor Z with strict normal crossings in a smooth K-variety X, and an integrable connection ∇ on $X \setminus Z$, regular along Z. We show that the connection admits a unique *extension with logarithmic poles* along Z and exponents in Im τ. Our proof is purely algebraic, in contrast to the Manin-Deligne construction [De, II, 5.4 and 5.7] which relies on monodromy and on the consideration of multivalued sections of moderate growth.

In Section 5, we introduce the global notion of regularity. We start from a *birational* definition, and give the following equivalent characterizations of a *regular* connection ∇ on X:

e) the restriction of ∇ to every smooth curve in X is regular in the classical sense,
f) ∇ is regular along the one-codimensional part of the boundary of X in some normal compactification $\overline{X}$.

Our proofs are *purely algebraic and do not use resolution of singularities.*

In Section 6, we define the *exponents* of a regular connection along a divisor and show that they coincide with the classical exponents of the differential module induced on any curve meeting the divisor transversally. We also introduce a global, birational notion of exponents of a regular connection, give several characterizations, and study its behaviour under inverse images and finite direct images.

Robba's result, reproduced in Appendix A, will be useful at several places in this book. It is a pleasure to publish without any change his letter from many years ago.

§1 Geometric models of divisorially valued function fields

1.1 Models

We set here:

$\mathcal{F}$ = a *function field* over K, *i.e.,* a finitely generated extension field of K;

$d = \operatorname{tr\,deg}_K \mathcal{F}$;

v = a *divisorial* valuation of $\mathcal{F}/K$, *i.e.* a discrete valuation of $\mathcal{F}$, trivial over K, normalized by the condition that its value group is *exactly* $\mathbf{Z}$, with valuation ring $R_v \supset K$, maximal ideal $\mathfrak{m}_v$ and residue field $k(v) \supset K$, such that

$$\operatorname{tr\,deg}_K k(v) = \operatorname{tr\,deg}_K \mathcal{F} - 1.$$

A uniformizing parameter of R_v will usually be denoted by t_v, or simply by t. We will denote by $(\hat{\mathcal{F}}, \hat{v})$ the completion of $(\mathcal{F}, v)$, and by $\hat{R}_v = R_{\hat{v}}$, $\hat{\mathfrak{m}}_v = \mathfrak{m}_{\hat{v}}$ the valuation ring and maximal ideal of $\hat{v}$. We will refer to $(\mathcal{F}, v)$ (resp. to $(\hat{\mathcal{F}}, \hat{v})$) as a *(divisorially) valued function field over* K (resp. a *field of formal functions in* d *variables over* K).

Definition 1.1.1 A *smooth (affine)* K*-model* of $(\mathcal{F}, v)$ is a pair (X, Z) consisting of a smooth (affine) connected K-variety X, with generic point η_X, and of a smooth closed irreducible subvariety Z, of generic point η_Z, together with isomorphisms of K-algebras $\kappa(\eta_X) \cong \mathcal{F}$, $\mathcal{O}_{X,\eta_Z} \cong R_v$, fitting in the natural commutative diagram

$$\begin{array}{ccc} \mathcal{O}_{X,\eta_Z} & \hookrightarrow & \kappa(\eta_X) \\ \cong\downarrow & & \downarrow\cong \\ R_v & \hookrightarrow & \mathcal{F}. \end{array}$$

We shall also simply write *model* instead of *smooth* K*-model.* In this situation, $\kappa(\eta_Z) = k(v)$. We will regard Z as a closed reduced subscheme of X, and denote its ideal sheaf by $\mathcal{J}_Z$.

Lemma 1.1.2 *Smooth K-models (X, Z) of $(\mathcal{F}, v)$ exist.*

Indeed, one may write R_v as a localization of a finitely generated integral K-algebra R; then v is the valuation associated to a prime divisor Z' of the normalization X' of Spec R. A sufficiently small open neighborhood X of $\eta_{Z'}$ in X', together with $Z = Z' \cap X$, fulfill the requirements.

Lemma 1.1.3 *Let (X, Z) and (X', Z') be two models of $(\mathcal{F}, v)$. Then there exist open neighborhoods U of η_Z in X and U' of $\eta_{Z'}$ in X', and an isomorphism $\epsilon : U \xrightarrow{\sim} U'$ that identifies $Z \cap U$ with the inverse image of $Z' \cap U'$ by ϵ.*

Proof: We may first replace X (resp. X') by an open affine neighborhood of η_Z (resp. $\eta_{Z'}$). Then we may assume that $\mathcal{O}(X)$ and $\mathcal{O}(X')$ are contained in $R_v \subset \mathcal{F}$. Let $\mathfrak{p}_Z$ (resp. $\mathfrak{p}_{Z'}$) be the prime ideal of $\mathcal{O}(X)$ (resp. $\mathcal{O}(X')$) corresponding to Z (resp. Z'), so that $R_v = \mathcal{O}(X)_{\mathfrak{p}_Z} = \mathcal{O}(X')_{\mathfrak{p}_{Z'}}$ and $\mathcal{O}(X) \cap \mathfrak{m}_v = \mathfrak{p}_Z$, $\mathcal{O}(X') \cap \mathfrak{m}_v = \mathfrak{p}_{Z'}$. Upon replacing X by a smaller open neighborhood of η_Z, we may then assume that there is a morphism $\phi : X \longrightarrow X'$, with $\phi(\eta_Z) = \eta_{Z'}$. Since $\phi^\sharp_{\eta_Z} : \mathcal{O}_{X',\eta_{Z'}} \xrightarrow{\sim} \mathcal{O}_{X,\eta_Z}$, we get the result by [EGA I, Proposition 5.6.4]. □

Definition 1.1.4 A *coordinatized K-model* $(X, Z, \underline{x})$ of $(\mathcal{F}, v)$ is a smooth affine K-model (X, Z) endowed with global étale coordinates $\underline{x} = (x_1, \dots, x_d)$ on X such that $\mathcal{J}_Z = x_1\mathcal{O}_X$. The étale coordinates $(t_v = x_1, x_2, \dots, x_d)$ on X are said to be *adapted* to Z.

Lemma 1.1.5 *Any model (X, Z) of $(\mathcal{F}, v)$ admits a Zariski covering by coordinatized models.*

Proof: Follows from [EGA IV, 4, Corollary 17.12.2, *d*]. □

Remark 1.1.6 In the situation of Definition 1.1.1 (resp. 1.1.4), we set $\hat{X}$ = the smooth formal K-scheme $\hat{X}_Z$, completion of X along Z. We say that $(\hat{X}, Z)$ (resp. $(\hat{X}, Z, \underline{x})$) is a *smooth (affine) formal model* (resp. a *coordinatized formal model*) of $(\hat{\mathcal{F}}, \hat{v})$.

1.2 Morphisms of models

We consider another divisorially valued function field $(\mathcal{G}, w)$ over K. Let (Y, W) be a model of $(\mathcal{G}, w)$.

Definition 1.2.1 A *morphism* $f : (X, Z) \longrightarrow (Y, W)$ is a morphism of K-schemes $\overset{\circ}{f} : X \longrightarrow Y$ (sometimes by an abuse of notation, denoted also by f) such that $\overset{\circ}{f}(X \setminus Z) \subset Y \setminus W$.

Let $(X, Z) \xrightarrow{f} (Y, W)$ be a morphism of models.

Definition 1.2.2 We say that f is a *closed immersion* (resp. *locally closed immersion*) if $\overset{\circ}{f} : X \longrightarrow Y$ is a closed immersion (resp. locally closed immersion) of

K-schemes, and X, W meet transversally along Z, *i.e.*, $Z = W \times_Y X$ (schematically). We say that f is *log dominant* if $\overset{\circ}{f}(\eta_Z) = \eta_W$. (This implies that $\overset{\circ}{f} : X \longrightarrow Y$ is dominant).

Remark 1.2.2.1 The generally accepted terminology for "log dominant" in the framework of log schemes is *vertical* at η_Z [N, 7.3]. More precisely, a log dominant morphism of models f as above, viewed as a morphism of log schemes, is vertical at all $z \in Z$ above η_W.

Each of the previous notions is stable under composition.

Remark 1.2.3 A morphism of K-models $f : (X, Z) \longrightarrow (Y, W)$ such that $\overset{\circ}{f}(Z) \subset W$, is log dominant if and only if $\overset{\circ}{f}$ is flat at η_Z.

Proof: Assume $\overset{\circ}{f}$ is flat at η_Z. Then $\overset{\circ}{f}$ is certainly dominant and $\dim \mathcal{O}_{X,\eta_Z} - \dim \mathcal{O}_{Y,\overset{\circ}{f}(\eta_Z)} = \dim \mathcal{O}_{\overset{\circ}{f}^{-1}(\overset{\circ}{f}(\eta_Z)),\eta_Z} \geq 0$. So, $\dim \mathcal{O}_{Y,\overset{\circ}{f}(\eta_Z)} \leq 1$, and since $\overset{\circ}{f}(\eta_Z) \in W$, $\overset{\circ}{f}(\eta_Z) = \eta_W$. Assume conversely that f is log dominant. Then $\mathcal{O}_{X,\eta_Z}$ is flat over the DVR $\mathcal{O}_{Y,\eta_W}$, since $\mathcal{O}_{Y,\eta_W} \hookrightarrow \mathcal{O}_{X,\eta_Z}$ is injective. □

Definition 1.2.4 Let $s = \overset{\circ}{f}(\eta_Z)$ and consider

$$\overset{\circ}{f}{}^{\sharp} : \mathcal{O}_{Y,s} \longrightarrow \mathcal{O}_{X,\eta_Z};$$

let $y \in \mathcal{O}_{Y,s}$ be a local equation of W in Y at s. The *ramification index* of f (at η_Z) is the positive integer $e_f = v(\overset{\circ}{f}{}^{\sharp}(y))$. If $e_f = 1$, we say that f is *unramified.* If f is log dominant and $\overset{\circ}{f}$ is quasi-finite at η_Z, we say that f is *Kummer étale* at η_Z.

Remark 1.2.5 The ramification index of f is then the multiplicity of the point η_Z in the fiber $\overset{\circ}{f}{}^{-1}(\overset{\circ}{f}(\eta_Z))$. So, f is unramified if and only if $\overset{\circ}{f}$ is transversal to W at η_Z [EGA IV, 17.13.3]. Notice that if f is unramified, then $\overset{\circ}{f}(Z) \subset W$.

Lemma 1.2.6

i) *f is unramified if $\overset{\circ}{f} : X \longrightarrow Y$ is smooth at η_Z and $\overset{\circ}{f}(\eta_Z) \in W$;*

ii) *if f is unramified and log dominant, then $\overset{\circ}{f} : X \longrightarrow Y$ is smooth at η_Z;*

iii) *a locally closed immersion of models is unramified;*

iv) *if $\overset{\circ}{f} : X \longrightarrow Y$ is a closed immersion and f is unramified, there are open neighborhoods U of η_Z in X and V of η_W in Y such that $(U, U \cap Z) \longrightarrow (V, V \cap W)$ is a closed immersion;*

v) *if f is log dominant, then, locally at η_Z, f factors as*

$$(X, Z) \xrightarrow{h} (Y', W') \xrightarrow{g} (Y, W)$$

where h and g are log dominant morphisms of models, $\overset{\circ}{h}$ is smooth at η_Z, g is Kummer étale at $\eta_{W'}$, and $e_f = e_g$.

Proof: i) and iii) follow from Remark 1.2.5. To prove ii), let $S = \operatorname{Spec} \mathcal{O}_{Y,\eta_W}$ and $\overset{\circ}{f}_S : X_S \longrightarrow S$ be the morphism obtained from $\overset{\circ}{f}$ by the base change $S \longrightarrow Y$. Let s be the closed point of S; Z_S is then a divisor of X_S. Under the assumptions of ii), $\overset{\circ}{f}_S$ induces an injective local homomorphism of discrete valuation rings $\mathcal{O}_{S,s} \longrightarrow \mathcal{O}_{X,\eta_{Z_S}}$, formally of the form $\kappa(W)[[t]] \hookrightarrow \kappa(Z)[[t]]$. This embedding (of t-adic rings) is clearly formally smooth. iv) follows from Remark 1.2.5 and [EGA IV, 9.9.2 (ii)], applied to $\mathcal{F} = \mathcal{O}_X$. The proof of v) is similar to that of ii): we take, locally at η_W, as Y' the normalization of Y in X and as W' the inverse image of W. The same base change used in ii) provides a local homomorphism of local rings of the form $\kappa(W)[[t]] \longrightarrow \kappa(Z)[[t]]$, $t \longmapsto t^{e_f}$. The conclusion follows. □

The following obvious lemma stated only is for ease of reference.

Lemma 1.2.7 *Let $(X, Z) \xrightarrow{f} (Y, W)$ be a log dominant morphism. We have a commutative diagram of inclusions*

$$\begin{array}{ccc} R_w & \longrightarrow & R_v \\ \downarrow & & \downarrow \\ \mathcal{G} & \longrightarrow & \mathcal{F} \end{array} \tag{1.2.7.1}$$

such that $\mathfrak{m}_v \cap R_w = \mathfrak{m}_w$ and $\mathfrak{m}_w R_v = \mathfrak{m}_v^{e_f}$. In particular, the K-embedding of fields $\mathcal{G} \hookrightarrow \mathcal{F}$ is compatible with the valuations, in the sense that $v_{|\mathcal{G}} = e_f w$, and induces an extension of residue fields $k(w) \hookrightarrow k(v)$; $e_f = e(v/w)$ is then the relative ramification index of v over w. Conversely, any diagram (1.2.7.1) *arises from a log dominant morphism of models.*

Proposition 1.2.8 *Let $f : (X, Z) \longrightarrow (Y, W)$ be a morphism of models with X a curve, such that $\overset{\circ}{f}(Z) \subset W$. Then there exists, étale-locally on X, a factorization*

$$(X, Z) \xrightarrow{i} (S, D) \xrightarrow{g} (Y, W),$$

where (S, D) is a model, i is a closed immersion, g is log dominant and $\overset{\circ}{g}_{|S \backslash D} : S \backslash D \longrightarrow Y \backslash W$ is smooth. One has $e_f = e_g$.

Proof: a) In the canonical graph factorization of f

$$(X, Z) \xrightarrow{(1, f)} (X \times Y, X \times W) \xrightarrow{p} (Y, W),$$

the second projection p is log dominant. We may replace f by $(1, f)$, hence assume that the underlying morphism $X \longrightarrow Y$ is a closed immersion, and that $X \times_Y W$ is the point Z with multiplicity e. The problem is local at Z (for the étale topology), both in X and in Y.

b) There exists a system of coordinates $(x_1, \ldots, x_n)$ on some étale neighborhood Y' of Z in Y, such that the inverse image W' of W is given by $x_1 = 0$ and the inverse image X' of X is given by $x_3 = \cdots = x_n = 0$, $x_1 = x_2^e$ (this is easily seen by choosing first $x_3, \ldots, x_n$ such that the surface $x_3 = \cdots = x_n = 0$ contains a Zariski neighborhood of Z in X and meets W transversally at Z, which reduces us to the familiar situation of the intersection of two smooth curves on a surface; see also part b) of the proof of (6.5.5) below). Since $(Y', W') \longrightarrow (Y, W)$ is log dominant and smooth outside W', we may replace (Y, W) by (Y', W').

c) Let us now consider the e-fold covering $\pi : Y' \longrightarrow Y$ ramified at $x_1 = 0$ defined by $x_1 \mapsto x_1^e$ (the other x_i's being unchanged). The inverse image of X is the union of the curves $X_\zeta : x_3 = \cdots = x_n = 0, x_1 = \zeta x_2$, where ζ runs among the e-th roots of unity. We may then replace (Y, W) by $(Y', W' : x_1 = 0)$ and (X, Z) by (X_ζ, Z). This yields the desired factorization, and the equality $e_f = e_g$ is immediate. □

Remark 1.2.9 The theory of log schemes allows us to prove Proposition 1.2.8 without the restriction that X is a curve (Appendix B), but we shall not need this stronger result.

Remark 1.2.10 Similar definitions are understood for formal schemes of the type considered in (1.1.6) where analogous results hold.

1.3 Tubular neighborhoods

Definition 1.3.1 Let X be a normal K-variety and Z be a closed subvariety of X of pure codimension 1. We say that X is a *tubular neighborhood* of Z, if there is a commutative diagram of K-varieties

(1.3.1.1)
$$\begin{array}{ccc} X & \overset{i}{\hookleftarrow} & Z \\ f\downarrow & \swarrow g & \\ S & & \end{array}$$

where

i) i is the canonical closed immersion;
ii) f is smooth of (pure) relative dimension 1;
iii) g is étale.

We will say that $(X \xrightarrow{f} S, t)$ is a *coordinatized tubular neighborhood* of Z in X (and t is a relative coordinate) if (1.3.1.1) can be completed into the commutative diagram

(1.3.1.2)
$$\begin{array}{ccccc} & & X & \overset{i}{\hookleftarrow} & Z \\ & \phi \swarrow & f\downarrow & \swarrow g & \\ \mathbf{A}^1_S & \longrightarrow & S & & \\ \mathrm{pr}_1 \downarrow & & \downarrow & & \\ \mathbf{A}^1_K & \longrightarrow & \mathrm{Spec}\, K & & \end{array}$$

where ϕ is étale, the square is fibered, $\mathrm{pr}_1 \circ \phi = t$, and $\mathcal{J}_Z = t\mathcal{O}_X$.

The (resp. coordinatized) tubular neighborhood (1.3.1.1) (resp. (1.3.1.2)) will be said to be *strict* if $S = Z$, g is the identity map, and the fibers of f are geometrically connected.

Lemma 1.3.2 *For any model (X, Z) of $(\mathcal{F}, v)$, there exists a Zariski (resp. étale) open cover of Z in X consisting of (resp. strict) coordinatized tubular neighborhoods $(U \xrightarrow{f} S, t)$ of the inverse image of Z in U.*

Proof: For any $x \in Z$, we may shrink (X, Z) to an affine *coordinatized* neighborhood of x as in (1.1.4). Then $f : X \longrightarrow S := \operatorname{Spec} K[x_2, \ldots, x_d]$ is smooth of pure relative dimension 1, and $g : Z \longrightarrow S$ is étale.

We now show that the statement for the étale topology follows from the one for the Zariski topology. Given the diagram (1.3.1.2) and a closed point $x \in Z$, there is an étale neighborhood $S' \longrightarrow S$ of $s = g(x) \in S$, and a closed point $s' \in S'$ above s such that in the diagram obtained by base change

$$
\text{(1.3.2.1)} \qquad
\begin{array}{ccccc}
 & & X_{S'} & \overset{i_{S'}}{\hookleftarrow} & Z_{S'} \\
 & \phi_{S'} \swarrow & f_{S'} \downarrow & \swarrow g_{S'} & \\
\mathbf{A}^1_{S'} & \longrightarrow & S' & & \\
\mathrm{pr}_1 \downarrow & & \downarrow & & \\
\mathbf{A}^1_K & \longrightarrow & \operatorname{Spec} K & & ,
\end{array}
$$

the morphism $g_{S'}$ admits a section $\psi : S' \longrightarrow Z_{S'}$, with $\psi(s')$ a closed point of $Z_{S'}$ lying above x [BLR, 2.2, Proposition 14]. We then replace $Z_{S'}$ by S', $g_{S'}$ by $1_{S'}$, and $i_{S'}$ by $i_{S'} \circ \psi$. So, we obtain a diagram

$$
\begin{array}{ccc}
X_{S'} & \overset{i'}{\hookleftarrow} & S' \\
f_{S'} \downarrow & \swarrow 1_{S'} & \\
S' & &
\end{array}
$$

which is a strict tubular neighborhood of S', but for the condition on (geometric) connectedness of the fibers of $f_{S'}$. To satisfy this, we replace $X_{S'}$ by the union of the connected components of $i'(y)$ in $f_{S'}^{-1}(y)$, for all points $y \in S'$, which is open in $X_{S'}$ [EGA IV, 15.6.5]. Since K is algebraically closed, we conclude, using [EGA IV, 9.7.7, i], that all fibers of $f_{S'}$ are in fact *geometrically* connected. □

Proposition 1.3.3 *Let $(C, P) \longrightarrow (X, Z)$ be a closed immersion of models, with C a curve. There exist an open neighborhood U of P in X and a coordinatized tubular neighborhood $(U \xrightarrow{f} S, t)$ of $Z \cap U$, such that, in the notation of* (1.3.1.2) *with (X, Z) replaced by $(U, Z \cap U)$, $C \cap U = f^{-1}(g(P))$.*

Proof: Since C and Z meet transversally at the closed point P, [EGA IV, 17.13.8.1] gives an isomorphism

$$\mathcal{J}_P/\mathcal{J}_P^2 \cong (\mathcal{J}_Z/\mathcal{J}_Z^2 \otimes_{\mathcal{O}_Z} \kappa(P)) \oplus (\mathcal{J}_C/\mathcal{J}_C^2 \otimes_{\mathcal{O}_C} \kappa(P))$$

where by $\mathcal{J}_Y$ we denote the $\mathcal{O}_X$-ideal corresponding to the closed subscheme Y of X. Since $\mathcal{J}_P/\mathcal{J}_P^2 \cong \Omega^1_{X/K} \otimes_{\mathcal{O}_X} \kappa(P)$, for any system of sections x_1 (resp. $x_2, \ldots, x_d$) of $\mathcal{O}_X$ in a neighborhood U of P, representing a system of generators of $\mathcal{J}_Z/\mathcal{J}_Z^2 \otimes_{\mathcal{O}_Z} \kappa(P)$ (resp. of $\mathcal{J}_C/\mathcal{J}_C^2 \otimes_{\mathcal{O}_C} \kappa(P)$), $(x_1, \ldots, x_d)$ are étale coordinates at P. We shrink U around P, so that the morphism $f : U \longrightarrow S := \mathrm{Spec}\, K[x_2, \ldots, x_d]$ is a tubular neighborhood of Z and fiber $f^{-1}(f(P))$ is connected. Certainly $C \cap U \subset f^{-1}(f(P))$, so by connectedness we have equality. □

These considerations extend without difficulty to the case of a non-connected (but still smooth) divisor Z.

1.4 Integral curves

1.4.1 In the situation of a tubular neighborhood, let $\partial \in \Gamma(X, \underline{Der}_{X/K})$ be a non-zero derivation such that

$$\langle \partial, f^*\omega \rangle = 0, \tag{1.4.1.1}$$

for any section ω of $\Omega^1_{S/K}$; we will say that (1.3.1.1), or simply $X \xrightarrow{f} S$, is a *family of integral curves of* ∂ *in* X *(based on* S*)*.
The commutative diagram (1.3.1.2) defines a splitting

$$\Omega^1_{X/K} = \mathcal{O}_X dt \oplus f^*\Omega^1_{S/K}. \tag{1.4.1.2}$$

If $(X \xrightarrow{f} S, t)$ is a coordinatized tubular neighborhood of Z and ∂, as in (1.4.1.1), also satisfies $\langle \partial, dt \rangle = 1$, we write $\partial = \partial_{t,f}$, or simply $\partial = \frac{\partial}{\partial t}$, when f is clearly understood.

Definition 1.4.2 Let (X, Z) be a model and $D \in \Gamma(X, \underline{Der}_{X/K})$. We say that D is *transversal to* Z *at* $\xi \in Z$ if for one (hence for any) local equation $l \in \mathcal{O}_{X,\xi}$ of Z at ξ, Dl is a unit in $\mathcal{O}_{X,\xi}$. In that case, we also say that lD has a *simple zero along* Z at ξ. We say that D is *transversal to* (resp. has a *simple zero along*) Z, if D is transversal to (resp. has a simple zero along) Z at every point $\xi \in Z$.

For instance, if $(X \xrightarrow{f} S, t)$ is a coordinatized tubular neighborhood of Z in X, $\partial_{t,f}$ is transversal to Z. Similar definitions are understood in case $(\hat{X}_Z, Z)$ is a smooth K-model of $(\hat{\mathcal{F}}, \hat{v})$ and $D \in \Gamma(\hat{X}_Z, \underline{Der}_{\hat{X}_Z/K})$.

Proposition 1.4.3 *Let* (X, Z) *be a model of* $(\mathcal{F}, v)$ *and* $D \in \Gamma(X, \underline{Der}_{X/K})$ *be transversal to (resp. have a simple zero along)* Z *at the point* $z \in Z$. *A necessary and* *sufficient condition for the existence of a family of integral curves* $(U \xrightarrow{f} S, t)$ *of* D *on some open neighborhood* U *of* z *in* X*, is that the field of constants* $\mathcal{F}^D$ *of* D *in* $\mathcal{F}$ *satisfy*

$$\mathrm{trdeg}_K \mathcal{F}^D = \mathrm{trdeg}_K \mathcal{F} - 1. \tag{1.4.3.1}$$

Proof: We first notice that $\mathcal{F}^D \backslash \{0\} \subset R_v^\times$. In fact, if x_1 is a parameter of R_v, and if $g = x_1^n u$, with $n > 0$ and $u \in R_v^\times$, were a non-zero element of $\mathcal{F}^D$, the equation $0 = Dg = n x_1^{n-1}(Dx_1)u + x_1^n Du$, would give a contradiction both in case D is transversal and in case D has a simple zero along v. So, by reduction modulo $\mathfrak{m}_v$, we get an injection $\pi : \mathcal{F}^D \hookrightarrow k(v) \cong \kappa(Z)$, which is an algebraic extension of fields, by (1.4.3.1). We then pick a local equation $x_1 \in \mathcal{O}_{X,z}$ for Z, and choose $x_2, \ldots, x_d$ in $\mathcal{F}^D \cap \pi^{-1}(\mathcal{O}_{Z,z})$, so that $(\pi(x_2), \ldots, \pi(x_d))$, is a system of parameters of $\mathcal{O}_{Z,z}$. By the argument used in the proof of *a)* $\Rightarrow$ *d)* in [EGA IV, Corollary 17.12.2], upon replacing X by a Zariski neighborhood of η_Z, we may assume that $x_1, \ldots, x_d$ are étale coordinates on X, adapted to Z. Upon further Zariski localization at z, we obtain a coordinatized tubular neighborhood $(X \xrightarrow{f} S, x_1)$ of Z in X as in (1.3.1.2), where $D = u \frac{\partial}{\partial x_1}$ (resp. $D = u x_1 \frac{\partial}{\partial x_1}$), with $u \in \Gamma(X, \mathcal{O}_X)$ never zero on Z if D is transversal to (resp. has a simple zero along) Z. □

Corollary 1.4.4 *Let* $(\mathcal{F}, v)$ *be a function field over* K *equipped with a divisorial valuation* v*. Let* $\partial \in Der(\mathcal{F}/K)$ *be such that* $\partial R_v \subset \mathfrak{m}_v$ *but* $\partial \mathfrak{m}_v \not\subset \mathfrak{m}_v^2$*, and assume* $\mathrm{trdeg}_K \mathcal{F}^\partial = \mathrm{trdeg}_K \mathcal{F} - 1$*. There exists a model* (X, Z) *of* $(\mathcal{F}, v)$ *such that* $\partial \in \Gamma(X, \underline{Der}_{X/K})$ *has a simple zero along* Z *and a coordinatized tubular neighborhood* $(X \xrightarrow{f} S, t)$ *of* Z *in* X*, which is a family of integral curves of* ∂ *in* X.

In the case of a smooth formal K-model $(\hat{X}_Z, Z)$ of $(\hat{\mathcal{F}}, \hat{v})$ one can be more precise.

Lemma 1.4.5 *Let* $(\hat{X}_Z, Z)$ *be a smooth* K*-model of* $(\hat{\mathcal{F}}, \hat{v})$.

i) *If there exists* $D \in \Gamma(\hat{X}_Z, \underline{Der}_{\hat{X}_Z/K})$ *transversal to* Z*, then there exists an isomorphism*

$$\hat{X}_Z \xrightarrow{\sim} \hat{\mathbf{A}}^1_K \times Z, \tag{1.4.5.1}$$

such that the composite map

$$Z \hookrightarrow \hat{X}_Z \xrightarrow{\sim} \hat{\mathbf{A}}^1_K \times Z \xrightarrow{\mathrm{pr}_2} Z \tag{1.4.5.2}$$

is the identity on the K*-scheme* Z.

ii) *Assume an isomorphism as in* i) *exists. Let* $D \in \Gamma(\hat{X}_Z, \underline{Der}_{\hat{X}_Z/K})$ *be transversal to (resp. have a simple zero along)* Z*. Then there exists a unique isomorphism as in* i) *in which* D *corresponds to the derivation* $(\frac{d}{dt}, 0)$ (resp. $(t\frac{d}{dt}, 0)$)*, where* t *denotes the canonical coordinate on* $\hat{\mathbf{A}}^1_K \cong \mathrm{Spf}\, K[[t]]$.

Proof: i) The existence of D transversal to Z implies the triviality of the conormal sheaf : $\mathcal{N}_{Z/X} = \mathcal{J}_Z/\mathcal{J}_Z^2 \cong \mathcal{O}_Z$. The sheaf $\mathcal{O}_{\hat{X}_Z}$ on Z coincides with the completion of the $\mathcal{O}_Z$-algebra $\mathbf{S}^{\cdot}_{\mathcal{O}_Z}(\mathcal{N}_{Z/X})$ in the topology defined by the ideal of augmentation. So, we get an isomorphism (1.4.5.1) satisfying the requirement in i).

ii) We treat the case of D transversal to Z first. We modify the given isomorphism (1.4.5.1) so as to satisfy the further condition on the derivation corresponding to D. Uniqueness of this modification allows one to assume that $\hat{X}_Z$ is affine. We set $Z = \operatorname{Spec} A$, $\hat{X}_Z = \operatorname{Spf} A[[x]]$ and $Dx = \sum_{j\geq 0} a_j x^j$, $a_j \in A$, $a_0 \in A^{\times}$. We first show that there is a unique map (of sets)

$$\begin{aligned} A &\longrightarrow A[[x]] \\ a &\longmapsto \varphi_a(x) \end{aligned}$$

such that $\varphi_a(0) = a$ and $D\varphi_a(x) = 0$. We write $\varphi_a(x) = a + \sum_{j>0} u_j x^j$, $Da = \sum_{h\geq 0} b_h x^h$, $b_h \in A$, and impose

$$0 = D\varphi_a(x) = Da + \sum_{j>0} j u_j x^{j-1} \sum_{i\geq 0} a_i x^i + \sum_{j>0} (Du_j) x^j .$$

This gives

$$u_1 = -b_0/a_0 \in A^{\times},$$

$$(h+1)a_0 u_{h+1} = -b_h - \sum_{\substack{i+j=h+1 \\ j\leq h}} j u_j a_i - \text{coefficient of } x^h \text{ in } \sum_{j=1}^{h} (Du_j) x^j ,$$

from which we see that $u_1, \ldots, u_h \in A$ uniquely determine $u_{h+1} \in A$. The map $a \longmapsto \varphi_a$ is necessarily a K-algebra homomorphism, and it is injective. If A' denotes the image of A in $A[[x]]$ via $a \longmapsto \varphi_a$, we have $A[[x]] = A'[[x]]$ and we see that we might have assumed from the beginning that $DA = 0$. We now look for $t = \sum_{j>0} u_j x^j$, $u_j \in A$, such that $Dt = 1$. This gives

$$1 = \sum_{j>0} j u_j x^{j-1} \sum_{i\geq 0} a_i x^i ,$$

$$u_1 = \frac{1}{a_0} \in A',$$

$$0 = \sum_{i+j=h+1} j u_j a_i \ , \ h > o,$$

which again shows that $u_1, \ldots, u_h \in A$ uniquely determine $u_{h+1} \in A$.

We now discuss the case of D having a simple zero along Z. We are given the isomorphism (1.4.5.1) (again we assume $Z = \operatorname{Spec} A$ affine) and the standard coordinate t on $\hat{\mathbf{A}}^1_K$. We apply the first part of this ii) paragraph to the derivation $t^{-1}D$.

We may then assume that $D = (u\frac{d}{dt}, 0)$, for a unit $u \in A[[t]]^\times$. We look for a parameter x of $A[[t]]$ such that $Dx = x$. We have $Dt = ut$. If $u = 1$, we take $x = t$. Otherwise, let $1 - u = \beta t^r$, with $\beta \in A[[t]]$, $r \in \mathbf{Z}_{>0}$. We set $x = \alpha t$, for a unit $\alpha \in A[[t]]^\times$. The condition $D(\alpha t) = \alpha t$ is equivalent to $D\alpha = \alpha(1-u) = \alpha\beta t^r$. So, $\alpha = \exp \beta t^{r+1}/(r+1)$ and $x = t \exp \beta t^{r+1}/(r+1)$ is the only solution. □

Corollary 1.4.6 *Let $(\hat{\mathcal{F}}, \hat{v})$ be a formal function field over K. Let $\partial \in Der(\hat{\mathcal{F}}/K)$ be such that $\partial R_{\hat{v}} \subset R_{\hat{v}}$ but $\partial \mathfrak{m}_{\hat{v}} \not\subset \mathfrak{m}_{\hat{v}}$ (resp. $\partial R_{\hat{v}} \subset \mathfrak{m}_{\hat{v}}$ but $\partial \mathfrak{m}_{\hat{v}} \not\subset \mathfrak{m}_{\hat{v}}^2$). Let $Z = \operatorname{Spec} A$ be any smooth affine K-model of $k(\hat{v})$. There exists an open dense subset U of Z and a smooth affine formal K-model $(\hat{X}_U, U)$ of $(\hat{\mathcal{F}}, \hat{v})$, with $\hat{X}_U = \hat{\mathbf{A}}^1_K \times U$, such that $\partial = (\frac{d}{dt}, 0)$(resp. $\partial = (t\frac{d}{dt}, 0)$), t denoting the canonical coordinate on $\hat{\mathbf{A}}^1_K \cong \operatorname{Spf} K[[t]]$.*

Definition 1.4.7 A derivation $\partial \in Der(\hat{\mathcal{F}}/K)$ such that $\partial R_{\hat{v}} \subset R_{\hat{v}}$ but $\partial \mathfrak{m}_{\hat{v}} \not\subset \mathfrak{m}_{\hat{v}}$ (resp. $\partial R_{\hat{v}} \subset \mathfrak{m}_{\hat{v}}$ but $\partial \mathfrak{m}_{\hat{v}} \not\subset \mathfrak{m}_{\hat{v}}^2$) is called *transversal to* $\hat{v}$ (resp. is said to have a *simple zero at* $\hat{v}$).

So, ∂ is transversal to $\hat{v}$ if and only if $t_{\hat{v}}\partial$ has a simple zero at $\hat{v}$.

§2 Logarithmic differential operators

2.1 Let (X, Z) be a model of the divisorially valued function field $(\mathcal{F}, v)$. We denote by j the immersion of $X \backslash Z$ in X.

We define $\underline{Der}_{X/K,Z}$ (resp. $\underline{Diff}^n_{X/K,Z}$, for any $n = 0, 1, \ldots$) as the subsheaf of $\underline{Der}_{X/K}$ (resp. $\underline{Diff}^n_{X/K}$) of the sections that preserve, as operators on $\mathcal{O}_X$, the $\mathcal{J}_Z$-adic filtration. Finally, $\underline{Diff}_{X/K,Z} = \cup_n \underline{Diff}^n_{X/K,Z}$. The stalk of $\underline{Der}_{X/K,Z}$ (resp. $\underline{Diff}^n_{X/K,Z}$) at η_Z is the R_v-module

(2.1.1)
$$Der_v(\mathcal{F}/K) = \{\partial \in Der(\mathcal{F}/K) | \partial \mathfrak{m}_v \subset \mathfrak{m}_v\}$$
$$(\text{resp. } Diff^n_v(\mathcal{F}/K) = \{L \in Diff^n(\mathcal{F}/K) | L\mathfrak{m}^i_v \subset \mathfrak{m}^i_v \,,\ \forall i = 0, 1, \ldots, n\}).$$

Lemma 2.1.2 *The $\mathcal{O}_X$-modules $\underline{Der}_{X/K,Z}$ and $\underline{Diff}^n_{X/K,Z}$ are locally free of finite type. The filtration of $\underline{Diff}^n_{X/K,Z}$ by the order of differential operators is canonically split. If $(t_v = x_1, x_2, \ldots, x_d)$ are étale coordinates on X adapted to Z, a basis of sections of $\underline{Der}_{X/K,Z}$ (resp. of $\underline{Diff}^n_{X/K,Z}$) is given by $\{x_1\frac{\partial}{\partial x_1}, \frac{\partial}{\partial x_2}, \ldots, \frac{\partial}{\partial x_d}\}$ (resp. by $\{(x_1\frac{\partial}{\partial x_1})^{\alpha_1}(\frac{\partial}{\partial x_2})^{\alpha_2}\ldots(\frac{\partial}{\partial x_d})^{\alpha_d} | \sum_{i=1}^d \alpha_i \le n\}$). Such a basis of $\underline{Der}_{X/K,Z}$ will be said to be* adapted *to Z.*

Proof: Let $X = \operatorname{Spec} A$ and $L \in \Gamma(X, \underline{Diff}^n_{X/K,Z})$. We can write $L = \sum_{i=0}^n P_i(\frac{\partial}{\partial x_1})^i$, with $P_i = \sum_{\underline{\beta} \in \mathbf{N}^{d-1}} a^{(i)}_{\underline{\beta}} (\frac{\partial}{\partial x_2})^{\beta_2} \ldots (\frac{\partial}{\partial x_d})^{\beta_d}$, and $a^{(i)}_{\underline{\beta}} \in A$. We first prove that $a^{(i)}_{\underline{\beta}} \in$

$x_1^i A, \forall i = 0, \dots, n$ and $\forall \underline{\beta} = (\beta_2, \dots, \beta_d) \in \mathbf{N}^{d-1}$. Assume $a_{\underline{\beta}}^{(i)} \in x_1^i A, \forall i < j \le n$ and $\forall \underline{\beta} \in \mathbf{N}^{d-1}$. Let $\underline{\beta} \in \mathbf{N}^{d-1}$; we compute

$$L(x_1^j \underline{x}^{\underline{\beta}}) = \sum_{i=0}^{j-1} j(j-1)\dots(j-i+1)x_1^{j-i} P_i(\underline{x}^{\underline{\beta}}) + j! P_j(\underline{x}^{\underline{\beta}}) \in x_1^j A.$$

It follows from the inductive assumption that $P_j(\underline{x}^{\underline{\beta}}) \in x_1^j A$. Therefore, $a_{\underline{\beta}}^{(j)} \in x_1^j A, \forall \underline{\beta} \in \mathbf{N}^{d-1}$. Hence, an A-basis of $\Gamma(X, \underline{Diff}^n_{X/K,Z})$ is given by the operators $\{x_1^{\alpha_1}(\frac{\partial}{\partial x_1})^{\alpha_1}(\frac{\partial}{\partial x_2})^{\alpha_2}\dots(\frac{\partial}{\partial x_d})^{\alpha_d} \mid \sum_{i=1}^d \alpha_i \le n\}$. Formulas

$$\begin{aligned} x^m \left(\frac{\partial}{\partial x}\right)^m &= x\frac{\partial}{\partial x}\left(x\frac{\partial}{\partial x} - 1\right)\dots\left(x\frac{\partial}{\partial x} - m + 1\right) \\ &= \left(x\frac{\partial}{\partial x}\right)^m + \sum_{j<m} b_{m,j}\left(x\frac{\partial}{\partial x}\right)^j \end{aligned} \tag{2.1.2.1}$$

and their reciprocals

$$\left(x\frac{\partial}{\partial x}\right)^m = x^m\left(\frac{\partial}{\partial x}\right)^m + \sum_{j<m} a_{m,j} x^j \left(\frac{\partial}{\partial x}\right)^j, \tag{2.1.2.2}$$

with $a_{m,j}, b_{m,j} \in \mathbf{Z}$, permit us to conclude. □

Remark 2.1.3 If X is a curve, the fiber of $\underline{Diff}^n_{X/K,Z}$ at the point Z has a *canonical* basis over K: $(1, x\frac{\partial}{\partial x}, \dots, (x\frac{\partial}{\partial x})^n)$ (which does not depend on the choice of the local coordinate x at Z).

2.1.4 Let (X, Z) be a smooth K-model and let $D \in \Gamma(X, \underline{Der}_{X/K})$. The following facts are obviously equivalent:

i) D satisfies (1.4.3.1) and there exists a local equation x of Z such that $Dx = x$;
ii) there exists an adapted basis $(D_1, \dots, D_d)$ of $(\underline{Der}_{X/K,Z})_{\eta_Z}$ with $D = D_1$.

If the previous conditions hold, then $D_j \mathcal{F}^D \subset \mathcal{F}^D$ and $D_j x = 0$, for $j = 2, \dots, d$.

2.1.5 One defines similarly the sheaves (resp. modules) $\underline{Der}_{\hat{X}_Z/K,Z}$ and $\underline{Diff}^n_{\hat{X}_Z/K,Z}$ (resp. $Der_{\hat{v}}(\hat{\mathcal{F}}/K)$ and ${}_{\hat{v}}(\hat{\mathcal{F}}/K)$) on a smooth formal model $(\hat{X}_Z, Z)$ of $(\hat{\mathcal{F}}, \hat{v})$. A system of étale coordinates $(x_1, \dots, x_d)$ on $\hat{X}_Z$ is *adapted* to Z if there is an isomorphism (1.4.5.1) such that x_1 corresponds to the standard coordinate on $\hat{\mathbf{A}}^1_K$ and $(x_2, \dots, x_d)$ to étale coordinates on Z. The basis $(x_1\frac{\partial}{\partial x_1}, \frac{\partial}{\partial x_2}, \dots, \frac{\partial}{\partial x_d})$ of $\underline{Der}_{\hat{X}_Z/K,Z}$ (and of $Der_{\hat{v}}(\hat{\mathcal{F}}/K)$) is then *adapted* to Z. According to (1.4.5), any $D \in Der_{\hat{v}}(\hat{\mathcal{F}}/K)$

having a simple zero at $\hat{v}$ is such that $Dx = x$ for some parameter x of $\hat{R}_v$ and is part of an adapted basis $(D = D_1, \ldots, D_d)$ of $Der_{\hat{v}}(\hat{\mathcal{F}}/K)$. Once again, $D_j \hat{\mathcal{F}}^D \subset \hat{\mathcal{F}}^D$ and $D_j x = 0$ for $j = 2, \ldots, d$. A result analogous to Lemma 2.1.2 holds. Notice that the ring $Diff_{cont}(\hat{\mathcal{F}}/K)$ of $\hat{v}$-adically continuous $\hat{\mathcal{F}}/K$-differential operators of $\hat{\mathcal{F}}$ is generated as a ("left") $\hat{\mathcal{F}}$-vector space by $Diff_{\hat{v}}(\hat{\mathcal{F}}/K) = \cup_n Diff^n_{\hat{v}}(\hat{\mathcal{F}}/K)$.

2.2 The sheaf $\Omega^1_{X/K}(\log Z)$ of differentials with logarithmic singularities at Z is the locally free $\mathcal{O}_X$-module locally generated by $\frac{dx_1}{x_1}, dx_2, \ldots, dx_d$ for any system $(t_v = x_1, x_2, \ldots, x_d)$ of étale coordinates on X adapted to Z. We view it as a subsheaf of $j_*\Omega^1_{X\backslash Z}$, and as a natural dual for $\underline{Der}_{X/K,Z}$. We define $\Omega^h_{X/K}(\log Z) = \bigwedge^h \Omega^1_{X/K}(\log Z)$, viewed as a subsheaf of $j_*\Omega^h_{X\backslash Z}$. It is clear that $\Omega^\bullet_{X/K}(\log Z)$ is a subcomplex of $j_*\Omega^\bullet_{X\backslash Z}$, called the *De Rham complex of X with logarithmic singularities at Z*.

Lemma 2.2.1 *$\Omega^\bullet_{X/K}(\log Z)$ is the smallest subcomplex of $j_*\Omega^\bullet_{X\backslash Z}$ containing $\Omega^\bullet_{X/K}$, stable under wedge product, and such that $\frac{df}{f}$ is a local section of $\Omega^1_{X/K}(\log Z)$ whenever f is a local section of $j_*\mathcal{O}_{X\backslash Z}$. A local section ω of $j_*\Omega^h_{X\backslash Z}$ is a section of $\Omega^h_{X/K}(\log Z)$ if and only if ω and $d\omega$ have a simple pole at Z.*

See [De, II, 3.2 and 3.3.1] (the proof is given in the analytic context but is easily translated into algebraic terms, working locally for the étale topology on X; it works more generally for divisors with strict normal crossings).

2.3 We consider a model (Y, W) of another divisorially valued function field $(\mathcal{G}, w)$ over K, and a morphism of models $f : (X, Z) \longrightarrow (Y, W)$. We denote by k the immersion of $Y\backslash W$ in Y. It follows from the description (2.2.1) that the functoriality map $f^*(k_*\Omega^\bullet_{Y\backslash W}) \longrightarrow j_*\Omega^\bullet_{X\backslash Z}$ induces a map

(2.3.1) $$f^*\Omega^\bullet_{Y/K}(\log W) \longrightarrow \Omega^\bullet_{X/K}(\log Z).$$

Using the fact that $\Omega^1_{X/K}(\log Z)$ is locally free, we see by duality that the functoriality map $\underline{Der}_{X/K} \longrightarrow f^*\underline{Der}_{Y/K}$ induces a map of $\mathcal{O}_X$-modules

(2.3.2) $$\mathcal{T}_f : \underline{Der}_{X/K,Z} \longrightarrow f^*\underline{Der}_{Y/K,W}.$$

Furthermore, using (2.1.2), we deduce that the functoriality map $\underline{Diff}^n_{X/K} \longrightarrow f^*\underline{Diff}^n_{Y/K}$ induces a map of $\mathcal{O}_X$-modules

(2.3.3) $$\mathcal{T}_f : \underline{Diff}^n_{X/K,Z} \longrightarrow f^*\underline{Diff}^n_{Y/K,W}.$$

If $g : (Y, W) \longrightarrow (T, D)$ is another morphism of models, we have

$$f^*(\mathcal{T}_g) \circ \mathcal{T}_f = \mathcal{T}_{g \circ f},$$

and $\mathcal{T}_1 = 1$. From these, we deduce maps of R_v-modules

$$\text{(2.3.4)} \qquad \begin{aligned} \mathcal{T}_f &: Der_v(\mathcal{F}/K) \longrightarrow Der_w(\mathcal{G}/K) \otimes_{R_w} R_v, \\ \mathcal{T}_f &: Diff^n_v(\mathcal{F}/K) \longrightarrow Diff^n_w(\mathcal{G}/K) \otimes_{R_w} R_v, \end{aligned}$$

and similar maps for completions.

Let us examine the special cases of closed immersions and log dominant $f : (X, Z) \longrightarrow (Y, W)$.

Proposition 2.4 i) *If f is a closed immersion, there is an exact sequence*

$$\text{(2.4.1)} \qquad 0 \longrightarrow \underline{Der}_{X/K,Z} \xrightarrow{\mathcal{T}_f} f^*\underline{Der}_{Y/K,W} \longrightarrow \mathcal{N}^\vee_{X/Y} \longrightarrow 0,$$

where $\mathcal{N}^\vee_{X/Y}$ denotes the normal sheaf.

ii) *If f is log dominant, there is an open neighborhood U of η_Z in X and an exact sequence*

$$\text{(2.4.2)} \qquad 0 \longrightarrow (\underline{Der}_{X/Y})_{|U} \longrightarrow (\underline{Der}_{X/K,Z})_{|U} \xrightarrow{\mathcal{T}_f} (f^*\underline{Der}_{Y/K,W})_{|U} \longrightarrow 0,$$

where $\underline{Der}_{X/Y}$ denotes the dual sheaf of $\Omega^1_{X/Y}$. In particular, if we set

$$Der_v(\mathcal{F}/\mathcal{G}) = Der_v(\mathcal{F}/K) \cap Der(\mathcal{F}/\mathcal{G}) ,$$

we may identify $\mathcal{T}_f(\partial)$ with the restriction of $\partial \in Der_v(\mathcal{F}/K)$ to an $\mathcal{F}$-valued derivation of $\mathcal{G}$, and we have an exact sequence

$$\text{(2.4.3)} \qquad 0 \longrightarrow Der_v(\mathcal{F}/\mathcal{G}) \longrightarrow Der_v(\mathcal{F}/K) \xrightarrow{\mathcal{T}_f} Der_w(\mathcal{G}/K) \otimes_{R_w} R_v \longrightarrow 0.$$

We may choose an adapted basis $(D_1, \dots , D_d)$ of $Der_v(\mathcal{F}/K)$, with $(D_{1|\mathcal{G}}, \dots , D_{r|\mathcal{G}})$ an adapted basis of $Der_w(\mathcal{G}/K)$, while $D_{i|\mathcal{G}} = 0$, for $i = r + 1, \dots , d$.

Proof: i) The immersion of X into Y is regular, so that there is an exact sequence $0 \longrightarrow \underline{Der}_{X/K} \longrightarrow f^*\underline{Der}_{Y/K} \longrightarrow \mathcal{N}^\vee_{X/Y} \longrightarrow 0$. Hence the problem is localized at Z. Since W and X meet transversally at Z, there exists in a neighborhood of every point of Z in Y a system of étale coordinates $(x_1, x_2, \dots , x_n)$ such that $x_1 = 0$ is a local equation for W, $(x_{r+1} = 0, \dots , x_n = 0)$ is a local equation for X, and $(\frac{\partial}{\partial x_{r+1}}, \dots , \frac{\partial}{\partial x_n})$ is a local basis of sections of $\mathcal{N}^\vee_{X/Y}$. Then according to (2.1.2), $(x_1 \frac{\partial}{\partial x_1}, \frac{\partial}{\partial x_2}, \dots , \frac{\partial}{\partial x_n})$ is a local basis of sections of $f^*\underline{Der}_{Y/K,W}$, $(x_1 \frac{\partial}{\partial x_1}, \frac{\partial}{\partial x_2}, \dots , \frac{\partial}{\partial x_r})$ is a local basis of sections of $\underline{Der}_{X/K,Z}$, and the result follows.

ii) There is an exact sequence $0 \longrightarrow \underline{Der}_{X/Y} \longrightarrow \underline{Der}_{X/K} \longrightarrow f^*\underline{Der}_{Y/K}$. We have to show that in some open neighborhood U of η_Z in X, $\mathrm{Im}(\underline{Der}_{X/Y} \longrightarrow \underline{Der}_{X/K}) \subset \underline{Der}_{X/K,Z}$, and $\mathcal{T}_f$ is surjective. We may replace X and Y by étale neighborhoods of η_Z and η_W respectively. In particular, we may assume that $W \longrightarrow Y$ is a section of a morphism $h : Y \longrightarrow S = W$ (1.3.2) with smooth connected one-dimensional fibers. We may assume that, if f_{η_S} denotes the fiber of the S-morphism f above η_S, f_{η_S} is smooth except above the point η_Z of the curve Y_{η_S}. Locally for the étale topology, there is a system of coordinates $(x_1, \dots, x_n)$ on X adapted to Z, a system of coordinates $(z_2, \dots, z_r)$ on S and a coordinate y on Y_{η_S} satisfying: $y \circ f = x_1^e$ on X_{η_S}, and $z_i \circ h \circ f = x_i$ for $i = 2, \dots, r$.

Then we have $x_1^{e-1}dx_1 = dx_2 = \cdots = dx_r = 0$ in $\Omega^1_{X/Y}$. It is now clear that $(\frac{\partial}{\partial x_{r+1}}, \dots, \frac{\partial}{\partial x_n})$ is a local basis of sections of $\underline{Der}_{X/Y}$, $(x_1\frac{\partial}{\partial x_1}, \frac{\partial}{\partial x_2}, \dots, \frac{\partial}{\partial x_n})$ is a local basis of sections of $\underline{Der}_{X/K,Z}$, $(x_1\frac{\partial}{\partial x_1}, \frac{\partial}{\partial x_2}, \dots, \frac{\partial}{\partial x_r})$ is a local basis of sections of $f^*\underline{Der}_{Y/K,W}$, and the result follows. □

An exact sequence similar to (2.4.3) holds for fields of formal functions.

Remark 2.5 In case ii), we see that $\underline{Der}_{X/Y}$ is flat at η_Z (in contrast to $\Omega^1_{X/Y}$ if $e \geq 2$). On U, the dual of $\underline{Der}_{X/Y}$ is $\Omega^1_{X/Y}(\log Z)$, the image of $\Omega^1_X(\log Z)$ in $j_*\Omega^1_{(X\setminus Z)/(Y\setminus W)}$. Notice that although $dx_1 \neq 0$ in $\Omega^1_{X/Y}$ if $e \geq 2$, we have $dx_1/x_1 = e^{-1}f^*dy/y = 0$ in $\Omega^1_{X/Y}(\log Z)$.

§3 Regularity along a divisor

3.1 Differential modules

3.1.1 Let $\mathcal{F}/K$ be a function field in d variables. An *$\mathcal{F}/K$-differential module* is a module over *Diff*$(\mathcal{F}/K)$ which is finitely dimensional as a vector space over $\mathcal{F}$; this dimension is called the *rank* of the differential module. Similarly, if $\hat{\mathcal{F}}/K$ is a formal function field, an $\hat{\mathcal{F}}/K$-differential module is a module over $\mathit{Diff}_{cont}(\hat{\mathcal{F}}/K)$ which is finitely dimensional as a vector space over $\hat{\mathcal{F}}$. (When using these definitions, it is convenient to drop our general assumption that K is algebraically closed).

If E_1 and E_2 are differential modules with respect to the same (formal) function field, there are canonically defined differential modules $E_1 \otimes_{\mathcal{F}} E_2$ and $Hom_{\mathcal{F}}(E_1, E_2)$.

3.1.2 Let V be a smooth connected variety over K, $\mathcal{F} = \kappa(V)$. A *connection* on a coherent $\mathcal{O}_V$-module $\mathcal{E}$ is an additive mapping

$$\nabla : \mathcal{E} \longrightarrow \mathcal{E} \otimes_{\mathcal{O}_V} \Omega^1_V \tag{3.1.2.1}$$

satisfying the Leibniz rule. It is called *integrable* if the induced $\mathcal{O}_V$-linear mapping

$$\nabla : \underline{Der}_V \longrightarrow \underline{End}_K \mathcal{E} \tag{3.1.2.2}$$

is compatible with the Lie bracket [,]; $\mathcal{E}$ is then necessarily locally free (*cf.* [Ka1] and (III.2.1) for more detail).

The integrability ensures that the generic fiber

$$\nabla : Der(\mathcal{F}/K) \longrightarrow End_K \mathcal{E}_{\eta_V} \tag{3.1.2.3}$$

extends to a structure of $\mathcal{F}/K$-differential module E on $\mathcal{E}_{\eta_V}$; we say that $(\mathcal{E}, \nabla)$ is a *model* of E on V.

If $(\mathcal{E}_1, \nabla_1)$ and $(\mathcal{E}_2, \nabla_2)$ are models of the $\mathcal{F}/K$-differential modules E_1 and E_2 respectively, there are canonically defined models $(\mathcal{E}_1, \nabla_1) \otimes (\mathcal{E}_2, \nabla_2)$ (connection: $\nabla_1 \otimes 1 + 1 \otimes \nabla_2$) and $\underline{Hom}((\mathcal{E}_1, \nabla_1), (\mathcal{E}_2, \nabla_2))$ (connection: $-(\nabla_1)^t \otimes 1 + 1 \otimes \nabla_2$) of $E_1 \otimes_{\mathcal{F}} E_2$ and $Hom_{\mathcal{F}}(E_1, E_2)$ respectively .

An $\mathcal{O}_V$-linear mapping $\mathcal{E}_1 \longrightarrow \mathcal{E}_2$ is called *horizontal* if it commutes with the connections.

3.1.3 Let us consider the situation of (1.2.7.1). Any $\mathcal{G}/K$-differential module E uniquely extends to an $\mathcal{F}/K$-differential module $E_{\mathcal{F}}$. Vice versa, if $[\mathcal{F} : \mathcal{G}] < \infty$, any $\mathcal{F}/K$-differential module E' may be regarded as a $\mathcal{G}/K$-differential module, which we denote by ${}_{\mathcal{G}}E'$ in that capacity. A similar notation will be used for $\hat{\mathcal{F}}/K$ or $\hat{\mathcal{G}}/K$-differential modules.

There is a canonical morphism $({}_{\mathcal{G}}E')\hat{} \longrightarrow_{\hat{\mathcal{G}}} (E'^{\wedge(v)})$, where $E'^{\wedge(v)}$ denotes the v-completion of E'. In fact, the canonical morphism $({}_{\mathcal{G}}E')\hat{} \longrightarrow \bigoplus_{v|w}\ {}_{\hat{\mathcal{G}}}(E'^{\wedge(v)})$ is an isomorphism.

3.1.4 Inverse and direct images of integrable connections will be discussed in detail in (III.2.1). It suffices here to say that if $(X, Z) \xrightarrow{f} (Y, W)$ is a log dominant morphism of models, corresponding to (1.2.7.1), then for any model $(\mathcal{E}, \nabla)$ of E on $Y \backslash W$, $f^*(\mathcal{E}, \nabla)$ is a model of $E_{\mathcal{F}}$ on $X \backslash Z$. If moreover f is finite étale on $X \backslash Z$, then for any model $(\mathcal{E}', \nabla')$ of E' on $X \backslash Z$, $f_*(\mathcal{E}', \nabla')$ is a model of ${}_{\mathcal{G}}E'$ on $Y \backslash W$.

3.2 Regularity: review of the one-variable case

Let C be a field of characteristic 0. We consider the field of formal Laurent series $C((x))$ with the x-adic valuation v.

Definition 3.2.1 A $C((x))/C$-differential module $\hat{E}$ is called *regular* if it contains a $C[[x]]$-lattice stable under $x\frac{d}{dx}$.

Let $\overline{C}$ be the algebraic closure of C and $C \subset C' \subset \overline{C}$ be an intermediate extension. For a $C((x))/C$-differential module $\hat{E}$ we set $\hat{E}_{C'} = C'((x)) \otimes_{C((x))} \hat{E}$, a $C'((x))/C'$-differential module.

The following results are classical (see [Man] or [DGS, III.8]).

Proposition 3.2.2 *A $C((x))/C$-differential module $\hat{E}$ of rank μ is regular if and only if the following equivalent conditions are satisfied:*

a) *for every $m \in \hat{E}$, the smallest $C[[x]]$-submodule of $\hat{E}$ containing m and stable under $x\frac{d}{dx}$ is finitely generated;*

b) *there is an integer c and a* $C[[x]]$*-lattice in* $\hat{E}$ *stable under* $x^c \mathit{Diff}_v(C((x))/C)$;

c) *for some (resp. any) cyclic vector* $m \in \hat{E}$*, the coefficients* b_j *of the associated differential equation* $(x\frac{d}{dx})^\mu m - \sum_{j<\mu} b_j (x\frac{d}{dx})^j m = 0$ *belong to* $C[[x]]$ (in this situation we also say that the differential equation is *regular*);

d) *there is a finite subset* $\mathcal{A}$ *of* $\overline{C}$ *such that intermediate extension* $C(\mathcal{A}) \subset C' \subset \overline{C}$ *there is a decomposition*

$$\hat{E}_{C'} \cong \oplus_{\alpha \in \mathcal{A}} C'((x)) \otimes_{C'} \mathrm{Ker}_{\hat{E}_{C'}} \left(x\frac{d}{dx} - \alpha \right)^\mu . \tag{3.2.2.1}$$

Moreover, if such a decomposition exists over C'*, it is canonical, and the factor* $C'((x)) \otimes_{C'} \mathrm{Ker}_{\hat{E}_{C'}} (x\frac{d}{dx} - \alpha)^\mu = C'((x)) \otimes_{C(\alpha)} \mathrm{Ker}_{\hat{E}_{C(\alpha)}} (x\frac{d}{dx} - \alpha)^\mu$ *does not change if* α *is replaced by* $\alpha + n, n \in \mathbf{Z}$ *.The subset* $\mathcal{A} + \mathbf{Z}$ *of* $\overline{C}$ *is uniquely determined by the condition*

$$\beta \in \mathcal{A} + \mathbf{Z} \text{ if and only if } \mathrm{Ker}_{\hat{E}_{\overline{C}}} \left(x\frac{d}{dx} - \beta \right)^\mu \neq 0. \tag{3.2.2.2}$$

Proposition 3.2.3 *Let* $\hat{E}_1$ *and* $\hat{E}_2$ *be* $C((x))/C$*-differential modules.*

i) *If* $\hat{E}$ *sits in an exact sequence of* $C((x))/C$*-differential modules*

$$0 \longrightarrow \hat{E}_1 \longrightarrow \hat{E} \longrightarrow \hat{E}_2 \longrightarrow 0$$

then $\hat{E}$ *is regular if and only if* $\hat{E}_1$ *and* $\hat{E}_2$ *are regular.*

ii) *If* $\hat{E}_1$ *and* $\hat{E}_2$ *are regular, so are* $\hat{E}_1 \otimes_{C((x))} \hat{E}_2$ *and* $Hom_{C((x))}(\hat{E}_1, \hat{E}_2)$.

3.2.4 Let $\mathcal{F}/K$ be a function field in one variable (K not necessarily algebraically closed, but closed in $\mathcal{F}$) and v be a discrete valuation of $\mathcal{F}$, trivial on K. Let ∂ be a K-linear derivation of $\hat{\mathcal{F}}$ transversal to v (1.4.7), and let $C = \hat{\mathcal{F}}^\partial$ be the constant subfield. Then $C \cong k(v)$ is a finite extension of K, and there is a canonical isomorphism $(\hat{R}_v, \hat{v}, \partial) \cong (C[[x]], x\text{-adic valuation}, \frac{d}{dx})$ (also (1.4.6)).

An $\mathcal{F}/K$-differential module E is called *regular at* v if the $\hat{\mathcal{F}}/C$-differential module $\hat{E} = E \otimes_{\mathcal{F}} \hat{\mathcal{F}}$ is regular.

Notice that if ∂ comes from a derivation of $\mathcal{F}$, then $\mathcal{F}^\partial = K$, which may be a proper subfield of $\hat{\mathcal{F}}^\partial$.

3.2.5 In the situation of (3.1.3), it is clear in the definition that $\hat{E}$ (resp. $\hat{E}'$ if $[\hat{\mathcal{F}} : \hat{\mathcal{G}}] < \infty$) is regular if and only if $\hat{E}_{\hat{\mathcal{F}}}$ (resp. ${}_{\hat{\mathcal{G}}}\hat{E}'$) is regular. Similarly, E (resp. E' if $[\mathcal{F} : \mathcal{G}] < \infty$) is regular at w (resp. at every $v \mid w$) if and only if $E_{\mathcal{F}}$ (resp. ${}_{\mathcal{G}}E'$) is regular at v (resp. w). Note that if a divisorial valuation v of $\mathcal{F}$ induces the trivial valuation on $\mathcal{G}$, then for any E, $E_{\mathcal{F}}$ is regular at v.

3.3 The case of a function field in several variables

3.3.1 Let $(\hat{\mathcal{F}}, v)$ be a field of formal functions over K (assumed again to be algebraically closed), and $(D_1, \ldots, D_d)$ an adapted basis of $Der_v(\hat{\mathcal{F}}/K)$ ((2.1.5)). We set $C = \hat{\mathcal{F}}^{D_1}$ so that $\hat{\mathcal{F}} = C((x))$, for a parameter x of $\hat{R}_v$ such that $D_1 x = x$, $D_j C \subset C$ and $D_j x = 0$, $j = 2, \ldots, d$.

Let $\hat{E}$ be an $\hat{\mathcal{F}}/K$-differential module of rank μ. Let us assume that the induced $C((x))/C$-differential module $\hat{E}_{/C}$ is regular and for $\alpha \in \mathcal{A}$, $\nu = 0, 1, \ldots,$ set $K_\alpha^{(\nu)} = \operatorname{Ker}_{\hat{E}}(D_1 - \alpha)^\nu$, a C-vector subspace of $\hat{E}$.

Lemma 3.3.2 *For all $j = 2, \ldots, d$, $\nu = 0, 1, \ldots,$ and $\alpha \in \mathcal{A}$*

$$D_j K_\alpha^{(\nu)} \subseteq K_\alpha^{(\nu+1)} .$$

Proof: Let $\partial \in \{D_2, \ldots, D_d\}$. The statement is trivial for $\nu = 0$. We proceed by induction: assume the statement holds for $\nu < \nu_0$ ($\nu_0 \geq 1$) and let $m \in K_\alpha^{(\nu_0)}$. So, $(D_1 - \alpha)^{\nu_0} m = 0$ and $(D_1 - \alpha) m \in K_\alpha^{(\nu_0 - 1)}$. Then $(D_1 - \alpha)^{\nu_0+1} \partial m = (D_1 - \alpha)^{\nu_0} \partial (D_1 - \alpha) m + (D_1 - \alpha)^{\nu_0} (\partial \alpha) m = 0$, since $\partial \alpha \in C$ and by the induction assumption. □

Proposition 3.3.3 i) *There exists an $\hat{\mathcal{F}}$-basis $\underline{e}$ of $\hat{E}$ such that*

$$D_j \underline{e} = \underline{e} G_j, \tag{3.3.3.1}$$

with $G_j \in M_{\mu \times \mu}(C)$ for all $j = 1, \ldots, d$.

ii) *For any $\hat{\mathcal{F}}$-basis $\underline{e}$ of $\hat{E}$ as in* i)*, the eigenvalues of G_1 are in K.*

Proof: After possibly replacing C by a finite extension, we may assume that C contains the set $\mathcal{A}$ in part d) of (3.2.2), so that $\hat{E}$ admits the decomposition (3.2.2.1). It follows from Lemma 3.3.2 that $D_j K_\alpha^{(\mu)} \subseteq K_\alpha^{(\mu)}$, for all $j = 2, \ldots, d$, and this implies i) under our assumption on C.

Let us turn to ii). For $j = 2, \ldots, d$, one has

$$D_j(G_1) = G_j G_1 - G_1 G_j,$$

whence $D_j(G_1^m) = G_j G_1^m - G_1^m G_j$, and $D_j \operatorname{Tr}(G_1^m) = \operatorname{Tr}(D_j(G_1^m)) = 0$. Therefore, the coefficients of the characteristic polynomial of G_1 are constant, i.e., in K. In particular $\mathcal{A}$ is a subset of K, so that there was in fact no need to enlarge C in i). □

Corollary 3.3.4 *For any $\hat{\mathcal{F}}/K$-differential module $\hat{E}$, the following conditions are equivalent:*

i) *there exists an $\hat{R}_v$-lattice $E_{\hat{R}_v}$ in $\hat{E}$ and an integer c such that $Diff_v(\hat{\mathcal{F}}/K) E_{\hat{R}_v} \subset x_1^{-c} E_{\hat{R}_v}$;*

ii) *for every $\hat{R}_v$-lattice $E_{\hat{R}_v}$ in $\hat{E}$, there is an integer c such that $Diff_v(\hat{\mathcal{F}}/K) E_{\hat{R}_v} \subset x_1^{-c} E_{\hat{R}_v}$;*

iii) *for some $D \in Der_v(\hat{\mathcal{F}}/K)$ with a simple zero at (or transversal to) v the induced $\hat{\mathcal{F}}/\hat{\mathcal{F}}^D$-differential module $\hat{E}_{/\hat{\mathcal{F}}^D}$ is regular.*

iv) *for every $D \in Der_v(\hat{\mathcal{F}}/K)$ with a simple zero at (or transversal to) v) the induced $\hat{\mathcal{F}}/\hat{\mathcal{F}}^D$-differential module $\hat{E}_{/\hat{\mathcal{F}}^D}$ is regular;*

v) *there exists an $\hat{R}_v$-lattice $E_{\hat{R}_v}$ in $\hat{E}$ such that $Diff_v(\hat{\mathcal{F}}/K)E_{\hat{R}_v} \subset E_{\hat{R}_v}$.*

Proof: i) $\Rightarrow$ iv) follows from the characterization b) of Proposition 3.2.2 of regularity in one variable. iv) $\Rightarrow$ iii) is trivial. iii) $\Rightarrow$ v) follows from i) of (3.3.3), using (2.1.5) to complete D into an adapted basis of $Der_v(\hat{\mathcal{F}}/K)$ and taking as lattice $E_{\hat{R}_v}$ the $\hat{R}_v$-span of the basis $\underline{e}$. v) $\Rightarrow$ i) is obvious. At last, if $E'_{\hat{R}_v}$ is another lattice, there exists a positive integer c' such that $x_1^{c'} E'_{\hat{R}_v} \subset E_{\hat{R}_v} \subset x_1^{-c'} E'_{\hat{R}_v}$. Then $Diff_v(\hat{\mathcal{F}}/K)E_{\hat{R}_v} \subset x_1^{-c-2c'} E_{\hat{R}_v}$. This shows that i) $\Leftrightarrow$ ii). □

Definition 3.3.5 An $\hat{\mathcal{F}}/K$-differential module $\hat{E}$ is called *regular* if it satisfies the equivalent conditions of (3.3.4). If $(\mathcal{F}, v)$ is a divisorially valued function field over K with completion $\hat{\mathcal{F}}$, an $\mathcal{F}/K$-differential module E is called *regular at* v if $\hat{E}$ is regular.

3.4 Regular connections along a divisor

3.4.1 Let (X, Z) be a model of a divisorially valued function field $(\mathcal{F}, v)$. Let E be an $\mathcal{F}/K$-differential module, and let $(\mathcal{E}, \nabla)$ be a model of E on $X \backslash Z$ (*cf.* (3.1)). The integrable connection ∇ is said to be *regular at* (or *along*) Z if and only if E is regular at v.

Proposition 3.4.2 *Assume that $(x_1, \ldots, x_d)$ is an adapted system of coordinates for (X, Z) (cf. (1.1.4)). Then ∇ is regular at Z if and only if there exists a coherent extension $\tilde{\mathcal{E}} \subset j_*\mathcal{E}$ of $\mathcal{E}$ on X and an integer c such that $\underline{Diff}_{X/K,Z}\tilde{\mathcal{E}} \subset x_1^{-c}\tilde{\mathcal{E}}$.*

Proof: Note that $\underline{Diff}_{X/K,Z}$ acts naturally on $j_*\mathcal{E}$ as a sub-K_X-algebra of $j_*(\underline{Diff}_{(X\backslash Z)/K})$. The "*if*"-part comes from (3.3.4, i). Conversely, let $\tilde{\mathcal{E}} \subset j_*\mathcal{E}$ be any coherent extension of $\mathcal{E}$ on X, and let us choose the integer c as in (3.3.4.*ii*) applied to the completion of the lattice $\tilde{\mathcal{E}}_{\eta_Z}$. Then there exists a closed subset $Z' \subset Z$ of codimension ≥ 2 in X such that $j'^*\underline{Diff}_{X/K,Z}\tilde{\mathcal{E}} \subset x_1^{-c} j'^*\tilde{\mathcal{E}}$, where j' denotes the embedding of $X\backslash Z'$ in X. It then suffices to replace $\tilde{\mathcal{E}}$ by $j'_* j'^*\tilde{\mathcal{E}}$, which is coherent according to [EGA IV, 5.11.4]. □

Proposition 3.4.3 *Let (X, Z) be a model and $(\mathcal{E}, \nabla)$, $(\mathcal{E}_1, \nabla_1)$, $(\mathcal{E}_2, \nabla_2)$ be coherent sheaves with integrable connections on $X\backslash Z$.*

i) *If $\mathcal{E}$ sits in a horizontal exact sequence $0 \longrightarrow \mathcal{E}_1 \longrightarrow \mathcal{E} \longrightarrow \mathcal{E}_2 \longrightarrow 0$, then ∇ is regular at Z if and only if ∇_1 and ∇_2 are regular at Z.*

ii) *If ∇_1 and ∇_2 are regular at Z, so are $(\mathcal{E}_1, \nabla_1) \otimes (\mathcal{E}_2, \nabla_2)$ and $\underline{Hom}((\mathcal{E}_1, \nabla_1), (\mathcal{E}_2, \nabla_2))$.*

Proof: This follows from (3.2.3). □

Proposition 3.4.4 *Let $(X, Z) \xrightarrow{f} (Y, W)$ be a morphism of models such that $f(Z) \subset W$. Let $(\mathcal{E}, \nabla)$ be a coherent sheaf with integrable connection on $Y \backslash W$. If ∇ is regular along W, then $f^*(\nabla)$ is regular along Z. Conversely, if $f^*(\nabla)$ is regular along Z and if f is log dominant, then ∇ is regular along W.*

Proof: We use the characterization (3.4.2) of regularity along a divisor and the functoriality (2.3.2) of logarithmic differential operators. By definition of the inverse image, we have the equality, for any $L \in \underline{Diff}_{X/K,Z}$ and any local section m of $\mathcal{E}$, $\nabla(L) f^*(m) = f^*(\nabla(\mathcal{T}_f L)m)$. Let $\tilde{\mathcal{E}} \subset k_* \mathcal{E}$ be a coherent extension of $\mathcal{E}$ on Y as in (3.4.2). We thus have

$$\underline{Diff}_{X/K,Z} f^* \tilde{\mathcal{E}} = f^*(\nabla(\mathcal{T}_f(\underline{Diff}_{X/K,Z}))\tilde{\mathcal{E}}) \subset f^*(\nabla(\underline{Diff}_{Y/K,W})\tilde{\mathcal{E}})$$

Let y be a local equation of W at $f(\eta_Z)$. If ∇ is regular at W, there is an integer c such that $\underline{Diff}_{Y/K,W} \tilde{\mathcal{E}} \subset y^{-c} \tilde{\mathcal{E}}$. We obtain that $\underline{Diff}_{X/K,Z} f^* \tilde{\mathcal{E}} \subset y^{-c} f^* \tilde{\mathcal{E}} \subset {x_1}^{-ce_f} f^* \tilde{\mathcal{E}}$, hence $f^*(\nabla)$ is regular along Z.

The second assertion follows from (3.1.4). and (3.2.5). □

Proposition 3.4.5 *The property of being regular at Z is local at η_Z for the* fppf *topology on X.*

Proof: If $X' \longrightarrow X$ is any *fppf* neighborhood of η_Z, we can extend it into a log dominant morphism $f : (X', Z') \longrightarrow (X, Z)$ of smooth K-models. If $f^*(\mathcal{E}, \nabla)$ is regular along Z', $(\mathcal{E}, \nabla)$ is regular along Z, by the previous statement. □

3.4.6 Examples

3.4.6.1 Take

$$\begin{array}{rcl} Y & = & \text{the } (x, y)\text{-plane with the } x\text{-axis } (y = 0) \text{ removed,} \\ W & = & y\text{-axis } (x = 0), \\ X & = & \text{the diagonal } (x = y) \text{ in the } (x, y)\text{-plane,} \\ Z & = & \text{the origin.} \end{array}$$

Let ∇ be the connection on $\mathcal{E} = \mathcal{O}_{Y \backslash W}$, whose formal solution is $e^{1/y}$. It is regular at W (in fact, it has no singularity at all along W), but its restriction to $X \backslash Z$ is not regular at Z. However the embedding of $X \backslash Z$ into $Y \backslash W$ does not extend to a morphism of *models* $(X, Z) \longrightarrow (Y, W)$.

3.4.6.2 Consider the morphism $f : (x, y) \longmapsto (u, v) = (xy, y)$ of $(X, Z) = (\mathbf{A}^2, y = 0)$ into $(Y, W) = (\mathbf{A}^2, v = 0)$. The connection ∇ on $\mathcal{E} = \mathcal{O}_{Y \backslash W}$ with formal solution $e^{u/v}$ pulls back to the connection $f^*\nabla$ killing e^x. So, $f^*\nabla$ is regular along Z, while ∇ is irregular along W. Notice that f is a birational map, but it is not flat at η_Z.

Theorem 3.4.7 *Let (X, Z) be a model and $(\mathcal{E}, \nabla)$ be a coherent sheaf with integrable connection on $X \setminus Z$. The following properties are equivalent:*

i) *∇ is regular at Z;*
ii) *for any locally closed immersion of models $(C, P) \overset{i}{\hookrightarrow} (X, Z)$, with C a curve, $i^*_{C \setminus P}(\nabla)$ is regular at P;*
iii) *there is an open neighborhood U of η_Z in X and a tubular neighborhood $U \overset{f}{\longrightarrow} S$ of $Z \cap U$, such that $\nabla_{|U_s}$ is regular at Z_s for a Zariski-dense set of closed points $s \in S$.*

Proof: The implication i) $\Rightarrow$ ii) is part of (3.4.4). The implication ii)$\Rightarrow$ iii) is immediate. Let us prove iii)$\Rightarrow$ i). We may replace U by any affine connected neighborhood of η_Z. This allows us to assume that

a) there is a transversal derivation $\partial = \partial_{t,f} \in \Gamma(U, \underline{Der}_{U/K})$ such that f is a family of integral curves of ∂; in particular, by (1.4.1.2), ∂ lies in the image of $\Gamma(U_s, \underline{Der}_{U_s/K})$ for every $s \in S$;
b) there is a section m of $\mathcal{E}$ over $U \setminus Z$ such that $\mathcal{E}$ is free with basis $(m, \nabla(t\partial)m, \ldots, \nabla(t\partial)^{\mu-1}m)$ (by the lemma of the cyclic vector).

Let us write the associated differential equation $(t\partial)^\mu m - \sum_{j<\mu} b_j (t\partial)^j m = 0$, with $b_j \in \mathcal{O}(U \setminus Z)$. By assumption the restriction of b_j to $U_s \setminus Z_s$ is holomorphic at every point of the finite set Z_s (Fuchs' criterion of regularity, c) of (3.2.2)). Since this holds for a dense set of points $s \in S$, this implies that b_j has no pole at η_Z, which means that $(\mathcal{E}, \nabla)$ is regular at Z. □

In (II.4), we shall show that in order to check the regularity of ∇, it suffices to find sufficiently many locally closed curves $C \subset X$, so that the points $C \cap Z$ are dense in Z, for which $\nabla_{|C \setminus C \cap Z}$ is regular.

3.4.8 (Counter) example Take $X =$ the (x, y)-plane, $Z = x$-axis $(y = 0)$, $C =$ any smooth curve cutting Z transversally (only) at the origin. (So $(C, P) \hookrightarrow (X, Z)$ is a closed immersion of models.) Let ∇_1 be the connection on $\mathcal{E} = \mathcal{O}_{X \setminus Z}$, whose formal solution is $e^{x/y}$ (*i.e.*, $\nabla_1(1) = -\frac{dx}{y} + x\frac{dy}{y^2}$). It is not regular at Z, but its restriction to $C \setminus P$ is regular at P. This shows the necessity of considering a dense set of points $P \in Z$.

3.4.9 Let X be a smooth connected variety over K, $\mathcal{F} = \kappa(X)$. Let $j : X \longrightarrow \overline{X}$ be a dominant open immersion of X in a normal K-variety $\overline{X}$, and let $\overline{Z}$ be a (not necessarily irreducible) subvariety of $\overline{X} \setminus X$ of pure codimension one in $\overline{X}$. If $(\mathcal{E}, \nabla)$ is a coherent sheaf with integrable connection on X, we say that ∇ is *regular* at $\overline{Z}$ if its generic fiber E is regular at the divisorial valuations of $\mathcal{F}$ defined by the irreducible components of $\overline{Z}$.

3.4.10 This definition applies especially in the case where $\overline{X}$ *and* $\overline{Z}$ *are smooth, but* $\overline{Z}$ *has several connected components*. This geometric setting is a useful generalization of the notion of smooth K-model. The discussion of logarithmic differential operators remains the same in this context. *All the results of this Section* 3.4 (except (3.4.5) which requires an obvious modification) *extend in a straightforward manner to this slightly more general setting.*

§4 Extensions with logarithmic poles

We fix here a section $\tau : K/\mathbf{Z} \longrightarrow K$ of the canonical projection $K \longrightarrow K/\mathbf{Z}$, such that $\tau(0) = 0$.

4.1 Let X be a smooth connected K-variety and $Z = \bigcup_{i=1}^{s} Z_i$ be a divisor with (strict) normal crossings in X, the sum of smooth connected divisors Z_i. Let $\epsilon : X' \hookrightarrow X$ be an open subvariety of X, containing $X \setminus Z$ and the generic point of each Z_i; we set $Z'_i = Z_i \cap X'$, $Z' = \bigcup_{i=1}^{s} Z'_i$. Then $X' = X \setminus T$, where T is a finite union $\bigcup T_\alpha$ of closed irreducible subvarieties T_α of some Z_i, with $\operatorname{codim}_{Z_i} T_\alpha \geq 1$, hence $\operatorname{codim}_X T_\alpha \geq 2, \forall \alpha$. Let us decompose j into $\epsilon \circ j'$ where $X \setminus Z = X' \setminus Z' \overset{j'}{\hookrightarrow} X' \overset{\epsilon}{\hookrightarrow} X$.

Lemma 4.2 *For any $h = 0, 1, \ldots,$ we have the equality of sheaves*

$$\epsilon^* j_* \Omega^h_{(X\setminus Z)/K} = j'_* \Omega^h_{(X'\setminus Z')/K},$$

and of their subsheaves

$$\epsilon^* \Omega^h_{X/K}(\log Z) = \Omega^h_{X'/K}(\log Z').$$

Moreover, the canonical morphism

$$\Omega^h_{X/K}(\log Z) \longrightarrow \epsilon_* \Omega^h_{X'/K}(\log Z')$$

[*EGA* 0_I, 4.4.3.2] *is an isomorphism.*

Proof: The equality $\epsilon^* j_* \Omega^h_{(X\setminus Z)/K} = j'_* \Omega^h_{(X'\setminus Z')/K}$ follows from [EGA IV, Proposition 5.9.4] applied to $\mathcal{F} = \Omega^h_{X/K}$. The second equality is clear from the explicit local description of $\Omega^h_{X/K}(\log Z)$ and $\Omega^h_{X'/K}(\log Z')$ respectively. Taking into account the fact that $\Omega^h_{X/K}(\log Z)$ is locally free, the third isomorphism follows according to [EGA IV, 5.10.2] ($\operatorname{prof}_T \Omega^h_{X/K}(\log Z) = \operatorname{prof}_T \mathcal{O}_X = \inf_{x\in T} \operatorname{prof}_{\mathcal{O}_{X,x}} \mathcal{O}_{X,x} = \inf_{x\in T} \dim \mathcal{O}_{X,x} \geq 2$). □

4.3 Let $\mathcal{M}$ be a torsion-free coherent $\mathcal{O}_X$-module, and let ∇ be an integrable X/K-connection with logarithmic singularities along Z

(4.3.1) $$\nabla : \mathcal{M} \longrightarrow \Omega^1_{X/K}(\log Z) \otimes \mathcal{M}.$$

Notice that the underlying coherent $\mathcal{O}_{X\setminus Z}$-module with connection $(\mathcal{M}, \nabla)_{|X\setminus Z}$ is regular along Z. In particular, $\mathcal{M}_{|X\setminus Z}$ is locally free. Since $\mathcal{M}$ is torsion-free, it follows that there exists an open subvariety X' as in (4.1) containing $X \setminus Z$ and the generic point of each Z_j, such that $\mathcal{M}_{|X'}$ is locally free. We denote by i'_j the locally closed embedding $Z'_j \hookrightarrow X$. In this situation, there is a well-defined endomorphism

$$\mathrm{Res}_{Z'_j} \nabla : {i'_j}^* \mathcal{M} \longrightarrow {i'_j}^* \mathcal{M},$$

called the *residue* of ∇ at Z'_j. It is induced, locally, by the action of $\nabla(x_j \frac{\partial}{\partial x_j})$, where $(x_1, \dots, x_d)$ denotes a system of local coordinates such that $Z_i = V(x_i)$ [De, II.3.8]. The *characteristic polynomial* $P_j(t) \in \mathcal{O}(Z'_j)[t]$ of $\mathrm{Res}'_{Z'_j} \nabla$ turns out to be *constant along* Z'_j, *i.e.* $P_j(t) \in K[t]$ (*cf.* [De, II.3.10]).

Definition 4.4 We denote by $\mathbf{MIC}^{\tau}(X(\log Z))$ the category of torsion-free coherent $\mathcal{O}_X$-modules with integrable X/K-connection with logarithmic singularities along Z, such that the roots of $P_j(t)$ $(j = 1, \dots, s)$ are in the image of τ. If $(\mathcal{E}, \nabla)$ is a coherent $\mathcal{O}_{X \setminus Z}$-module with integrable connection, and if $(\tilde{\mathcal{E}}, \tilde{\nabla})$ is an object of $\mathbf{MIC}^{\tau}(X(\log Z))$ such that $\tilde{\mathcal{E}}$ is locally free and $(\tilde{\mathcal{E}}, \tilde{\nabla})_{|X \setminus Z} \cong (\mathcal{E}, \nabla)$, we say that $(\tilde{\mathcal{E}}, \tilde{\nabla})$ is a τ*-extension* of $(\mathcal{E}, \nabla)$ on X (with logarithmic poles on Z).

The condition that $\tilde{\mathcal{E}}$ is locally free is not automatic[(1)]: indeed if $X =$ the x, y plane, and $Z : xy = 0$, the ideal of the origin carries a natural connection with logarithmic singularities at Z, with residue $= 0$; its restriction outside the origin is the trivial connection.

Before we discuss the existence of τ-extensions, let us recall a few general facts about extensions of locally free sheaves. Let Y be a normal variety, $T \subset Y$ a closed subvariety of codimension ≥ 2, $\epsilon : Y \setminus T \hookrightarrow Y$ the open immersion, $\mathcal{F}$ a locally free $\mathcal{O}_{Y \setminus T}$-module of finite rank. Then

i) $\epsilon_* \mathcal{F}$ is a reflexive[(2)] coherent $\mathcal{O}_Y$-module, *cf.* [EGA IV, Corollary 5.11.4] and [Se, Proposition 7],

ii) $\epsilon_* \mathcal{F}$ is the unique reflexive extension of $\mathcal{F}$, up to unique isomorphism [Se, Proposition 7 and §3, Remark 2]; the point is that if $\tilde{\mathcal{F}}$ is a reflexive coherent $\mathcal{O}_Y$-module, then

$$\begin{aligned} \epsilon_* \epsilon^* \tilde{\mathcal{F}} &= \epsilon_* \underline{Hom}(\epsilon^* \tilde{\mathcal{F}}^{\vee}, \mathcal{O}_{Y \setminus T}) = \underline{Hom}(\tilde{\mathcal{F}}^{\vee}, \epsilon_* \mathcal{O}_{Y \setminus T}) \\ &= \underline{Hom}(\tilde{\mathcal{F}}^{\vee}, \mathcal{O}_Y) = \tilde{\mathcal{F}}; \end{aligned}$$

(applying this to $\tilde{\mathcal{F}} = (\epsilon_* \mathcal{F})^{\vee\vee}$, one deduces that $(\epsilon_* \mathcal{F})^{\vee\vee} = \epsilon_*(\epsilon^* \tilde{\mathcal{F}}) = \epsilon_* \mathcal{F}$, *i.e.* the reflexivity of $\epsilon_* \mathcal{F}$ stated in i)).

[(1)] We are grateful to H. Esnault for pointing out this subtlety to us.

[(2)] *i.e.* isomorphic to its bidual. Any locally free module of finite rank is obviously reflexive. The converse is not true if the rank is > 1.

iii) $\epsilon_* \mathcal{F}$ need not be locally free if rk $\mathcal{F} > 1$; for example, if $Y = \mathbf{A}^3$, $T =$ the origin, $\pi : Y \backslash T \to \mathbf{P}^2$ the natural projection, then $\mathcal{F} = \pi^* \Omega^1_{\mathbf{P}^2}$ does not extend to a locally free $\mathcal{O}_Y$-module [Se, 5.(a)].
We now revisit a result of Gérard and Levelt [GL].

Proposition 4.5 *For any object* $(\mathcal{M}, \nabla)$ *of* $\mathbf{MIC}^\tau(X(\log Z))$, $\mathcal{M}$ *is locally free if and only if it is reflexive.*

Due to the previous remarks (and Lemma 4.2), this is equivalent to

Proposition 4.6 *Let* $(\mathcal{E}, \nabla)$ *be a coherent* $\mathcal{O}_{X \backslash Z}$*-module with integrable connection, and let* $(\tilde{\mathcal{E}}', \tilde{\nabla}') \in \mathbf{MIC}^\tau(X'(\log Z'))$ *be a* τ*-extension of* $(\mathcal{E}, \nabla)$ *on* X' (as in (4.3)). *Then* $(\tilde{\mathcal{E}}, \tilde{\nabla}) := (\epsilon_* \tilde{\mathcal{E}}', \epsilon_* \tilde{\nabla}') \in \mathbf{MIC}^\tau(X(\log Z))$ *is a* τ*-extension of* $(\mathcal{E}, \nabla)$ *on* X.

Proof: Using Lemma 4.2, one sees that $(\tilde{\mathcal{E}}, \tilde{\nabla})$ is equipped with an integrable connection with logarithmic singularity as in (4.3.1), and it is clear that $(\tilde{\mathcal{E}}, \tilde{\nabla})$ is an object of $\mathbf{MIC}^\tau(X(\log Z))$. The problem is to show that the (reflexive coherent) extension $\tilde{\mathcal{E}}$ of the locally free $\mathcal{O}_{X'}$-module $\tilde{\mathcal{E}}'$ is locally free. For any closed point $x \in X$, we denote by $(\hat{\mathcal{O}}_{X,x}, \hat{\mathfrak{m}}_{X,x})$ the completion of the local ring $(\mathcal{O}_{X,x}, \mathfrak{m}_{X,x})$. It is enough to prove that, for any closed point $z \in Z$, $\widehat{\tilde{\mathcal{E}}}_z := \tilde{\mathcal{E}}_z \otimes \hat{\mathcal{O}}_{X,z}$ is a free $\hat{\mathcal{O}}_{X,z}$-module. We recall that $\hat{\mathcal{O}}_{X,z} \cong K[[x_1, \ldots, x_d]]$, and that we may assume that in this isomorphism, $x_1, \ldots, x_d$ are étale coordinates in X at z, and $x_1 \ldots x_s = 0$, $s \geq 2$, is a local equation at z of the divisor Z in X. We now apply the theory of [GL, §1]: the $\hat{\mathcal{O}}_{X,z}$-module $\widehat{\tilde{\mathcal{E}}}_z$ is torsion free, and $\nabla(x_j \frac{\partial}{\partial x_j})$ are commuting K-derivations of $\widehat{\tilde{\mathcal{E}}}_z$. Let ${}^s\nabla_j$ be the semisimple part of the action of $\nabla(x_j \frac{\partial}{\partial x_j})$ on the filtered K-vector space $(\widehat{\tilde{\mathcal{E}}}_z, (\hat{\mathfrak{m}}^n_{X,z} \widehat{\tilde{\mathcal{E}}}_z)_n)$, in the sense of *loc. cit.*; these operators determine a new integrable connection ${}^s\nabla$ on $\widehat{\tilde{\mathcal{E}}}_z$, with logarithmic poles along $x_1 \ldots x_s = 0$, by ${}^s\nabla(x_j \frac{\partial}{\partial x_j}) = {}^s\nabla_j$; in particular, the operators ${}^s\nabla_j$ mutually commute. There is a finite-dimensional K-vector space $V \subset \widehat{\tilde{\mathcal{E}}}_z$ stable under the ${}^s\nabla_j$, and such that $V \oplus \hat{\mathfrak{m}}_{X,z} \widehat{\tilde{\mathcal{E}}}_z = \widehat{\tilde{\mathcal{E}}}_z$. We have $\nu := \dim V \geq \mu :=$ generic rank of $\mathcal{E}$, and the problem is to prove $\nu = \mu$.

Let $\underline{e} = (e_1, \ldots, e_\nu)$ be a common basis of eigenvectors: ${}^s\nabla_j e_k = \alpha_{jk} e_k$. We notice that α_{jk} is a root of P_j whenever the image of e_k in $\tilde{\mathcal{E}}_z \otimes \hat{\mathcal{O}}_{Z_j,z}$ does not vanish.

Lemma 4.6.1 *The following conditions are equivalent:*

i) *for every* (j, k), α_{jk} *is a root of* P_j,
ii) *every* α_{jk} *lies in the image of* τ,
iii) $\widehat{\tilde{\mathcal{E}}}_z$ *is free.*

It is clear that i) implies ii), and the last remark shows that iii) implies i), since $\underline{e}$ is a basis of $\widehat{\tilde{\mathcal{E}}}_z$ if $\widehat{\tilde{\mathcal{E}}}_z$ is free. The implication ii) $\Rightarrow$ iii) is due to Gérard-Levelt

[GL, Theorem 3.4.i]. For convenience, let us sketch the argument. If $\nu \neq \mu$, we can write $e_{r+1} = \lambda_1 e_1 + \cdots + \lambda_r e_r$, with $\lambda_i \in \operatorname{Frac} \hat{\mathcal{O}}_{X,z}$ (not all in K) and r minimal. Applying ${}^s\nabla_j$, we get $\alpha_{j,r+1} e_{r+1} = \sum_{k \leq r} (x_j \frac{\partial}{\partial x_j} \lambda_k + \alpha_{jk} \lambda_k) e_k$, and on combining these equalities:

$$x_j \frac{\partial}{\partial x_j} \lambda_1 + (\alpha_{j1} - \alpha_{j,r+1}) \lambda_1 = \cdots = x_j \frac{\partial}{\partial x_j} \lambda_r + (\alpha_{jr} - \alpha_{j,r+1}) \lambda_r = 0.$$

Let us remark that for any $\lambda \in \operatorname{Frac} \hat{\mathcal{O}}_{X,z}$, $\lambda \neq 0$, $\beta_j \in K$, the simultaneous equations $x_j \frac{\partial}{\partial x_j} \lambda = \beta_j \lambda$ imply that all β_j are integers. We conclude that if some $\lambda_k \neq 0$, then $\alpha_{jk} = \alpha_{j,r+1}$ for all j, so that $\lambda_k \in K$. This shows $\lambda_k \in K$ for *all* k, a contradiction.

Let us show that in our situation, condition i) of the lemma is satisfied. We fix (j, k). If e_k does not vanish in $\tilde{\mathcal{E}}_z \otimes \hat{\mathcal{O}}_{Z_j,z}$, we are done. Otherwise, there exists $i \neq j$ such that e_k does not vanish in $\tilde{\mathcal{E}}_z \otimes \hat{\mathcal{O}}_{Z_i,z}$ (by construction of V). In order to show that α_{jk} is a root of P_j, we may localize $\hat{\mathcal{O}}_{X,z}$ at the prime (x_i, x_j) and thus reduce to the case of the completed local ring $\hat{\mathcal{O}}_{X,\eta_{i,j}}$ of the generic point $\eta_{i,j}$ of $Z_i \cap Z_j$. This is a complete regular local ring of dimension 2. It is known that any reflexive module of finite type over such a ring is free. In particular, $\tilde{\mathcal{E}}_z \otimes \hat{\mathcal{O}}_{X,\eta_{i,j}}$ is free. As in the implication iii) $\Rightarrow$ i) of the lemma, we conclude that α_{jk} is a root of P_j. This achieves the proof of (4.5) and (4.6).

Proposition 4.7 *Let $(\mathcal{M}_n, \nabla_n)$, for $n = 1, 2$, be objects of* $\mathbf{MIC}^\tau(X(\log Z))$, *with $\mathcal{M}_n$ locally free, and let $\varphi : \mathcal{M}_{1|X\setminus Z} \longrightarrow \mathcal{M}_{2|X\setminus Z}$ be a horizontal isomorphism on $X \setminus Z$. Then φ extends (uniquely) to an isomorphism $\overline{\varphi} : \mathcal{M}_1 \longrightarrow \mathcal{M}_2$ in* $\mathbf{MIC}^\tau(X(\log Z))$.

Proof: Let z be a closed point of Z. Upon replacing X by an affine open neighborhood of z, we may assume $X = \operatorname{Spec} A$ affine and étale over $\mathbf{A}^d_K$ with coordinates $(x_1, \ldots, x_d)$, $Z = V(x_1 \cdots x_s)$, and for $n = 1, 2$, $\mathcal{M}_n = M_n \otimes \mathcal{O}_X$, with M_n a free A-module of rank μ, $M_1 = \oplus_{j=1}^{\mu} A e_j$, $M_2 = \oplus_{j=1}^{\mu} A e'_j$. We have

$$\text{(4.7.1)} \qquad \begin{aligned} \nabla_1 \left(x_i \frac{\partial}{\partial x_i} \right) \underline{e} &= \underline{e} G^{(i)}, \\ \nabla_2 \left(x_i \frac{\partial}{\partial x_i} \right) \underline{e}' &= \underline{e}' H^{(i)} \end{aligned}$$

for $i = 1, \ldots, d$, where $G^{(i)}, H^{(i)} \in M_\mu(A)$, and there exists $B \in GL(\mu, A[\frac{1}{x_1 \cdots x_s}])$, namely the matrix of φ w.r.t. the bases $\underline{e}$, $\underline{e}'$, such that

$$\text{(4.7.2)} \qquad x_i \frac{\partial B}{\partial x_i} = B G^{(i)} - H^{(i)} B$$

for $i = 1, \ldots, d$. We prove that $B \in GL(\mu, A)$. It is sufficient to prove that for $i = 1, \ldots, s$, $B \in M_\mu(A_{(x_i)})$. Let $C_i = A_{(x_i)} / x_i A_{(x_i)}$ be the residue field of the DVR

$A_{(x_i)}$ and assume $N_i \geq 0$ is the order of the pole of B at $Z_i = V(x_i)$. Let $B_{-N_i} \neq 0$, $G_0^{(i)}$, $H_0^{(i)} \in M_\mu(C_i)$ be the reductions of $x_i^{N_i} B$, $G^{(i)}$, $H^{(i)}$ modulo $x_i A_{(x_i)}$. From

$$x_i \frac{\partial (x_i^{N_i} B)}{\partial x_i} = x_i^{N_i} B(G^{(i)} + N_i) - x_i^{N_i} H^{(i)} B \tag{4.7.3}$$

we deduce in $M_\mu(C_i)$

$$N_i B_{-N_i} = H_0^{(i)} B_{-N_i} - B_{-N_i} G_0^{(i)} \tag{4.7.4}$$

for all $i = 1, \ldots, s$. But $\det(tI_\mu - G_0^{(i)})$, $\det(tI_\mu - H_0^{(i)})$ are the characteristic polynomials of $\mathrm{Res}_{Z_i} \nabla$, $\mathrm{Res}_{Z_i} \nabla'$, respectively, hence their roots are in the image of τ. The linear transformation

$$X \longmapsto H_0^{(i)} X - X G_0^{(i)}$$

of $M_\mu(C_i)$ into itself cannot have integral non-zero characteristic roots. Hence $N_i > 0$ is impossible. □

Proposition 4.8 *Let (X, Z) be a model of a divisorially valued function field $(\mathcal{F}, v)$, and let $\mathcal{E}$ be a coherent $\mathcal{O}_{X\setminus Z}$-module with integrable connection ∇, regular along Z. Then there exists an affine neighborhood U of η_Z such that the restriction of $(\mathcal{E}, \nabla)$ to $U \setminus (Z \cap U)$ admits a τ-extension on U.*

Proof: We may restrict X around the generic point of Z, hence assume that $X = \mathrm{Spec} A$ is étale over $\mathbf{A}_K^d$ with coordinates $(x_1, \ldots, x_d)$, $Z = V(x_1)$ irreducible, and that $\mathcal{E}_{|X\setminus Z} \cong \Gamma(X \setminus Z, \mathcal{E}) \otimes \mathcal{O}_{X\setminus Z}$, with $\Gamma(X \setminus Z, \mathcal{E}) = \oplus_{j=1}^{\mu} A[\frac{1}{x_1}] e_j$ a free $A[\frac{1}{x_1}]$-module of rank μ. The valuation ring R_v of $(\mathcal{F}, v)$ is the localized ring $A_{(x_1)}$. The completion $(\hat{\mathcal{F}}, \hat{v})$ of $(\mathcal{F}, v)$ is isomorphic to $\kappa(Z)((x_1))$ equipped with the x_1-adic valuation, and $R_{\hat{v}}$ corresponds to $\kappa(Z)[[x_1]]$.

Let us consider the $\hat{\mathcal{F}}/K$-differential module $\hat{E} = \Gamma(X \setminus Z, \mathcal{E}) \otimes \hat{\mathcal{F}}$. According to Proposition 3.3.3, the assumption of regularity implies the existence of an $\hat{\mathcal{F}}$-basis $\underline{\hat{e}}$ of $\hat{E}$ such that

$$\hat{\nabla}\left(x_1 \frac{\partial}{\partial x_1}\right) \underline{\hat{e}} = \underline{\hat{e}} T, \quad \hat{\nabla}\left(\frac{\partial}{\partial x_i}\right) \underline{\hat{e}} = \underline{\hat{e}} H_i, \quad \textit{for } i = 2, \ldots, d, \tag{4.8.1}$$

with $T \in M_\mu(K)$ the eigenvalues of T being in the image of τ, and $H_i \in M_\mu(\kappa(Z))$. Let us write $\underline{\hat{e}} = \underline{e} P$, with $P \in GL(\mu, \hat{\mathcal{F}})$. Let $Q \in GL(\mu, \mathcal{F})$ be a matrix sufficiently close $\hat{v}$-adically to P so that

$$\begin{aligned} &\hat{v}(P - Q) > -\hat{v}(P^{-1}), \\ &\hat{v}(I_\mu - P^{-1} Q) > -\hat{v}(P) - \hat{v}(P^{-1}) \, (\geq 0), \\ &\hat{v}(I_\mu - Q^{-1} P) > \max(0, \hat{v}(P)). \end{aligned} \tag{4.8.2}$$

We then set $\underline{\tilde{e}} = \underline{e} Q = \underline{\hat{e}} P^{-1} Q$, and calculate

$$\begin{aligned}
\nabla\left(x_1\frac{\partial}{\partial x_1}\right)\underline{\tilde{e}} &= \underline{\hat{e}}\left(TP^{-1}Q - P^{-1}\frac{x_1\partial P}{\partial x_1}P^{-1}Q + P^{-1}\frac{x_1\partial Q}{\partial x_1}\right)\\
&= \underline{\tilde{e}}\left(Q^{-1}PTP^{-1}Q - Q^{-1}\frac{x_1\partial P}{\partial x_1}P^{-1}Q + Q^{-1}\frac{x_1\partial Q}{\partial x_1}\right)\\
&= \underline{\tilde{e}}\left(Q^{-1}PTP^{-1}Q - Q^{-1}\frac{x_1\partial P}{\partial x_1}(P^{-1}Q - I_\mu) - Q^{-1}\frac{x_1\partial(P-Q)}{\partial x_1}\right)\\
&= \underline{\tilde{e}}\Big(T + (Q^{-1}P - I_\mu)T + T(P^{-1}Q - I_\mu)\\
&\quad + (Q^{-1}P - I_\mu)T(P^{-1}Q - I_\mu) + P^{-1}\frac{x_1\partial P}{\partial x_1}(P^{-1}Q - I_\mu)\\
&\quad + (P^{-1} - Q^{-1})\frac{x_1\partial(P^{-1}Q - I_\mu)}{\partial x_1} - P^{-1}\frac{x_1\partial(P-Q)}{\partial x_1}\\
&\quad + (P^{-1} - Q^{-1})\frac{x_1\partial(P-Q)}{\partial x_1}\Big).
\end{aligned}$$

By (4.8.2), we see that

$$\nabla\left(x_1\frac{\partial}{\partial x_1}\right)\underline{\tilde{e}} \equiv \underline{\tilde{e}}T \text{ modulo } x_1\sum_{i=1}^{\mu} R_{\hat{v}}\underline{\tilde{e}}_i.$$

Therefore,

$$\nabla\left(x_1\frac{\partial}{\partial x_1}\right)\underline{\tilde{e}} = \underline{\tilde{e}}\tilde{G}^{(1)}, \tag{4.8.3}$$

with $\tilde{G}^{(1)} \in M_\mu(R_v)$, and the eigenvalues of the image of the matrix $\tilde{G}^{(1)}$ in $M_\mu(\kappa(Z))$ modulo x_1R_v belong to the image of τ.[(1)] A similar calculation for $i = 2, \ldots, d$, shows that

$$\nabla\left(\frac{\partial}{\partial x_i}\right)\underline{\tilde{e}} = \underline{\tilde{e}}\tilde{G}^{(i)}, \tag{4.8.4}$$

[(1)] As for (4.8.2), we may be a little more explicit by noticing that if $Q \in M_\mu(\mathcal{F})$ is such that

$$\hat{v}(P-Q) = h - (\mu-1)\hat{v}(P) + \max(\hat{v}(\det P), -\mu\hat{v}(P)),$$

with $h > 0$, then:

(1) $\hat{v}(Q) = \hat{v}(P)$;
(2) $\hat{v}(\det Q - \det P) \geq h + \max(\hat{v}(\det P), -\mu\hat{v}(P))$;
(3) $\hat{v}(\det Q) = \hat{v}(\det P)$, hence $Q \in GL(\mu, \mathcal{F})$;
(4) $\hat{v}(Q^{-1}) = \hat{v}(P^{-1})$;
(5) $\hat{v}(Q^{-1} - P^{-1}) \geq h - \hat{v}(P) \geq h + \hat{v}(P^{-1})$;
(6) $\hat{v}(I_\mu - Q^{-1}P)$, $\hat{v}(I_\mu - P^{-1}Q) \geq h$.

with $\tilde{G}^{(i)} \in M_\mu(R_v)$. Since $\underline{\tilde{e}} = \underline{e}Q$, with $Q \in GL(\mu, \mathcal{F})$, $\underline{\tilde{e}}$ represents a basis of sections of $\Gamma(V, \mathcal{E})$, on a dense open subset V of $X \setminus Z$ of the form $V = X \setminus (Z \cup Y)$, where $Y = \bigcup_\alpha Y_\alpha$ is a finite union of irreducible *divisors* Y_α of X, such that $Y_\alpha \not\subset Z, \forall \alpha$. We may also assume that V is so small that the matrices $\tilde{G}^{(i)}$ are holomorphic on V. We claim that the proposition holds with $U = X \setminus Y$. Indeed, the matrices $G^{(i)}$ are in $M_\mu(A)$, and the eigenvalues of the image of $G^{(1)}$ modulo $x_1 A$ in $M_\mu(\kappa(Z))$ belong to the image of τ. We then extend trivially $\mathcal{E}$ to X, by putting $\tilde{\mathcal{E}} = \sum_{i=1}^{\mu} \mathcal{O}_X j_* e_i$. The connection ∇ then obviously extends to a connection $\tilde{\nabla}$ on $\tilde{\mathcal{E}}$ with logarithmic singularities along Z. The residue of $\tilde{\nabla}$ along Z is represented in the basis $(i_Z^*(j_* e_1), \ldots, i_Z^*(j_* e_\mu))$ by the matrix $G^{(1)}$ modulo $x_1 A$, hence $(\tilde{\mathcal{E}}, \tilde{\nabla})$ is an object of $\mathbf{MIC}^\tau(X(\log Z))$. □

Theorem 4.9 *In the general situation* (4.1), *any coherent $\mathcal{O}_{X \setminus Z}$-module with integrable connection, regular along Z, admits a τ-extension on X, unique up to unique isomorphism.*

Proof: Let $(\mathcal{E}, \nabla)$ be a coherent $\mathcal{O}_{X \setminus Z}$-module with integrable connection, regular along Z. We already proved that a τ-extension $(\tilde{\mathcal{E}}, \tilde{\nabla})$ of a coherent $\mathcal{O}_{X \setminus Z}$-module $(\mathcal{E}, \nabla)$ with integrable connection is unique if it exists. By (4.8), there exists an affine neighborhood U_i of η_{Z_i} such that the restriction of $(\mathcal{E}, \nabla)$ to $U_i \setminus (Z \cap U_i)$ admits a τ-extension $(\mathcal{E}, \nabla)$ on U_i. But $(\mathcal{E}, \nabla)$ itself and these τ-extensions glue together, so we get a τ-extension of $(\mathcal{E}, \nabla)$ on $X' = (X \setminus Z) \cup \bigcup U_i$. We apply at last Proposition 4.6 and get a τ-extension of $(\mathcal{E}, \nabla)$ on X. □

Remark 4.10 Since τ is not additive, the formation of the τ-extension is not compatible with $\otimes$, $\underline{Hom}$, and is not functorial in general. However:

Proposition 4.11 *Let* $(X, Z) \xrightarrow{f} (Y, W)$ *be an* unramified *morphism of models. Let* $(\mathcal{E}, \nabla)$ *be a coherent sheaf with integrable connection on* $Y \setminus W$, *regular along* W, *and let* $(\tilde{\mathcal{E}}, \tilde{\nabla})$ *be its τ-extension on* Y. *Then* $f^*(\tilde{\mathcal{E}}, \tilde{\nabla})$ *is the τ-extension of* $f^*(\mathcal{E}, \nabla)$ *on* X.

Proof: $f^*(\tilde{\mathcal{E}}, \tilde{\nabla})$ is an extension with logarithmic poles along Z and the problem is to show that the eigenvalues of $Res_Z f^*(\nabla)$ are in the image of τ. It is enough to show that the characteristic polynomials P_W and P_Z of $Res_W \nabla$ and $Res_Z f^*(\nabla)$ respectively coincide in $K[t]$. This is clear if f is a closed immersion of models, since P_Z is a specialization of P_W in this case. In general, in the canonical graph factorization of f

$$(X, Z) \xrightarrow{i = (1, f)} (X \times Y, X \times W) \xrightarrow{p} (Y, W),$$

i is unramified, hence a closed immersion of models in some neighborhood of $i(\eta_Z)$ (1.2.5). We are reduced to the trivial case where $f = p$ is a projection. □

§5 Regular connections: the global case

5.1 Let X be a smooth connected variety over K, $\mathcal{F} = \kappa(X)$. Let E be an $\mathcal{F}/K$-differential module, and let $(\mathcal{E}, \nabla)$ be a model of E on X. Then $(\mathcal{E}, \nabla)$ or (E, ∇) (or simply ∇) is said to be *regular* or *fuchsian* if (E, ∇) is regular at every divisorial valuation v of $\mathcal{F}$. For instance, the trivial connection $(\mathcal{O}_X, d_{X/K})$ is regular. Since any such v admits a smooth K-model (X, Z), this notion of (global) regularity reduces to the notion of regularity along a divisor (previous section), by using various models. One should however keep in mind the birational character of this notion.

Proposition 5.2 *Let $(\mathcal{E}, \nabla)$, $(\mathcal{E}_1, \nabla_1)$, $(\mathcal{E}_2, \nabla_2)$ be coherent sheaves with integrable connections on X.*

i) *If $\mathcal{E}$ sits in a horizontal exact sequence $0 \longrightarrow \mathcal{E}_1 \longrightarrow \mathcal{E} \longrightarrow \mathcal{E}_2 \longrightarrow 0$, then ∇ is regular if and only if ∇_1 and ∇_2 are regular.*
ii) *If ∇_1 and ∇_2 are regular, so are $(\mathcal{E}_1, \nabla_1)\otimes(\mathcal{E}_2, \nabla_2)$ and $\underline{Hom}((\mathcal{E}_1, \nabla_1), (\mathcal{E}_2, \nabla_2))$.*

Proof: This follows from (3.2.3) (or (3.4.3)). □

Proposition 5.3 *The property of being regular is invariant under inverse image by dominant morphisms and under direct image by finite étale morphisms. In particular, for an object $(\mathcal{E}, \nabla)$ of* **MIC**(X)*, the property of being regular is local at η_X for the* fppf *(hence for the étale) topology on X.*

Proof: This follows from (3.1.4) and (3.2.5) (for the case of a dominant morphism f, notice that f induces an embedding of function fields $\kappa(Y) \hookrightarrow \kappa(X)$, and that any divisorial valuation of $\kappa(Y)$ extends to a divisorial valuation of $\kappa(X)$). □

Proposition 5.4 *Let X' be a normal K-variety, and let $(\mathcal{E}, \nabla)$ be a coherent sheaf with integrable connection on a smooth open dense subset $X \subset X'$. Let P be a closed point of $\partial X = X' \backslash X$. Assume that ∇ is regular at every component of ∂X of codimension 1 in X' passing through P. If $\overline{C} \xrightarrow{h} X'$ is any morphism from a smooth curve $\overline{C}$, such that $h(\overline{C}) \not\subset \partial X$, and Q is a point of $\overline{C}$ such that $h(Q) = P$, then, for $C = h^{-1}(X)$, $\nabla_{|C}$ is regular at Q.*

Proof: We may assume that $\overline{C}$ is the disjoint union of C and $\{Q\}$. We may also replace X' by any affine neighborhood of P, and assume that $\partial X \cap h(\overline{C}) = \{P\}$. Let $\{x_1, x_2 \dots\}$ be a finite set of generators of the ideal of ∂X in $\mathcal{O}(X')$ such that $x_i \neq 0$ on C. We may and shall replace ∂X by the reduced divisor defined by the ideal $\sqrt{(\Pi x_i)}$. There is a closed subset $T \subset \partial X$ such that $\tilde{Z} = \partial X \backslash T$ is a disjoint union of smooth divisors Z_i in $\tilde{X} = X' \backslash T$ defined by the equation $x_i = 0$. We denote by j' (resp. $\tilde{j}$, resp. k) the open immersion of $\tilde{X}$ into X' (resp. X into $\tilde{X}$, resp. C into $\overline{C}$), and set $j = j' \circ \tilde{j}$. If $P \notin T$, the result follows from (3.4.4) (which relies upon (3.4.2)). In order to settle the case $P \in T$, we use the existence of an extension $(\tilde{\mathcal{E}} \subset \tilde{j}_*\mathcal{E}, \tilde{\nabla})$ of $(\mathcal{E}, \nabla)$ on $\tilde{X}$ with logarithmic poles along $\tilde{Z}$ (4.9). We need the

Lemma 5.5 *The functoriality map* $h^*(j_*\Omega^1_{X/K}) \longrightarrow k_*\Omega^1_{C/K}$ *induces a map*

$$h^* j'_* \Omega^1_{\tilde{X}/K}(\log \tilde{Z}) \longrightarrow \Omega^1_{\overline{C}/K}(\log Q). \tag{5.5.1}$$

Indeed, $j'_*\Omega^1_{\tilde{X}/K}(\log \tilde{Z})$ is generated by $\Omega^1_{X'/K}$ and by the sections $\frac{dx_i}{x_i}$ (taking into account the fact that $j'_*\mathcal{O}_{\tilde{X}} = \mathcal{O}_{X'}$), and $\Omega^1_{\overline{C}/K}(\log Q)$ contains $h^*(\frac{dx_i}{x_i}) = \frac{d(x_i \circ h)}{x_i \circ h}$ ((2.2.1)).

Let us turn back to the proof of (5.4). Due to the functoriality (5.5), the mapping $\tilde{\nabla}$ gives rise to a composite mapping

$$\begin{aligned} h^* j'_* \tilde{\nabla} : h^* j'_* \tilde{\mathcal{E}} &\longrightarrow h^* j'_* \tilde{\mathcal{E}} \otimes_{\mathcal{O}_{\overline{C}}} h^* j'_* \Omega^1_{\tilde{X}/K}(\log \tilde{Z}) \\ &\longrightarrow h^* j'_* \tilde{\mathcal{E}} \otimes_{\mathcal{O}_{\overline{C}}} \Omega^1_{\overline{C}/K}(\log Q) \end{aligned}$$

which is an extension of $\nabla_{|C}$ with logarithmic poles at Q. Hence $\nabla_{|C}$ is regular at Q. □

Remark 5.6 The use of extensions with logarithmic poles is not essential in the proof of (5.4). We sketch a variant, which uses instead jets with logarithmic poles (*cf.* Appendix B.6). The sheaf of $\mathcal{O}_{\tilde{X}}$-algebras $\mathcal{P}^n_{\tilde{X}/K,\tilde{Z}}$ is generated by the $\frac{j_n x_i}{x_i}$ (*cf.* (B.6.2)); as a left $\mathcal{O}_{\tilde{X}}$-module, it is the dual of $\underline{Diff}^n_{\tilde{X}/K,\tilde{Z}}$. The argument of (5.5) gives a map of functoriality

$$h^* j'_* \mathcal{P}^n_{\tilde{X}/K,\tilde{Z}} \longrightarrow \mathcal{P}^n_{\overline{C}/K,Q}. \tag{5.6.1}$$

We can then imitate (3.4.5): let $\tilde{\mathcal{E}} \subset k_*\mathcal{E}$ be a coherent extension of $\mathcal{E}$ on $\tilde{X}$ as in (3.4.2). Then for all n, the mapping $\tilde{\nabla}$ gives rise to a composite mapping

$$\begin{aligned} h^* j'_* \tilde{\nabla} : h^* j'_* \tilde{\mathcal{E}} &\longrightarrow (\Pi x_i)^{-c} h^* j'_* \tilde{\mathcal{E}} \otimes_{\mathcal{O}_{\overline{C}}} h^* j'_* \mathcal{P}^n_{\tilde{X}/K,\tilde{Z}} \\ &\longrightarrow x^{-c'} h^* j'_* \tilde{\mathcal{E}} \otimes_{\mathcal{O}_{\overline{C}}} \mathcal{P}^n_{\overline{C}/K,Q} \end{aligned}$$

for some c' independent of n; hence $\nabla_{|C}$ is regular at Q. (Note however that since $j'_*\Omega^1_{\tilde{X}/K}(\log \tilde{Z})$ is not flat in general, it is not clear from (5.5.1) whether $\underline{Der}_{\overline{C}/K} \longrightarrow h^*\underline{Der}_{X'/K}$ induces maps of $\mathcal{O}_{X'}$-modules

$$\mathcal{T}_h : \underline{Der}_{\overline{C}/K,Q} \longrightarrow h^* j'_*(\underline{Der}_{\tilde{X}/K,\tilde{Z}}) \text{ and } \mathcal{T}_h : \underline{Diff}^n_{\overline{C}/K,Q} \longrightarrow h^* j'_*(\underline{Diff}^n_{\tilde{X}/K,\tilde{Z}}).)$$

We shall give still another proof of (5.4) in (6.5.6).

Theorem 5.7 *Let* $(\mathcal{E}, \nabla)$ *be a coherent sheaf with integrable connection on a smooth connected* K*-variety* X. *The following properties are equivalent:*

i) ∇ *is regular;*

ii) *for any morphism* $C \xrightarrow{h} X$, *with* C *a smooth* K*-curve,* $h^*(\nabla)$ *is regular;*

iii) *for any locally closed immersion of a smooth* K*-curve* $C \overset{i}{\hookrightarrow} X$, $i^*(\nabla)$ *is regular.*

Proof: i) $\Rightarrow$ ii): we may extend X to a proper normal variety $\overline{X}$ and i to a morphism $\overline{C} \xrightarrow{h} \overline{X}$, where $\overline{C}$ is a smooth projective model of C. If the image of h is contained in X, the result is trivial ($h^*(\nabla)$ has no singular points). Otherwise, we conclude by (5.4).

ii) $\Rightarrow$ iii) is obvious.

iii) $\Rightarrow$ i) : let v be a divisorial valuation of $\kappa(X)$, and let (X', Z') be a model of $(\kappa(X), v)$. We may shrink X' and X, and assume that $X = X' \setminus Z'$. Then iii) implies that ∇ is regular at Z', according to (3.4.7). □

Corollary 5.8 *Let* $f : X \longrightarrow Y$ *be a morphism of smooth varieties, and* $(\mathcal{E}, \nabla)$ *be a coherent sheaf with regular connection on* Y. *Then* $f^*(\nabla)$ *is regular. The converse is true if* f *is dominant.*

Proof: Immediate from (5.7). □

Remark 5.9 The proofs of (5.7) and (5.8) found in the literature are usually much more complicated and use Hironaka's resolution of singularities. The proof in [De] is incomplete, since it uses an assertion about non-existence of confluences which is false as it stands [*loc. cit.* 4.1.2, 1.24]; this point is corrected in the erratum (April 1971). The treatment in [Bn] of the problem of inverse images relies on the test of curves which is stated in a wrong way: in [*loc. cit.* 4 (iii)], it is asserted that regularity at a divisor Z may be checked by considering *one* curve which cuts Z transversally (*cf.* counterexample (3.4.8)).

We now show that the birational notion of global regularity actually involves only *finitely many* divisorial valuations of the function field.

Theorem 5.10 *Let* X *be a smooth connected* K*-variety, and let* $j : X \longrightarrow \overline{X}$ *be an open dominant immersion of* X *in a proper normal* K*-variety* $\overline{X}$; *we denote by* $\overline{X}^{sm}$ *the smooth part of* $\overline{X}$. *Let* $(\mathcal{E}, \nabla)$ *be a coherent sheaf with integrable connection on* X. *The following properties are equivalent:*

i) ∇ *is regular;*

ii) ∇ *is regular at every irreducible component* $\overline{Z}$ *of* $\overline{X} \setminus j(X)$ *of codimension* 1 *(*see (3.4.9)*);*

iii) *for every irreducible component* $\overline{Z}$ *of* $\overline{X} \setminus j(X)$ *of codimension* 1, *there is an open neighborhood* U *of* η_Z *in* $\overline{X}^{sm}$ *and a tubular neighborhood* $U \xrightarrow{f} S$ *of* $Z \cap U$, *such that* $\nabla_{|U_s}$ *is regular at* Z_s *for a Zariski-dense set of closed points* $s \in S$.

Proof: The implication i) $\Rightarrow$ iii) follows from (5.8), while iii) $\Rightarrow$ ii) follows from (3.4.7) (since $\overline{X}$ is normal, the smooth part $\overline{Z}^{sm}$ of Z is open dense). Let us prove ii) $\Rightarrow$ (∇ is regular at every divisorial valuation v of $\kappa(\overline{X})$).

Since $\overline{X}$ is proper, v is centered at some point $x \in \overline{X}$ of codimension ≥ 1. More precisely, there is a model (X', Z') of $(\kappa(\overline{X}), v)$ and a birational map $X' \xrightarrow{\epsilon} \overline{X}$ such that $\epsilon(\eta_{Z'}) = x$. We have to check the regularity of $\epsilon^*\nabla$ on the inverse image of X, along Z'. We may assume $x \in \overline{X}\backslash X$ and $\epsilon^{-1}(\overline{X} \setminus X) \subset Z'$. Let P be any closed point in $\epsilon(Z') \subset \overline{\{x\}}_{\overline{X}}$. We see via (3.4.7) that it is enough to show that for any morphism $\overline{C} \xrightarrow{h} \overline{X}$ of a smooth connected curve $\overline{C}$ to $\overline{X}$ such that $h(\overline{C}) \not\subset \overline{X}\backslash X$ and for any closed point $Q \in \overline{C}$ with $h(Q) = P$, $\nabla_{|C}$, for $C = h^{-1}(X)$ is regular at Q. Since ∇ is regular at every component $\overline{Z}$ of $\overline{X}\backslash X$ containing P and of codimension 1 in $\overline{X}$, this holds by (5.4). □

§6 Exponents

6.1 Indicial polynomials

Definition 6.1.1 Let C be a field of characteristic 0, and let $\Lambda \in C((x))[\frac{xd}{dx}]$ be a scalar differential operator. The *indicial polynomial* of Λ at $x = 0$ is the unique $\phi = \phi_{\Lambda,0} \in C[s]$ such that for every $s \in \mathbf{Z}$,

$$\Lambda x^s = \phi(s)x^{s+r} + \text{higher order terms in } x.$$

The roots (in $\overline{C}$) of $\phi_{\Lambda,0}$ are called the (*Fuchs*) *exponents* of Λ at $x = 0$.

For $\Lambda \in C(x)[\frac{d}{dx}]$, and $\theta \in C$ (resp. $\theta = \infty$), one defines in the same way the indicial polynomial and the exponents $\phi_{\Lambda,\theta}$ of Λ at $x = \theta$.

Let us write Λ, as in the previous definition, in the form $\sum_{i=0}^{\mu} \gamma_i (x\frac{d}{dx})^i$, $\gamma_i \in C((x))$, and set

$$(6.1.1.1) \qquad \begin{aligned} r &= \min_i \operatorname{ord}_0 \gamma_i, \\ \nu &= \max\{i \text{ such that } \operatorname{ord}_0 \gamma_i = r\}. \end{aligned}$$

Clearly, $\phi(s) = \sum_{i=0}^{\nu} (x^{-r}\gamma_i)_{|x=0} s^i$.

6.1.2 We now assume that C is a function field over an algebraically closed field K.

Proposition 6.1.3 *Let Λ be as before, and assume $C((x))[\frac{d}{dx}]/C((x))[\frac{d}{dx}]\Lambda$ underlies a $C((x))/K$-differential module. Then the roots of ϕ are in K.*

This is a consequence of Robba's result reproduced in Appendix A; however, we present here a proof which avoids Turrittin's theorem.

Proof: i) Let us write $\Lambda_f = \sum_{i=0}^{\nu} (\frac{\gamma_i}{\gamma_\nu})(x\frac{d}{dx})^i$. Then Λ_f is regular, and $\Lambda = \gamma_\nu(\Lambda_f + x\Lambda')$, where $\Lambda' \in C[[x]][x\frac{d}{dx}]$. Clearly, $\phi_{\Lambda,0}(s) = (x^{-r}\gamma_\nu)(0) \cdot \phi_{\Lambda_f,0}(s)$, and $\phi_{\Lambda_f,0}(s)$ is the (usual) indicial polynomial at 0 of the regular differential operator Λ_f. Hence, we are reduced to the regular case.

ii) For a regular Λ, we may change the cyclic basis of $C((x))[\frac{d}{dx}]/C((x))[\frac{d}{dx}]\Lambda$ into a basis in which $\nabla(x\frac{d}{dx})$ is represented by a constant matrix in Jordan normal form (after replacing C by a finite extension if necessary). The roots of ϕ are the diagonal entries of this matrix. Then Robba's argument goes through, and shows that the exponents belong to K. □

Corollary 6.1.4 *Let $\Lambda \in C(x)[\frac{d}{dx}]$ be such that $C(x)[\frac{d}{dx}]/C(x)[\frac{d}{dx}]\Lambda$ underlies a $C(x)/K$-differential module. Then the roots of the indicial polynomial of Λ at any $\theta \in C \cup \infty$, are in K.*

Proof: The indicial polynomial (and the exponents) of Λ at θ, resp. at ∞, may be defined using a change of variable $x \longmapsto x' = x - \theta$ $(\theta \in C)$, resp. $x \longrightarrow x' = \frac{1}{x}$. One can apply (6.1.3) to Λ viewed as an element of $C((x'))[\frac{d}{dx'}]$. □

6.2 Exponents: review of the one variable regular case

Let again C be a field of characteristic 0 and consider a regular $C((x))/C$-differential module $\hat{E}$ as in (3.2.2). We apply the notation of Section 1 to $(\hat{\mathcal{F}}, v) = (C((x)), x$-adic valuation), despite the fact that C is not necessarily algebraically closed.

Definition 6.2.1 The elements of $\mathcal{A}$, considered up to addition by an integer, are called the *exponents* of $\hat{E}$. The subset $\mathcal{A} + \mathbf{Z}$ of $\overline{C}$ is denoted by $Exp(\hat{E})$. Following the last assertion in (3.2.2), $Exp(\hat{E})$ is canonically attached to $\hat{E}$.

The following results are classical (see [Man] or [DGS, III.8]).

Proposition 6.2.2

i) $Exp(\hat{E})$ coincides, modulo $\mathbf{Z}$, with the roots of the indicial polynomial of the differential operator associated with any cyclic vector m as in c) of (3.2.2).
ii) Let $E_{\hat{R}_v}$ be an $\hat{R}_v$-lattice of $\hat{E}$ stable by $Diff_v(\hat{\mathcal{F}}/K)$, as in v) of (3.3.4). Let $(D_1, \ldots, D_d)$ be an adapted basis of $Der_v(\hat{\mathcal{F}}/K)$; $\nabla(D_1)$ induces a K-linear endomorphism $Res_v\nabla$ of the finite dimensional C-vector space $E_{\hat{R}_v}/\hat{\mathfrak{m}}_v E_{\hat{R}_v}$. Then $Exp(\hat{E})$ coincides, modulo $\mathbf{Z}$, with the eigenvalues of $Res_v\nabla$.

Proposition 6.2.3 *Let $\hat{E}_1$ and $\hat{E}_2$ be regular $C((x))/C$-differential modules.*

i) *If $\hat{E}$ sits in an exact sequence of $C((x))/C$-differential modules*

$$0 \longrightarrow \hat{E}_1 \longrightarrow \hat{E} \longrightarrow \hat{E}_2 \longrightarrow 0$$

then $Exp(\hat{E}) = Exp(\hat{E}_1) \cup Exp(\hat{E}_2)$.

ii) *$Exp(\hat{E}_1 \otimes_{C((x))} \hat{E}_2) = Exp(\hat{E}_1) + Exp(\hat{E}_2)$ and $Exp(Hom_{C((x))}(\hat{E}_1, \hat{E}_2)) = -Exp(\hat{E}_1) + Exp(\hat{E}_2)$.*

6.2.4 Let $\mathcal{F}/K$ be a function field in one variable (K not necessarily algebraically closed), and let v be a non-trivial valuation of $\mathcal{F}$ trivial on K. Then $C := k(v)$

is a finite extension of K, and there is an element x of $\mathfrak{m}_v$ such that $(\hat{\mathcal{F}}, \hat{v}) = (C((x)), x\text{−adic valuation})$, $\hat{R}_v = C[[x]]$.

If E is an $\mathcal{F}/C$-differential module, regular at v, we set $Exp_v(E) = Exp(\hat{E})$.

Lemma 6.2.5 *In the situation of* (3.1.3), *where* $\hat{E}$ *is a* $\hat{\mathcal{G}}/K$*-differential module, we have* $Exp(\hat{E}_{\hat{\mathcal{F}}}) = e(v/w)Exp(\hat{E}) + \mathbf{Z}$. *If* $[\hat{\mathcal{F}} : \hat{\mathcal{G}}] < \infty$, *we have* $Exp(\hat{E}') = e(v/w)Exp({}_{\hat{\mathcal{G}}}\hat{E}') + \mathbf{Z}$ *for any* $\hat{\mathcal{F}}/K$*-differential module* $\hat{E}'$.

Proof: Let x be a generator of $\mathfrak{m}_v$ such that $y = x^{e(v/w)}$ is a generator of $\mathfrak{m}_w$, so that $\hat{\mathcal{F}} \cong C((x))$, $\hat{\mathcal{G}} \cong C'((y))$, with $C' = C \cap \hat{\mathcal{G}}$. Using i) of (6.2.3), it suffices to consider the case when $\hat{E} \cong C'((y)) \otimes_{C'} \mathrm{Ker}_{\hat{E}}(y\frac{d}{dy} - \alpha)^{\mu}$. By reason of dimension, we then have $\hat{E}_{\hat{\mathcal{F}}} \cong C((x)) \otimes_C \mathrm{Ker}_{\hat{E}_{\hat{\mathcal{F}}}}(y\frac{d}{dy} - \alpha)^{\mu}$. Since $x\frac{d}{dx} = e(v/w)y\frac{d}{dy}$, this implies the first assertion. To deduce from it (with $\hat{E} = {}_{\hat{\mathcal{G}}}\hat{E}'$) the second one, it suffices to show that $({}_{\hat{\mathcal{G}}}\hat{E}')_{\hat{\mathcal{F}}} \cong (\hat{E}')^{[\hat{\mathcal{F}}:\hat{\mathcal{G}}]}$. As before, we may assume that $\hat{E}' \cong C((x)) \otimes_C \mathrm{Ker}_{\hat{E}'}(x\frac{d}{dx} - \alpha)^{\mu}$, and we are reduced to the case of the trivial $C((x))/C$-differential module $C((x))$, for which it is immediate that $({}_{C'((y))}C((x)))_{C((x))} \cong C((x))^{[C((x)):C'((y))]}$.

Notice that if v is a divisorial valuation of a function field $\mathcal{F}$ over K which induces the trivial valuation on the subfield $\mathcal{G} \subset K$, then for any $\mathcal{G}/K$-differential module E, $Exp_v(E_{\mathcal{F}}) = \mathbf{Z}$. □

6.3 The case of a function field in several variables

Let $(\hat{\mathcal{F}}, v)$ be a field of formal functions over K, and $(D_1, \ldots, D_d)$ an adapted basis of $Der_v(\hat{\mathcal{F}}/K)$. We set $C = \hat{\mathcal{F}}^{D_1}$ so that $\hat{\mathcal{F}} = C((x))$, for a parameter x of $\hat{R}_v$ such that $D_1 x = x$, $D_j C \subset C$ and $D_j x = 0$, for $j = 2, \ldots, d$.

Definition 6.3.1 The *exponents* (*at* v) of a regular $\hat{\mathcal{F}}/K$-differential module $\hat{E}$ are the exponents of the induced $\hat{\mathcal{F}}/C$-differential module (in one variable x) $\hat{E}_{/C}$. We set $Exp_v(\hat{E}) = Exp(\hat{E}_{/C}) = \mathcal{A} + \mathbf{Z}$, where $\mathcal{A}$ is the set of eigenvalues of the matrix G_1 appearing in (3.3.3).

Proposition 6.3.2 $Exp_v(\hat{E}) \subset K$.

This follows from (3.3.3) as well as from (6.1.3), (6.2.2).

6.4 Exponents of a regular connection along a divisor

Let (X, Z) be a model of a divisorially valued function field $(\mathcal{F}, v)$. Let E be an $\mathcal{F}/K$-differential module regular at v, and let $(\mathcal{E}, \nabla)$ be a model of E on $X \backslash Z$ ((3.1)).

Definition 6.4.1 The *exponents* of ∇ *along* Z are the exponents of $\hat{E}$ at v. We set $Exp_Z(\nabla) = Exp_v(\hat{E})$.

Proposition 6.4.2 *Let* (X, Z) *be a model and* $(\mathcal{E}, \nabla)$, $(\mathcal{E}_1, \nabla_1)$, $(\mathcal{E}_2, \nabla_2)$ *be coherent sheaves with integrable regular connections on* $X \backslash Z$.

i) *If $\mathcal{E}$ sits in a horizontal exact sequence* $0 \longrightarrow \mathcal{E}_1 \longrightarrow \mathcal{E} \longrightarrow \mathcal{E}_2 \longrightarrow 0$, *then* $Exp_Z(\nabla) = Exp_Z(\nabla_1) \cup Exp_Z(\nabla_2)$.

ii) $Exp_Z((\mathcal{E}_1, \nabla_1) \otimes (\mathcal{E}_2, \nabla_2)) = Exp_Z(\nabla_1) + Exp_Z(\nabla_2)$, $Exp_Z(\underline{Hom}((\mathcal{E}_1, \nabla_1), (\mathcal{E}_2, \nabla_2))) = -Exp_Z(\nabla_1) + Exp_Z(\nabla_2)$.

Proof: This follows from (6.2.3). □

Proposition 6.4.3 *Let* $(X, Z) \xrightarrow{f} (Y, W)$ *be a morphism of models such that* $f(Z) \subset W$. *Let* $(\mathcal{E}, \nabla)$ *be a coherent sheaf with integrable connection on* $Y \setminus W$, *regular along* W. *Then* $Exp_Z(f^*(\nabla)) = e_f Exp_W(\nabla) + \mathbf{Z}$.

Proof: i) If f is log dominant, the result follows from (6.2.5).

ii) Let us next assume that X is a curve. We may replace X by any connected étale neighborhood of Z. Using the factorization (1.2.7), and the previous case, we are reduced to the case where f is a closed immersion of models. In this case, the result follows from (4.11).

iii) In the general case, we can find a closed immersion $(C, P) \xrightarrow{i} (X, Z)$ with C a curve. Then $e_f = e_{i \circ f}$ and the result follows from ii). □

6.4.5 Example As in (3.4.6), take

$$\begin{aligned} Y &= \text{the } (x, y)\text{-plane with the } x\text{-axis } (y = 0) \text{ removed}, \\ W &= y\text{-axis}(x = 0), \\ X &= \text{the diagonal } (x = y) \text{ in the } (x, y)\text{-plane}, \\ Z &= \text{the origin.} \end{aligned}$$

For $a, b \in K$, let $\nabla_{a,b}$ be the connection on $\mathcal{E} = \mathcal{O}_{Y \setminus W}$, whose formal solution is $x^a y^b$. It is regular at W with $Exp_W(\nabla_{a,b}) = a + \mathbf{Z}$; its restriction to $X \setminus Z$ is also regular at Z, but has $Exp_Z = a + b + \mathbf{Z}$. Again, the point is that the embedding of $X \setminus Z$ into $Y \setminus W$ does not extend to a morphism of models $(X, Z) \longrightarrow (Y, W)$.

6.4.6 The definition of exponents along a divisor generalizes in an obvious way to the situation of (3.4.9): let X be a smooth connected variety over K, $\mathcal{F} = \kappa(X)$. Let $j : X \longrightarrow \overline{X}$ be a dominant open immersion of X in a normal K-variety $\overline{X}$, and let $\overline{Z}$ be a (not necessarily irreducible) subvariety of $\overline{X} \setminus X$. If $(\mathcal{E}, \nabla)$ is a coherent sheaf with integrable connection on X, we denote by $Exp_{\overline{Z}}(\nabla)$ the set of exponents of the generic fiber E at the divisorial valuations of $\mathcal{F}$ defined by the irreducible components of $\overline{Z}$ *of pure codimension one in* $\overline{X}$.

6.5 Global exponents

6.5.1 Let X be a smooth connected variety over K, $\mathcal{F} = \kappa(X)$. Let E be a regular $\mathcal{F}/K$-differential module, and let $(\mathcal{E}, \nabla)$ be a model of E on X. The *exponents* of E or ∇ are the exponents of E at any divisorial valuation of $\mathcal{F}$.

Notation 6.5.2 We denote by $\mathbf{Z}Exp(\nabla)$ (resp. $\mathbf{Q}Exp(\nabla)$) the additive subgroup (resp. sub-$\mathbf{Q}$-vector space) of K *generated* by the exponents of ∇. Notice that $\mathbf{Z}Exp(\nabla)$ contains 1, and that $\mathbf{Q}Exp(\nabla) = \mathbf{Q} \otimes_{\mathbf{Z}} \mathbf{Z}Exp(\nabla)$.

Proposition 6.5.3 *Let $(\mathcal{E}, \nabla)$, $(\mathcal{E}_1, \nabla_1)$, $(\mathcal{E}_2, \nabla_2)$ be coherent sheaves with regular connections on X.*

i) *If $\mathcal{E}$ sits in a horizontal exact sequence* $0 \longrightarrow \mathcal{E}_1 \longrightarrow \mathcal{E} \longrightarrow \mathcal{E}_2 \longrightarrow 0$, *then* $\mathbf{Z}Exp(\nabla) = \mathbf{Z}Exp(\nabla_1) + \mathbf{Z}Exp(\nabla_1)$.
ii) $\mathbf{Z}Exp((\mathcal{E}_1, \nabla_1) \otimes (\mathcal{E}_2, \nabla_2)) \subset \mathbf{Z}Exp(\nabla_1) + \mathbf{Z}Exp(\nabla_2)$, $\mathbf{Z}Exp(\underline{Hom}((\mathcal{E}_1, \nabla_1), (\mathcal{E}_2, \nabla_2))) \subset \mathbf{Z}Exp(\nabla_1) + \mathbf{Z}Exp(\nabla_2)$.

Proof: This follows from (6.4.2). □

Proposition 6.5.4 *$\mathbf{Q}Exp(\nabla)$ is invariant under inverse image by dominant morphisms and under direct image by finite étale morphisms. In particular, its formation is local at η_X for the fppf (hence for the étale) topology on X.*

Proof: This follows from (6.2.5). □

Proposition 6.5.5 *Let X' be a normal K-variety, and let $(\mathcal{E}, \nabla)$ be a coherent sheaf with regular connection on an open dense subset $X \subset X'$. Let P be a closed point of $\partial X = X' \backslash X$. Let $\overline{C} \xrightarrow{h} X'$ be a morphism from a smooth curve $\overline{C}$ such that $h(\overline{C}) \not\subset \partial X$ and assume there is a point $Q \in \overline{C}$ with $h(Q) = P$. Then, for $C = h^{-1}(X)$, $Exp_Q(\nabla_{|C}) \subset \mathbf{Z}Exp(\nabla)$. If moreover X' is smooth and $\partial X = \cup Z_i$ is a divisor with strict normal crossings, then $Exp_Q(\nabla_{|C})$ is contained in the subgroup of K generated by 1 and the sets $Exp_{Z_i}(\nabla)$.*

Proof: We may assume that C is the complement of $\{Q\}$ in $\overline{C}$.

a) Let us first prove the second assertion. In this situation, we have the τ-extension $(\tilde{\mathcal{E}}, \tilde{\nabla})$ on X' at our disposal (4.9). We may assume that $\tilde{\mathcal{E}}$ is free, that each component Z_i is globally described by an equation $x_i = 0$, and that $\Omega^1_{\overline{C}/K}(\log Q)$ is free with basis $\frac{dx}{x}$ for some global coordinate x on $\overline{C}$. The functoriality $h^*\Omega^1_{X'/K}(\log \partial X) \longrightarrow \Omega^1_{\overline{C}/K}(\log Q)$ induces a linear map

$$T_h^* : \frac{h^*\Omega^1_{X'/K}(\log \partial X)}{h^*\Omega^1_{X'/K}} \longrightarrow \frac{\Omega^1_{\overline{C}/K}(\log Q)}{\Omega^1_{\overline{C}/K}}$$

of skyscraper sheaves of K-vector spaces (concentrated at Q). A basis of $\frac{h^*\Omega^1_{X'/K}(\log \partial X)}{h^*\Omega^1_{X'/K}}$ (resp. $\frac{\Omega^1_{\overline{C}/K}(\log Q)}{\Omega^1_{\overline{C}/K}}$) at Q is given by the classes of the $h^*(\frac{dx_i}{x_i})$ (resp. by the class of $\frac{dx}{x}$), and one has $T_h^*(h^*(\frac{dx_i}{x_i})) \equiv e_i \frac{dx}{x}$ mod. $\Omega^1_{\overline{C}/K}$ for some intersection multiplicities $e_i \in \mathbf{Z}_{\geq 0}$ (e_i is the valuation of the image in $\mathcal{O}_{\overline{C},Q}$ of a local equation of Z_i in $\mathcal{O}_{X',P}$).

Now $\tilde{\nabla}$ induces a composite map

$$(h^*\tilde{\mathcal{E}})_Q \otimes \kappa(Q) \overset{u=(\ldots,u_i,\ldots)}{\longrightarrow} (h^*\tilde{\mathcal{E}})_Q \otimes \frac{h^*\Omega^1_{X'/K}(\log \partial X)}{h^*\Omega^1_{X'/K}}$$
$$\overset{v=1\otimes T_h^*}{\longrightarrow} (h^*\tilde{\mathcal{E}})_Q \otimes \frac{\Omega^1_{\overline{C}/K}(\log Q)}{\Omega^1_{\overline{C}/K}} \cong (h^*\tilde{\mathcal{E}})_Q \otimes \kappa(Q).$$

The map u_i is nothing but the specialization of $Res_{Z_i}\nabla$ at P, and $v \circ u$ is nothing but $Res_Q h^*\nabla$. Since the $Res_{Z_i}\nabla$ mutually commute, the endomorphisms $v \circ u_i = e_i u_i$ of the finite dimensional $K = \kappa(Q)$-vector space $(h^*\tilde{\mathcal{E}})_Q \otimes \kappa(Q)$ also commute and the eigenvalues of $\sum_i v \circ u_i$ are of the form $\sum_i e_i\lambda_i$, for eigenvalues λ_i of u_i. We conclude that $Exp_Q(h^*\nabla) \subset (\sum e_i \; Exp_{Z_i}(\nabla)) + \mathbf{Z}$.

b) Let us now turn to the first assertion. By factoring h via its graph and localizing around P, we reduce to the case where h is a closed immersion, $P = Q$. Let $\overline{X} \subset \mathbf{P}^N$ be a normal projective completion of X', $D = \overline{X}\backslash X$, $\overline{\overline{C}} :=$ closure of C in $\mathbf{P}^N$. Let $d = \dim X$. Then, for $\delta >> 0$, a system $\{\mathcal{Y}_1, \ldots, \mathcal{Y}_{d-2}\}$ of $d-2$ hypersurfaces of degree δ passing through $\overline{\overline{C}}$ and sufficiently general, verifies:

(i) $\overline{S} = \mathcal{Y}_1 \cap \cdots \cap \mathcal{Y}_{d-2} \cap \overline{X}$ is a proper normal surface with (isolated) singularities contained in $\mathcal{Y}_1 \cap \cdots \cap \mathcal{Y}_{d-2} \cap \overline{X}^{sing}$,
(ii) $S' = \mathcal{Y}_1 \cap \cdots \cap \mathcal{Y}_{d-2} \cap X'$ contains $\overline{C}$,
(iii) $S = \mathcal{Y}_1 \cap \cdots \cap \mathcal{Y}_{d-2} \cap X$ is a smooth open subset of S' containing C,
(iv) S' cuts D transversally except at a finite set of points.

For lack of reference, we give some detail. Let $\epsilon : Bl \longrightarrow \overline{X}$ be the blowing-up of $\overline{X}$ centered at $\overline{\overline{C}}$, with exceptional divisor E. We set $\mathcal{O}_{Bl}(1) = \epsilon^*\mathrm{Ker}(\mathcal{O}_{\overline{X}} \longrightarrow \mathcal{O}_{\overline{\overline{C}}})$. Then for $\delta >> 0$, $\epsilon^*\mathcal{O}_{\overline{X}}(\delta) \otimes \mathcal{O}_{Bl}(1)$ is very ample; a basis of sections defines an embedding of Bl into $\mathbf{P}^M$. Since

$$\begin{aligned}\epsilon_*(\epsilon^*\mathcal{O}_{\overline{X}}(\delta) \otimes \mathcal{O}_{Bl}(1)) &\cong \mathcal{O}_{\overline{X}}(\delta) \otimes \mathrm{Ker}(\mathcal{O}_{\overline{X}} \longrightarrow \mathcal{O}_{\overline{\overline{C}}}) \\ &\cong \mathrm{Ker}(\mathcal{O}_{\overline{X}}(\delta) \longrightarrow \mathcal{O}_{\overline{\overline{C}}}(\delta)),\end{aligned}$$

the linear system of hypersurfaces of degree δ in $\overline{X}$ passing through $\overline{\overline{C}}$ gives rise to an embedding

$$\overline{X}\backslash\overline{\overline{C}} \cong Bl\backslash E \hookrightarrow \mathbf{P}^M.$$

Besides, we note that $\epsilon^{-1}(\overline{X}^{sing})$ and $\epsilon^{-1}(D^{sing})$ are of codimension ≥ 2 in Bl. The above properties then follow from Bertini's theorem applied to Bl [Kl].

The local theory of the embedded resolution of singularities of curves on a surface shows that there is a smooth surface S'', a smooth curve $\overline{C}'$ and a birational morphism

$\epsilon : S'' \longrightarrow S'$ inducing an isomorphism $\overline{C}' \cong \overline{C}$, such that $(\epsilon^{-1}(\partial S))^{red}$ is a divisor with normal crossings and that $\overline{C}'$ cuts $(\epsilon^{-1}(\partial S))^{red}$ transversally. In fact, S'' is obtained by performing a sequence of monoidal transformations (centered at closed points), $\overline{C}'$ is the iterated strict transform of $\overline{C}$ and cuts $(\epsilon^{-1}(\partial S))^{red}$ only at some point Q' of the exceptional divisor E' of the last monoidal transformation. If we perform the corresponding sequence of monoidal transformations (centered at the same points) on X' instead of S', we get a commutative diagram

$$\begin{array}{ccc} S'' & \stackrel{\epsilon}{\longrightarrow} & S' \\ \iota' \downarrow & & \iota \downarrow \\ X'' & \stackrel{\epsilon'}{\longrightarrow} & X' \end{array}$$

where ι' is a closed immersion; the exceptional divisor E'' of the last monoidal transformation restricts to E' on S'', hence meets S'' transversally in a neighborhood of Q' in E'. Applying (6.4.3), we get

$$\begin{aligned} Exp_Q(\nabla_{|C}) &= Exp_{Q'}(\epsilon^*\iota^*\nabla_{|\overline{C}'\setminus Q'}) = Exp_{E'}(\epsilon^*\iota^*\nabla) \\ &= Exp_{E'}(\iota'^*\epsilon'^*\nabla) = Exp_{E''}(\epsilon'^*\nabla) \subset \mathbf{Z}Exp(\epsilon'^*\nabla) \\ &= \mathbf{Z}Exp(\nabla). \end{aligned}$$

□

Remark 6.5.6 The argument in part b) also provides another proof of (5.4), replacing the reference to (6.4.3) by a reference to (3.4.4).

Open question 6.5.7 Is it possible, in the situation of (6.5.5) (without assumption of normal crossings), that $Exp_Q(\nabla_{|C})$ is always contained in the subgroup of K generated by 1 and the set $Exp_{\partial X}(\nabla)$? In fact, the argument in part *b*) reduces the question to the case when X is a surface.

Theorem 6.5.8 *We fix an additive subgroup Σ of K containing 1. Let $(\mathcal{E}, \nabla)$ be a coherent sheaf with regular connection on a smooth connected K-variety X. The following properties are equivalent:*

i) *the exponents of ∇ are in Σ;*
ii) *for any morphism $C \stackrel{h}{\longrightarrow} X$, with C a smooth K-curve, the exponents of $h^*(\nabla)$ are in Σ;*
iii) *for any locally closed immersion of a smooth K-curve $C \stackrel{i}{\hookrightarrow} X$, the exponents of $i^*(\nabla)$ are in Σ.*

Proof: (Parallel to the proof of (5.7)).

i) $\Rightarrow$ ii) We may extend X to a proper normal variety $\overline{X}$ and i to a morphism $\overline{C} \stackrel{h}{\longrightarrow} \overline{X}$, where $\overline{C}$ is a smooth projective model of C. If the image of h is contained in X, the result is trivial ($h^*(\nabla)$ has no singular points). Otherwise, we conclude by the first assertion in (6.5.5).

ii) $\Rightarrow$ iii) is obvious.

iii) $\Rightarrow$ i) Let v be a divisorial valuation of $\kappa(X)$, and let (X', Z') be a model of $(\kappa(X), v)$. We may shrink X' and X, and assume that $X = X' \setminus Z'$. The result follows from (6.4.3). □

Corollary 6.5.9 *Let $f : X \longrightarrow Y$ be a morphism of smooth varieties, and $(\mathcal{E}, \nabla)$ be a coherent sheaf with regular connection on Y. Then $\mathbf{Z}Exp f^*(\nabla) \subset \mathbf{Z}Exp(\nabla)$.*

Proof: Immediate from (6.5.8). □

Open question 6.5.10 Let $j : X \longrightarrow \overline{X}$ be an open dominant immersion of X in a proper normal K-variety $\overline{X}$. Is it true that $Exp(\nabla)$ is the subgroup of K generated by 1 and the sets $Exp_{\overline{Z}}(\nabla)$, for every irreducible component $\overline{Z}$ of $\overline{X}\backslash j(X)$, of codimension 1 in $\overline{X}$? This would be true if the previous open question has a positive answer. In this direction, we have:

Theorem 6.5.11 *Assume that $\overline{X}$ is smooth (and proper) and that X is the complement of a divisor $\overline{Z}$ with strict normal crossings in $\overline{X}$. Then $Exp(\nabla)$ is the subgroup of K generated by* 1 *and the set $Exp_{\overline{Z}}(\nabla)$.*

Proof: Let us prove that for any divisorial valuation v of $\kappa(\overline{X})$, the exponents of the generic fiber of ∇ are in the subgroup Σ of K generated by 1 and the set $Exp_{\overline{Z}}(\nabla)$. Since $\overline{X}$ is proper, v is centered at some point $x \in \overline{X}$. More precisely, there is a model (X', Z') of $(\kappa(\overline{X}), v)$ and a birational map $X' \xrightarrow{\epsilon} \overline{X}$ such that $\epsilon(\eta_{Z'}) = x$. We have to show that $Exp_{Z'}(\epsilon^*\nabla) \subset \Sigma$. Let P be any closed point in $\epsilon(Z') \subset \{x\}^{-}_{\overline{X}}$. By (6.5.8), it suffices to show that for any morphism $\overline{C} \xrightarrow{h} \overline{X}$ from a smooth curve $\overline{C}$ such that $h(\overline{C}) \not\subset \partial X$ and for any point $Q \in \overline{C}$ with $h(Q) = P$, $Exp_Q(\nabla_{|C}) \subset \Sigma$. This follows from the second assertion of (6.5.5). □

Appendix A: A letter of Philippe Robba (Nov. 2, 1984)

Mon cher Francesco,

Soit $y = (y_1, \ldots, y_n)$. Soit M un $K(x, y)$-module différentiel intégrable libre de rang s.

Considérons y comme des paramètres et M comme un $K(y)(x)$-module différentiel. Alors

Théorème: Soit $a \in K$. Les exposants de ce module différentiel en a sont indépendants de y.

Démonstration. Prenons $a = 0$. Soit $F = \begin{pmatrix} F_1 \\ \vdots \\ F_s \end{pmatrix}$ une base de M. Les matrices de dérivation sont

$$\frac{\partial}{\partial x} F = HF, \qquad \frac{\partial}{\partial y_i} F = G_i F$$

où $H, G_i \in Mat(s, K(x, y))$. L'hypothèse que M est intégrable se traduit par

$$\frac{\partial}{\partial y_i} H + HG_i = \frac{\partial}{\partial x} G_i + G_i H$$

pour $1 \leq i \leq n$. D'après le théorème de Turrittin, il existe m tel que si $z = x^m$ et si l'on considère M comme $K(y)^{alg}((z))$-module différentiel, il existe une base $\hat{F}$ telle que la matrice de dérivation $\hat{H}$ associée à $\frac{\partial}{\partial z}$,

$$\frac{\partial}{\partial z} \hat{F} = \hat{H} \hat{F}$$

soit sous forme de Jordan. Les quantités $Res_0(\hat{H}_{jj}) \in K(y)^{alg}$ sont m fois les exposants de M. On doit donc montrer que les $Res_0(\frac{\partial}{\partial y_i} \hat{H}_{jj})$ sont nuls.

On a

$$\frac{\partial}{\partial y_i} \hat{H} = \frac{\partial}{\partial z} \hat{G}_i + \hat{G}_i \hat{H} - \hat{H} \hat{G}_i.$$

On sait que $Res_0(\frac{\partial}{\partial z} \hat{G}_i) = 0$. Il reste donc à considérer les termes diagonaux de $\hat{G}_i \hat{H} - \hat{H} \hat{G}_i$.

Considérons la décomposition de $\hat{H}$ en ses sous-blocs

$$\hat{H} = \begin{pmatrix} \hat{H}^1 & 0 & 0 & \ldots & 0 \\ 0 & \hat{H}^2 & 0 & \ldots & 0 \\ \ldots & \ldots & \ldots & \ldots & \ldots \\ 0 & 0 & 0 & \ldots & 0 \\ 0 & 0 & 0 & \ldots & \hat{H}^r \end{pmatrix}$$

où $\hat{H}^j$ est de la forme

$$\hat{H}^j = \begin{pmatrix} \eta_j & 1 & 0 & \dots & 0 & 0 \\ 0 & \eta_j & 1 & 0 & \dots & 0 \\ \dots & \dots & \dots & \dots & \dots & \dots \\ \dots & \dots & \dots & \dots & \dots & \dots \\ & & & \dots & \eta_j & 1 \\ 0 & 0 & 0 & \dots & 0 & \eta_j \\ . & & & & & \end{pmatrix}$$

Si l'on considère la décomposition par blocs correspondante de $\hat{G}_i$

$$\hat{G}_i = \begin{pmatrix} \hat{G}_i^{11} & \hat{G}_i^{12} & \dots & \dots \\ \hat{G}_i^{21} & \hat{G}_i^{22} & \dots & \dots \\ \dots & \dots & \dots & \dots \\ \dots & \dots & \dots & \hat{G}_i^{rr} \end{pmatrix}$$

les blocs diagonaux de $\hat{G}_i\hat{H} - \hat{H}\hat{G}_i$ seront $\hat{G}_i^{jj}\hat{H}^j - \hat{H}^j\hat{G}_i^{jj}$.

Si $\hat{H}^j$ est de dimension s_j et si $\alpha_j = Res_0(\eta_j)$, on aura

$$\mathrm{Tr} \circ Res_0 \left(\frac{\partial}{\partial y_i} \hat{H}^j \right) = s_j \frac{\partial}{\partial y_i} \alpha_j = Res_0 \circ \mathrm{Tr}(\hat{G}_i^{jj}\hat{H}^j - \hat{H}^j\hat{G}_i^{jj}) = 0,$$

et donc $\frac{\partial}{\partial y_i}\alpha_j = 0$. **C.Q.F.D.**

Philippe

Copie à Bernie.

Appendix B: Models and log schemes

B.1 The notion of smooth model of a divisorially valued function field is a somewhat *ad hoc* elementary notion, not too far from the birational point of view. A broader and more conceptual setting is provided by the theory of log schemes ([KK], [I1, 2]). In this appendix, we recast some of our results of §§1, 2 in this framework.

B.2 We identify a smooth K-model (X, Z) (and, more generally, any K-variety X carrying a marked divisor Z) with the log scheme (X, M_Z), whose log structure is given in terms of the inclusion $j : X \setminus Z \longrightarrow X$ by $M_Z = \mathcal{O}_X \cap j_* \mathcal{O}^\times_{X\setminus Z} \overset{\alpha_Z}{\hookrightarrow} \mathcal{O}_X$. A morphism of two such structures $f : (X, Z) \longrightarrow (Y, W)$ in the sense of log schemes, is a morphism of K-schemes $\overset{\circ}{f} : X \longrightarrow Y$ such that the set-theoretic inverse image of W by $\overset{\circ}{f}$ *is contained in* Z. This is compatible with our Definition 1.2.1 of a morphism of models.

If X is smooth and Z is a divisor with reduced normal crossings[(1)] in X, (X, M_Z) is a fine saturated log scheme, smooth over K, *i.e.* over the trivial log scheme $(\mathrm{Spec}K, K^\times)$ associated to $\mathrm{Spec}K$. More precisely, we consider the category of fine saturated log schemes over K. We denote it $\mathcal{L}V_K$ for short. The category of K-schemes may be regarded as a full subcategory of $\mathcal{L}V_K$, by identifying a K-scheme X with the log scheme $(X, \mathcal{O}_X^\times)$. In that sense, for any object (X, M) of $\mathcal{L}V_K$, we have a canonical morphism $(X, M) \longrightarrow X$ of log schemes over K. It will be convenient to regard $\mathrm{Spec}K$ as a smooth K-model.

B.2.1 Example Let $X = \mathrm{Spec}K$ and $M = K^\times \oplus \mathbf{N} \longrightarrow K$ be the log structure $(x, n) \longmapsto x$ if $n = 0$ and $(x, n) \longmapsto 0$ otherwise. Then the canonical morphism $(X, M) \longrightarrow X$ is not log smooth. In fact, its relative dimension is 0, while the rank of $\omega^1_{(X,M)/X}$ is 1 [I1, Exmaple II, Corollary 3.15].

It may be easily checked that any morphism $f : (X, Z) \longrightarrow (Y, W)$ of smooth K-models is exact [KK, §4]. It is in fact integral [KK, Corollary 4.4 (ii)]. In particular, a morphism of smooth K-models is a closed immersion in the sense of log schemes if and only if it is an exact closed immersion. If f is smooth, the underlying morphism of schemes is flat [KK, Corollary 4.5].

We recall from [I1] that the *characteristic monoid* (resp. *abelian sheaf*) $\mathcal{C}_X$ (resp. $\mathcal{C}_X^{gp}$) of a log scheme (X, M) is the étale sheaf of monoids $(M/\mathcal{O}_X^\times)$ (resp. the abelian sheaf $(M/\mathcal{O}_X^\times)^{gp}$).

Proposition B.3 *Let (X, M) be an objet of $\mathcal{L}V_K$ log smooth over K, such that the rank of $\mathcal{C}_X^{gp}$ is everywhere ≤ 1. Then X is étale locally a smooth K-model.*

---

(1) In this appendix we think in terms of the étale topology.

Proof: We first prove that étale locally at any closed point x of X, there exist strict (log) smooth morphisms $(X, M) \xrightarrow{\varphi} \mathbf{A}_K^{1\,log} = \mathrm{Spec}(K[t]) = (\mathbf{A}_K^1, M_{t=0})$. We use [KK, Example 3.7]. We may assume that there exists a finitely generated integral monoid P and an étale morphism $X \longrightarrow \mathrm{Spec}K[P]$ such that $M = P\mathcal{O}_X^\times$. The map $P \longrightarrow \mathcal{C}_{X,x}$ is then surjective. We have to consider two cases:

1) The rank of $\mathcal{C}_{X,x}^{gp}$ is 0. Then also $\mathcal{C}_{X,x}^{gp} = \mathcal{C}_{X,x} = 0$.

2) The rank of $\mathcal{C}_{X,x}^{gp}$ is 1, so that $\mathcal{C}_{X,x} \cong \mathbf{N}$.

In case 1), the structural morphism $(X, M) \longrightarrow K$ is strict. Therefore x is a smooth point of X and any smooth morphism of a neighborhood of x to $\mathbf{A}_K^1 \backslash \{t = 0\}$ gives what we need. In case 2), we pick an element $p \in P$ such that $p \longmapsto 1$, under the canonical map $P \longrightarrow \mathcal{C}_{X,x} \cong \mathbf{N}$. The corresponding morphism $\mathrm{Spec}K[P] \longrightarrow \mathrm{Spec}K[\mathbf{N}]$ is then log smooth while the composition $X \longrightarrow \mathrm{Spec}K[P] \longrightarrow \mathrm{Spec}K[\mathbf{N}]$ is strict.

Now, étale locally, the log structure of (X, M) is the inverse image of the log structure of $\mathbf{A}_K^{1\,log}$, hence it coincides with the log structure M_Z, where Z is the smooth divisor $\varphi^{-1}(t = 0)$. □

Lemma B.4.1 *Let $f : (X, Z) \longrightarrow (Y, N)$ be a morphism of smooth objects in $\mathcal{LV}_K$, with (X, Z) a smooth K-model.*

(i) *f is an exact closed immersion if and only if étale locally on Y it is a closed immersion of smooth K-models as defined in* (1.2.2).

(ii) *Assume $(Y, N) = (Y, W)$ is also a smooth K-model. Then f is a closed immersion of log schemes if and only if it is a closed immersion of K-models as defined in* (1.2.2).

Proof: We first prove the second part.

(ii) By definition f is a closed immersion of usual schemes. Let z be any closed point of Z and let t be a local equation of W in Y at $f(z)$. Then $M_{W,f(z)} = \mathcal{O}_{Y,f(z)}^\times \oplus t^{\mathbf{N}}$ and $M_{Z,z} = f^*(M_W)_z = \mathcal{O}_{X,z}^\times \oplus (f^*t)^{\mathbf{N}}$. So, f^*t is a local equation for Z at z in X.

(i) Since f is strict, and the rank of $\mathcal{C}_X^{gp}$ is at most 1, $f(X)$ is contained in the open subset of Y where the rank of $\mathcal{C}_Y^{gp}$ is at most 1. By Proposition B.3, this open set is covered by étale neighborhoods that are smooth K-models when equipped with the inverse image log structure. By [I1, Exp. II, 2.1] a closed (resp. an exact closed) immersion of log schemes remains such after *strict* base change in $\mathcal{LV}_K$. After strict étale localization to the previous étale neighborhoods, we get exact closed immersions of log schemes, that are smooth K-models, hence the result according to (ii). □

Lemma B.4.2 *Let $f : (X, M) \longrightarrow (Y, N)$ be a morphism locally of finite type (as a morphism of schemes) in $\mathcal{LV}_K$. Then there exists étale locally on X a factorization in $\mathcal{LV}_K$*

$$(X, M) \xrightarrow{i} (S, L) \xrightarrow{g} (Y, N),$$

with i an exact closed immersion and g smooth. If f is a closed immersion, g can be taken to be étale.

Proof: This is proven in [HK, (2.9.2), (2.9.3)], except for the condition that (S, L) should be saturated. So now, assume (S, L) is only fine and $(X, M)\xrightarrow{i}(S, L)$ is an exact closed immersion. We follow the argument of [KK, Proposition 4.10 (1)]. Etale locally on X, there is a chart $(P_X \longrightarrow M, Q_S \longrightarrow L, h : Q \longrightarrow P)$ of i with h surjective. Let $Q' = (h^{gp})^{-1}(P)$ where $h^{gp} : Q^{gp} \longrightarrow P^{gp}$; since P is saturated, so is Q'. Let then

$$S' = S \times_{\mathrm{Spec}(\mathbf{Z}[Q])} \mathrm{Spec}(\mathbf{Z}[Q']),$$

and endow S' with the inverse image L' of the canonical log structure of $\mathrm{Spec}(\mathbf{Z}[Q'])$. Then, as in *loc.cit.*, we see that i factors into

$$(X, M)\xrightarrow{i''}(S', L')\xrightarrow{i'}(S, L),$$

with i'' an exact closed immersion and i' étale. □

B.4.3 Example Let $(X, Z) = (\mathrm{Spec} K[t], t = 0)$, and let $f : (X, Z) \longrightarrow X$ be the canonical morphism. Then f is not smooth, but it decomposes into the diagonal embedding $i : (X, Z) \longrightarrow (X \times_K X = \mathrm{Spec} K[x, y], y = 0)$, $i(t) = (t, t)$, which is a closed immersion, followed by the projection $p : (\mathrm{Spec} K[x, y], y = 0) \longrightarrow X$, $(x, y) \longmapsto x$, which is smooth.

Corollary B.4.4 *Let $f : (X, Z) \longrightarrow (Y, W)$ be a morphism of smooth K-models. Then there exists étale locally on X a factorization in $\mathcal{L}V_K$*

$$(X, Z)\xrightarrow{i}(S, D)\xrightarrow{g}(Y, W),$$

where (S, D) is a smooth K-model, i is a closed immersion and g smooth.

Proof: We first decompose f, étale locally on X, into

$$(X, Z)\xrightarrow{i}(S, L)\xrightarrow{g}(Y, W),$$

with i an exact closed immersion and g smooth. We then observe that (S, L) is smooth, and apply Lemma B.4.1 to further reduce i, étale locally on S, to an exact closed immersion of smooth K-models. □

B.5 We now characterize geometric smoothness for a morphism of K-models.

Proposition B.5.1 *Let $(X, Z)\xrightarrow{f}(Y, W)$ be a morphism of smooth K-models such that $f(\overset{\circ}{Z}) \subset W$. Then the following are equivalent:*

i) f *is smooth at* η_Z,

ii) $\overset{\circ}{f}$ *is flat at* η_Z,

iii) f *is log dominant in the sense of* (1.2.2).

Proof: i) $\Rightarrow$ ii) by [KK, Corollary 4.5]. ii) $\Leftrightarrow$ iii) by Remark 1.2.3. We prove iii) $\Rightarrow$ i) We must prove two things: that the canonical map $(f^*\omega^1_{(Y,W)/K})_{\eta_Z} \longrightarrow \omega^1_{(X,Z)/K,\eta_Z}$ is injective and that the $\mathcal{O}_{X,\eta_Z}$-module $\omega^1_{(X,Z)/(Y,W),\eta_Z}$ is free. The first statement holds because both the source and the target are free $\mathcal{O}_{X,\eta_Z}$-modules that inject into $\kappa(X) \otimes_{\kappa(Y)} \Omega^1_{\kappa(Y)/K}$ and $\Omega^1_{\kappa(X)/K}$, respectively. On the other hand, as all extensions are separably generated, the sequence of $\kappa(X)$-vector spaces

$$0 \longrightarrow \kappa(X) \otimes_{\kappa(Y)} \Omega^1_{\kappa(Y)/K} \longrightarrow \Omega^1_{\kappa(X)/K} \longrightarrow \Omega^1_{\kappa(X)/\kappa(Y)} \longrightarrow 0$$

is exact. As for the second statement, it is clear, by definition, that

$$\omega^1_{(X,Z)/(Y,W),\eta_Z}/torsion = \Omega^1_{X/Y,\eta_Z}/torsion. \tag{B.5.1.1}$$

By [H3, Chapter II, Lemma 8.9] it suffices to prove that

$$\dim_{\kappa(X)}\omega^1_{(X,Z)/(Y,W),\eta_Z} \otimes_{\mathcal{O}_{X,\eta_Z}} \kappa(X) = \dim_{\kappa(Z)}\omega^1_{(X,Z)/(Y,W),\eta_Z} \otimes_{\mathcal{O}_{X,\eta_Z}} \kappa(Z).$$

Now the *l.h.s.* equals $\dim_{\kappa(X)}\Omega^1_{\kappa(X)/\kappa(Y)} = \dim X - \dim Y$, since the log structure is trivial at the generic point of X. The *r.h.s.* equals $\Omega^1_{X/Y,\eta_Z} \otimes_{\mathcal{O}_{X,\eta_Z}} \kappa(Z) = \Omega^1_{\kappa(Z)/\kappa(W)}$ by (B.5.1.1). But $\dim_{\kappa(Z)}\Omega^1_{\kappa(Z)/\kappa(W)} = \dim Z - \dim W$. Hence they are equal. □

B.5.1.2 Remark It follows from the proposition that if the morphism of models f is Kummer étale at η_Z in the sense of (1.2.4), then f is an étale morphism of log schemes.

Corollary B.5.1.3 *Let* $S := \mathrm{Spec}\mathcal{O}_{W,f(\eta_Z)}$ *be endowed with the log structure* M_S *induced by* M_Z *via the canonical map* $S \longrightarrow X$ *and let* $f_S := f \times_{(Y,W)} (S, M_S)$: $(X, Z) \times_{(Y,W)} (S, M_S) \longrightarrow (S, M_S)$ *be the map of fs log schemes deduced by base change. The conditions of the proposition are also equivalent to the following.*

iv) f_S *is locally a product of a classically smooth morphism followed by a Kummer étale one* [I2].

v) f_S *is saturated* [T].

We skip the proof of this statement.

B.5.2 Examples

B.5.2.1 Let X be the affine (x, y)-plane deprived of the y-axis $(x = 0)$, and Z be the divisor $y = 0$ in X. Let Y be the affine (x, y)-plane and W be its y-axis. The morphism $(X, Z)\overset{f}{\longrightarrow}(Y, W)$, $(x, y) \longmapsto (xy, y)$, is not smooth since it is not flat.

B.5.2.2 Let X be the affine (x, y)-plane and let Z be the divisor $y = 0$ in X. Let Y be the projective x-line and W be its point at infinity. The projection $(X, Z) \xrightarrow{f} (Y, W)$, $(x, y) \longmapsto x$, is smooth, but $f(Z) \subset Y \setminus W$.

B.6 Logarithmic differential operators Let $f : (X, M) \longrightarrow (Y, N)$ be a morphism in $\mathcal{LV}_K$. We refer to [KK, Remark 5.8] and to [Og, 1.1] for the description of how the Grothendieck theory of differential operators may be developed for the morphism f. Namely, let

$$\Delta : (X, M) \longrightarrow (X, M) \times_{(Y,N)} (X, M)$$

be the diagonal embedding in $\mathcal{LV}_K$, which is a closed immersion. One applies Lemma B.4.2 to decompose Δ étale locally on X into an exact closed immersion

$$\Delta' : (X, M) \longrightarrow (X', M'),$$

of sheaf of ideals $\mathcal{J}_{X'}$, followed by an étale morphism. Let $P^n_{(X,M)/(Y,N)} = (P^n_{\Delta'}, M_n)$ be the n-th infinitesimal neighborhood of X in X' in the scheme theoretic sense, equipped with the log structure M_n inverse image of M'. One checks that this local definition makes global sense and provides a canonical nilpotent exact closed immersion $\Delta^{(n)} : (X, M) \longrightarrow P^n_{(X,M)/(Y,N)}$ and two canonical projections $p_1^{(n)}$ and $p_2^{(n)} : P^n_{(X,M)/(Y,N)} \longrightarrow (X, M)$ in $\mathcal{LV}_K$, such that $p_i^{(n)} \circ \Delta^{(n)} = 1_{(X,M)}$, for $i = 1, 2$. The *sheaf of jets of order n of f* is the augmented $\mathcal{O}_X$-algebra

$$\mathcal{P}^n_{(X,M)/(Y,N)} = p^{(n)}_{1*}(\mathcal{O}_{P^n_{(X,M)/(Y,N)}}) \cong \Delta'^{-1}(\mathcal{O}_{X'}/\mathcal{J}^{n+1}_{X'}), \tag{B.6.1}$$

equipped with the canonical ring homorphisms

$$\mathcal{O}_X \xrightarrow{j_n = p_2^{(n)*}} \mathcal{P}^n_{(X,M)/(Y,N)} \xrightarrow{\Delta^{(n)*}} \mathcal{O}_X. \tag{B.6.2}$$

It is filtered by the powers of the augmentation ideal $\mathrm{Ker}\Delta^{(n)*}$, and these data fit in an inverse system via the natural projections

$$\mathcal{P}^n_{(X,M)/(Y,N)} \longrightarrow \mathcal{P}^{n-1}_{(X,M)/(Y,N)}, \tag{B.6.3}$$

with kernel $(\mathrm{Ker}\Delta^{(n)*})^n$. One finds

$$\omega^1_{(X,M)/(Y,N)} = \mathrm{Ker}\Delta^{(n)*}/(\mathrm{Ker}\Delta^{(n)*})^2 \cong \Delta'^{-1}(\mathcal{J}_{X'}/\mathcal{J}^2_{X'}), \tag{B.6.4}$$

for any $n \geq 1$.

Definition B.6.5 For a morphism f as before, we define the sheaf of *relative differential operators* of order n as

$$\underline{Diff}^{n}_{(X,M)/(Y,N)} = \underline{Hom}_{\mathcal{O}_X}(\mathcal{P}^{n}_{(X,M)/(Y,N)}, \mathcal{O}_X), \tag{B.6.5.1}$$

and the sheaf of *relative derivations* as

$$\underline{Der}_{(X,M)/(Y,N)} = \underline{Hom}_{\mathcal{O}_X}(\omega^{1}_{(X,M)/(Y,N)}, \mathcal{O}_X). \tag{B.6.5.2}$$

There are natural morphisms

$$\mathcal{P}^{n}_{X/Y} \longrightarrow \mathcal{P}^{n}_{(X,M)/(Y,N)} \tag{B.6.5.3}$$

and

$$\Omega^{1}_{X/Y} \longrightarrow \omega^{1}_{(X,M)/(Y,N)}. \tag{B.6.5.4}$$

One defines as in the classical case a ring structure in

$$\underline{Diff}_{(X,M)/(Y,N)} = \cup_{n\in\mathbf{N}} \underline{Diff}^{n}_{(X,M)/(Y,N)}, \tag{B.6.5.5}$$

a natural injective morphism

$$\underline{Der}_{(X,M)/(Y,N)} \hookrightarrow \underline{Diff}^{1}_{(X,M)/(Y,N)}, \tag{B.6.5.6}$$

and a left $\underline{Diff}_{(X,M)/(Y,N)}$-structure on $\mathcal{O}_X$, compatible with the classical action of $\underline{Diff}_{X/Y}$, via the natural morphism, dual to (B.6.5.3),

$$\underline{Diff}_{(X,M)/(Y,N)} \longrightarrow \underline{Diff}_{X/Y}. \tag{B.6.5.7}$$

and inducing, dually to (B.6.5.4),

$$\underline{Der}_{(X,M)/(Y,N)} \longrightarrow \underline{Der}_{X/Y}. \tag{B.6.5.8}$$

Proposition B.6.6 *Let* (X, Z) *be a smooth* K*-model. Then* $\omega^{1}_{(X,Z)/K}$ *(resp.* $\underline{Der}_{(X,Z)/K}$*) identifies via* (B.6.5.4) *(resp.* (B.6.5.8)*) to* $\Omega^{1}_{X/K}(\log Z)$ *(resp.* $\underline{Der}_{X/K,Z}$*). Furthermore,* (B.6.5.7) *is injective and identifies* $\underline{Diff}_{(X,Z)/K}$ *with* $\underline{Diff}_{X/K,Z}$*.*

Proof: We may assume that (X, Z) is coordinatized with adapted coordinates $(x_1, \ldots, x_d)$. Then both $\omega^{1}_{(X,Z)/K}$ and $\Omega^{1}_{X/K}(\log Z)$ are freely generated by $\{\frac{dx_1}{x_1}, dx_2, \ldots, dx_d\}$, so the first assertion is clear.

The polynomial structure of $\underline{Diff}_{(X,Z)/K}$ is proved as in [Og, 1.1], with the simplification due to the assumption that K is a field of characteristic zero, so that the divided power calculus is not needed. Injectivity of $\underline{Diff}^n_{(X,Z)/K} \longrightarrow \underline{Diff}^n_{X/K}$, and therefore of (B.6.5.5), follows by lack of torsion. The identification of $\underline{Diff}_{(X,Z)/K}$ with $\underline{Diff}_{X/K,Z}$ then follows from Lemma 2.1.2.

With this definition, the functoriality of $\underline{Diff}^n_{(X,M)/(Y,N)}$ is formal. □

Proposition B.7 *Under the equivalent conditions of proposition* (B.5.1), *there is an exact sequence of free $\mathcal{O}_{X,\eta_Z}$-modules of finite type:*

(B.7.1)
$$\begin{aligned} 0 \longrightarrow \omega^1_{(Y,W)/K,\eta_W} \otimes_{\mathcal{O}_{X,\eta_W}} \mathcal{O}_{X,\eta_Z} &\longrightarrow \omega^1_{(X,Z)/K,\eta_Z} \\ \longrightarrow \omega^1_{(X,Z)/(Y,W),\eta_Z} &\longrightarrow 0. \end{aligned}$$

In the situation of ii) of Proposition 2.4,

$$\omega^1_{(X,Z)/(Y,W),\eta_Z} = \Omega_{X/Y,\eta_Z}/torsion,$$

hence

$$\begin{aligned} \underline{Der}_{(X,Z)/(Y,W),\eta_Z} &= Hom_{\mathcal{O}_{X,\eta_Z}}(\omega^1_{(X,Z)/(Y,W),\eta_Z}, \mathcal{O}_{X,\eta_Z}) \\ &= Hom_{\mathcal{O}_{X,\eta_Z}}(\Omega_{X/Y,\eta_Z}, \mathcal{O}_{X,\eta_Z}) \\ &= \underline{Der}_{X/Y,\eta_Z}. \end{aligned}$$

Then

$$Der_v(\mathcal{F}/\mathcal{G}) = \underline{Der}_{X/Y,\eta_Z} = \underline{Der}_{(X,Z)/(Y,W),\eta_Z}$$

and the exact sequence (2.4.3) is dual to (B.7.1).

2 Irregularity in several variables

Introduction

In this chapter, we tackle the study of irregularity in several variables. This domain is far less explored than the island of regularity.

A systematic study has been undertaken by Y. Laurent and Z. Mebkhout [L], [Me4], [LMe]. Their main theme is the geometric construction, based upon the idea of the V-filtration, of some algebraic cycles which turn out to be positive and to correspond, through the "comparison theorem", to the perverse sheaf of irregularity and its Gevrey filtration introduced in [Me2]; in the process, they attach a Newton polygon to any holonomic $\mathcal{D}$-module along a hypersurface, and even a collection of Newton polygons indexed by the components of the characteristic variety of the sheaf of irregularity.

Our approach is rather more algebraic, with a flavor of ultrametric analysis. It is close in spirit to Chapter 1.

In that chapter, we saw that one of the characterizations of regularity along a divisor is the condition that logarithmic differential operators of increasing order act with poles of bounded order at the generic point of Z. The consideration of this order of growth leads, in the general case, to the notion of the Poincaré-Katz rank $\rho_Z(\nabla)$, which may be viewed as the coarsest measure of irregularity of an integrable connection ∇ at a singular divisor Z.

We introduce and study a finer invariant, the Newton polygon $NP_Z(\nabla)$. It has the following properties:

a) there is a dense open subset U of Z such that the Newton polygon (in the usual sense) of the restriction of ∇ to any curve meeting Z transversally inside U coincides with $NP_Z(\nabla)$,
b) the maximal slope of $NP_Z(\nabla)$ is equal to $\rho_Z(\nabla)$.

More precisely, we construct a partition of the singular locus Z by constructible subsets, the *Newton polygon partition*, together with a bunch of secondary Newton polygons, with the property that a polygon N is the Newton polygon of the restriction of the connection to "almost every curve" passing through the stratum attached to N. Moreover, we prove that the associated function on Z (assumed to be smooth), with polygonal values, is *semicontinuous*. One of our main technical tools is a version of the Turrittin-Levelt-Hukuhara decomposition in the relative setting. Unfortunately, the relation between the Laurent-Mebkhout polygons and ours remains unclear for us (except for the "principal" one). It also remains to understand the link with C. Sabbah's analytic study of irregularity in two variables [S].

Let us now describe in more detail the content of each section.

In Section 1 we recall some properties of the spectral norm of a non-archimedean Banach algebra A over the field K endowed with the trivial valuation (so, the topology of K is discrete), in some special situations. Namely, we are given a complete discretely valued extension field $(\hat{\mathcal{F}}, \hat{v})$ of K and a finite dimensional vector space M over $\hat{\mathcal{F}}$: on arbitrarily choosing the absolute value $a^{-1} \in (0, 1)$ of a uniformizer of $\hat{\mathcal{F}}$ (resp. an $\hat{\mathcal{F}}$-basis of M to whose elements we give norm one) both $\hat{\mathcal{F}}$ and M may be naturally regarded as K-Banach spaces. Typically $A = \mathcal{L}_K(M)$, the K-algebra of continuous K-linear endomorphisms of M, endowed with the operator norm $|| \ ||_{\mathcal{L}_K(M)}$.

In Section 2, we define a measure of the defect of regularity along an irreducible divisor, the *Poincaré-Katz rank* of a differential module $(\mathcal{M}, \nabla)$ on $X \setminus Z$, for a smooth model (X, Z) of $(\mathcal{F}, v)$. To do this, we first take the formal v-completions $(\hat{\mathcal{F}}, \hat{v})$ of $(\mathcal{F}, v)$ and $(\hat{M}, \hat{\nabla})$ of the generic fiber (M, ∇) of $(\mathcal{M}, \nabla)$. We define the Poincaré-Katz rank of $(\hat{M}, \hat{\nabla})$ along $\hat{v}$ (and of (M, ∇) along Z) in terms of the spectral norms of all covariant derivations preserving the ideal of the valuation $\hat{v}$ of $\hat{\mathcal{F}}$ (in one variable, the use of spectral norms in this context already appeared in the work of G. Christol and B. Dwork [CD]). We then give an asymptotic formula for the rank, involving the action of logarithmic differential operators of increasing order at the generic point of the singular divisor. We examine the behaviour of the rank under inverse (resp. finite direct) images, and some exactness properties related to the notion of rank. We also examine the case of a confluence (2.5.8).

From this down-to-earth, birational and ultrametric point of view, the Poincaré-Katz rank appears as a more elementary invariant than the irregularity (Fuchs number). In the sequel of the chapter, we shall in fact reprove in a different (indirect) way most of these properties of the rank, as consequences of results on Newton polygons.

In Section 3, using Turrittin's theorem, we show that the rank may be read algebraically in terms of the iterated action of *any* vector field transversal to the singular divisor, and we define the *slope filtration* of $\hat{M}$.

In Section 4, we review Newton polygons in one variable, study their variation in a family of integral curves (constructibility, semicontinuity), and clarify some aspects of the phenomenon of confluence. Using the slope filtration, we then introduce the Newton polygon of $\hat{M}$; its maximal slope is the Poincaré-Katz rank. We examine some functorialities.

In the last three sections, which are of a less elementary nature, we give up the birational point of view. We then study the variation of Newton polygons when one restricts the connection to various curves transversal to Z, not necessarily belonging to a family of integral curves. Using Gauss maps, we establish property a) stated above (5.3.1). This allows us to give complements to some of our results of (I) about regular connections (5.3.3), (5.3.5). We introduce the (principal) Newton polygon of an integrable connection at a singular divisor and study its behaviour under inverse image. We then introduce secondary Newton polygons (5.4) and prove the semicontinuity theorem (5.4.4).

In Section 6, we establish, under a technical assumption, a formal decomposition of integrable connections in several variables at a singular divisor.

The last section contains technical results about cyclic vectors and indicial polynomials which will be crucial in the last chapter. In particular, we give sufficient (and essentially necessary) conditions for the existence, in the neighborhood of a singular divisor, of a cyclic vector with respect to a transversal derivation.

§1 Spectral norms

1.1 Let $\mathcal{F}$ be a function field over a field K of characteristic 0, and let v be a divisorial valuation of $\mathcal{F}/K$. Let $(\hat{\mathcal{F}}, v = \hat{v})$ denote the completion. We fix a real constant $a > 1$, and consider the non-archimedean absolute value $|\ |_{\hat{\mathcal{F}},a} = |\ |_a$ on $\hat{\mathcal{F}}$ defined by $|y|_a = a^{-v(y)}$. For any matrix $A = (a_{i,j}) \in M_{\nu\times\nu'}(\hat{\mathcal{F}})$, we set $||A||_{\hat{\mathcal{F}},a} = ||A||_a := \sup_{i,j} |a_{i,j}|_a$, a norm of $\hat{\mathcal{F}}$-Banach space on $M_{\nu\times\nu'}(\hat{\mathcal{F}})$ such that

$$(1.1.1) \qquad ||AB||_a \leq ||A||_a ||B||_a$$

whenever multiplication is possible.

Let $(\mathcal{F}, v)/(\mathcal{G}, w)$ be an extension of divisorially valued function fields (or fields of formal functions); this means that we have an extension of fields $\mathcal{G} \subset \mathcal{F}$, that $\mathfrak{m}_v \cap R_w = \mathfrak{m}_w$ and $\mathfrak{m}_w R_v = \mathfrak{m}_v^e$, where $e = e(v/w)$ is the *relative ramification index*. Then for $y \in \hat{\mathcal{G}}$, we have

$$(1.1.2) \qquad |y|_{\hat{\mathcal{G}},a}^{e(v/w)} = |y|_{\hat{\mathcal{F}},a},$$

and similarly for matrix-norms.

Let M be a finite dimensional $\hat{\mathcal{F}}$-vector space. Any choice of a basis $\underline{e}$ of M over $\hat{\mathcal{F}}$ determines on M a norm $||\sum a_i e_i||_{M,a,\underline{e}} = \sup_i |a_i|_a$ of Banach space, both over $\hat{\mathcal{F}}$ and over K endowed with the trivial absolute value (inducing the discrete topology on K). Two norms $|\ |_1, |\ |_2$ on a vector space V are *strongly equivalent* if there exist positive constants C_1, C_2 such that $C_1|y|_1 \leq |y|_2 \leq C_2|y|_1$, for all $y \in V$; we write $|\ |_1 \sim |\ |_2$ if that is the case. If $\underline{f} = \underline{e}A$ is another basis of $\hat{M}$, the norms $||\ ||_{M,a,\underline{e}}$ and $||\ ||_{M,a,\underline{f}}$ are strongly equivalent, since for any $y \in M$

$$(1.1.3) \qquad ||A||_a^{-1} ||y||_{M,a,\underline{e}} \leq ||y||_{M,a,\underline{f}} \leq ||A^{-1}||_a ||y||_{M,a,\underline{e}}\ .$$

In particular, $(\hat{\mathcal{F}}, |\ |_a)$ is a K-Banach space. Since $\hat{\mathcal{F}} = C((x))$, for a uniformizer x of R_v and a function field $C \subset R_v$ over K, $\{x^i\}_{i\in\mathbf{Z}}$ is an orthogonal (topological) basis of $(\hat{\mathcal{F}}, |\ |_a)$ over K.

In the situation of formula 1.1.2, for a $\hat{\mathcal{G}}$-vector space N freely generated by $\underline{g}$, and $N_{\hat{\mathcal{F}}} := N \otimes_{\hat{\mathcal{G}}} \hat{\mathcal{F}}$,

$$(1.1.4) \qquad ||\ ||_{N_{\hat{\mathcal{F}}},a,\underline{g}\otimes 1} = ||\ ||_{N,a,\underline{g}}^{e(v/w)}\ .$$

Under the same assumptions, the $\hat{\mathcal{G}}$-vector space obtained by restriction of scalars from a finite-dimensional $\hat{\mathcal{F}}$-vector space M, will be denoted by ${}_{\hat{\mathcal{G}}}M$. If $\underline{e}$ (resp. $\underline{f}$)

is a basis of the $\hat{\mathcal{F}}$-vector space M (resp. of the $\hat{\mathcal{G}}$-vector space $\hat{\mathcal{F}}$), $\underline{f}\underline{e} := (f_i e_j)$ is a basis of ${}_{\hat{\mathcal{G}}}M$. Under the extra assumption that $\hat{\mathcal{G}}$ is of finite index in $\hat{\mathcal{F}}$,

$$|| \; ||_{M,a,\underline{e}} \sim || \; ||^{e(v/w)}_{\hat{\mathcal{G}}M,a,\underline{f}\underline{e}} \; . \tag{1.1.5}$$

1.2 Let $(M, || \; ||_M)$, $(N, || \; ||_N)$ be two K-Banach spaces. The K-vector space $\mathcal{L}_K(M, N)$ of continuous K-linear maps from M to N is a K-Banach space in the usual norm

$$||L||_{\mathcal{L}_K(M,N)} := \sup\{||L(y)||_N/||y||_M | y \in M \; , \; y \neq 0 \},$$

for $L \in \mathcal{L}_K(M, N)$. The ring $\mathcal{L}_K(M) := \mathcal{L}_K(M, M)$ of continuous K-linear endomorphisms of M becomes then a K-Banach algebra with respect to the norm $|| \; ||_{\mathcal{L}_K(M)}$. Strongly equivalent norms on M and N induce strongly equivalent norms on $\mathcal{L}_K(M, N)$.

We denote by $(M\hat{\otimes}N, || \; ||_{M\hat{\otimes}N})$ the topological tensor product of $(M, || \; ||_M)$ and $(N, || \; ||_N)$: it is the completion of $(M \otimes N, || \; ||_{M\otimes N})$, where

$$||x||_{M\otimes N} = \inf \max_i ||m_i||_M ||n_i||_N \; ,$$

over all representations of $x \in M \otimes N$ as $x = \sum_i m_i \otimes n_i$. If M, N are finite dimensional $\hat{\mathcal{F}}$-vector spaces, $|| \; ||_M = || \; ||_{M,a,\underline{e}}$, $|| \; ||_N = || \; ||_{N,a,\underline{f}}$, for an $\hat{\mathcal{F}}$-basis $\underline{e}$ of M (resp. $\underline{f}$ of N), we have a canonical projection

$$M\hat{\otimes}_K N \longrightarrow M \otimes_{\hat{\mathcal{F}}} N \; ,$$

and a canonical injection

$$\mathcal{L}_{\hat{\mathcal{F}}}(M, N) \hookrightarrow \mathcal{L}_K(M, N) \; .$$

One easily checks, using orthogonal bases, that

$$||x||_{M\otimes_{\hat{\mathcal{F}}}N,a,\underline{e}\otimes\underline{f}} = \inf_{x' \longmapsto x} ||x'||_{M\hat{\otimes}_K N} = \min_{x' \longmapsto x} ||x'||_{M\hat{\otimes}_K N} \; , \tag{1.2.1}$$

and

$$||x||_{\mathcal{L}_{\hat{\mathcal{F}}}(M,N),a,\underline{e}^\vee\otimes\underline{f}} = ||x||_{\mathcal{L}_K(M,N)} \; . \tag{1.2.2}$$

For $L \in \mathcal{L}_K(M)$, and M as in the previous formulas, the norm $||L||_{\mathcal{L}_K(M)}$ will be also denoted by $||L||_{M,a,\underline{e}}$.

1.3 Let $(A, || \ ||_A)$ be a Banach algebra with identity over K. As in the classical case and with the same proof [T.S., 1.2.3], one shows that for any $f \in A$, the limit $\lim_{n\to\infty}(||f^n||_A)^{1/n}$ exists in $\mathbf{R}_{\geq 0}$; it is called the *spectral norm* $||f||_{A,\mathrm{sp}}$ of f. We have

$$(1.3.1) \qquad ||f||_{A,\mathrm{sp}} := \lim_{n\to\infty}(||f^n||_A)^{1/n} = \inf_{n\geq 1}(||f^n||_A)^{1/n} .$$

The theory of analytic spaces of Berkovich [Bk] offers the possibility of dealing with the spectrum $\Sigma_f \subset \mathbf{A}^1$ of f as in the classical case [*loc. cit.*, Chap. 7]. In particular, [*loc. cit.*, 7.1.2] shows that $||f||_{A,\mathrm{sp}}$ is determined by Σ_f.

Proposition 1.3.2 *Let $\mathcal{J}$ be a square-zero bilateral ideal of A and u be an element of $\mathcal{J}$. The spectral norm of $f + u$ coincides with the one of f.*

The proposition is a consequence of the following

Lemma 1.3.3 *Let notation be as in* (1.3.2), $(\overline{A} = A/\mathcal{J}, || \ ||_{\overline{A}})$ *be the quotient K-Banach algebra, and $f \longmapsto \overline{f}$ be the canonical projection. For any $f \in A$*

$$||f||_{A,\mathrm{sp}} = ||\overline{f}||_{\overline{A},\mathrm{sp}}.$$

Proof: We first observe that if z is in J, then $(1+z)(1-z) = (1-z)(1+z) = 1$ in A. Clearly, $\Sigma_{f'} \subset \Sigma_f$. We have to show that $\mathbf{A}^1 \setminus \Sigma_{f'} \subset \mathbf{A}^1 \setminus \Sigma_f$. For a (Berkovich) point $x \in \mathbf{A}^1$, $\mathcal{H}(x)$ denotes the corresponding complete valuation field (*cf. loc.cit.*): we have to prove that if x is such that $1 \otimes T(x) - f' \otimes 1$ is invertible in $B\hat{\otimes}\mathcal{H}(x)$, then $1 \otimes T(x) - f \otimes 1$ is invertible in $A\hat{\otimes}\mathcal{H}(x)$. There is a $y \in A\hat{\otimes}\mathcal{H}(x)$ such that $u = 1 - (1 \otimes T(x) - f \otimes 1)y \in J$ and $v = 1 - y(1 \otimes T(x) - f \otimes 1) \in J$. But then $1-u$ and $1-v$ are invertible in A and $(1 \otimes T(x) - f \otimes 1)y(1+u) = (1+v)y(1\otimes T(x) - f\otimes 1) = 1$. Therefore, $y(1+u) = (1+v)y$ and $1\otimes T(x) - f\otimes 1$ is invertible in $A\hat{\otimes}\mathcal{H}(x)$. Therefore, $x \notin \Sigma_f$. □

In the particular case of $(A, || \ ||_A) = (\mathcal{L}_K(M), || \ ||_{\mathcal{L}_K(M)})$, where $(M, || \ ||_M) = (M, || \ ||_{M,a,\underline{e}})$, with M a finite dimensional $\hat{\mathcal{F}}$-vector space and $\underline{e}$ a basis of it, $||L||_{A,\mathrm{sp}}$, will be denoted by $||L||_{M,\mathrm{sp}}$. It is in fact clearly independent of the choice of the basis $\underline{e}$.

Lemma 1.4 *Let $(M, || \ ||_M)$ and $(N, || \ ||_N)$ be Banach spaces over K.*

i) *If D (resp. D') is an element of $\mathcal{L}_K(M)$ (resp. of $\mathcal{L}_K(N)$),*

$$||D\hat{\otimes}_K D'||_{M\hat{\otimes}_K N,\mathrm{sp}} = ||D||_{M,\mathrm{sp}}||D'||_{N,\mathrm{sp}},$$

$$||\mathcal{L}_K(D, D')||_{\mathcal{L}_K(M,N),\mathrm{sp}} = ||D||_{M,\mathrm{sp}}||D'||_{N,\mathrm{sp}},$$

$$||D\hat{\otimes}_K 1_N + 1_M\hat{\otimes}_K D'||_{M\hat{\otimes}_K N,\mathrm{sp}} \leq \max(||D||_{M,\mathrm{sp}}, ||D'||_{N,\mathrm{sp}}),$$

$$||\mathcal{L}_K(1_M, D') - \mathcal{L}_K(D, 1_N)||_{\mathcal{L}_K(M,N),\mathrm{sp}} \leq \max(||D||_{M,\mathrm{sp}}, ||D'||_{N,\mathrm{sp}}).$$

If $||D||_{M,\mathrm{sp}} \neq ||D'||_{N,\mathrm{sp}}$, equality holds in the previous formulas.

ii) *Assume that in* i) M *(resp.* N*) is a* $\hat{\mathcal{F}}$*-vector space of finite dimension and that* $\|\ \|_M$ *(resp.* $\|\ \|_N$*) is the norm associated to a* $\hat{\mathcal{F}}$*-basis of* M *(resp.* N*), and assume that*

$$(1.4.1)\qquad \begin{aligned}&D\hat{\otimes}_K 1_N + 1_M\hat{\otimes}_K D' \text{ kills the closed } K\text{-vector subspace}\\ &\mathrm{Ker}(M\hat{\otimes}_K N \longrightarrow M\otimes_{\hat{\mathcal{F}}} N) \text{ of } M\hat{\otimes}_K N.\end{aligned}$$

Then $D\hat{\otimes}_K 1_N + 1_M\hat{\otimes}_K D'$ *induces a continuous* K*-linear operator* $F \in \mathcal{L}_K(M\otimes_{\hat{\mathcal{F}}} N)$*. We have*

$$\begin{aligned}\|F\|_{M\otimes_{\hat{\mathcal{F}}}N,\mathrm{sp}} &= \|D\hat{\otimes}_K 1_N + 1_M\hat{\otimes}_K D'\|_{M\hat{\otimes}_K N,\mathrm{sp}}\\ &\le \max(\|D\|_{M,\mathrm{sp}}, \|D'\|_{N,\mathrm{sp}})\ ,\end{aligned}$$

with equality holding if $\|D\|_{M,\mathrm{sp}} \ne \|D'\|_{N,\mathrm{sp}}$.

iii) *Assumptions are as in* ii)*, with* (1.4.1) *replaced by*

$$(1.4.2)\qquad \begin{aligned}&\mathcal{L}_K(1_M, D') - \mathcal{L}_K(D, 1_N) \text{ is stable on the closed } K\text{-vector subspace}\\ &Hom_{\hat{\mathcal{F}}}(M, N) \text{ of } \mathcal{L}_K(M, N).\end{aligned}$$

Then $\mathcal{L}_K(1_M, D') - \mathcal{L}_K(D, 1_N)$ *induces a continuous* K*-linear operator* $G \in \mathcal{L}_K(Hom_{\hat{\mathcal{F}}}(M, N))$*. We have*

$$\begin{aligned}\|G\|_{Hom_{\hat{\mathcal{F}}}(M,N),\mathrm{sp}} &= \|\mathcal{L}_K(1_M, D') - \mathcal{L}_K(D, 1_N)\|_{\mathcal{L}_K(M,N),\mathrm{sp}}\\ &\le \max(\|D\|_{M,\mathrm{sp}}, \|D'\|_{N,\mathrm{sp}})\ ,\end{aligned}$$

with equality holding if $\|D\|_{M,\mathrm{sp}} \ne \|D'\|_{N,\mathrm{sp}}$.

Proof: To prove i), we first note that the operators $D\hat{\otimes}_K 1_N$ and $1_M\hat{\otimes}_K D'$ on $M\hat{\otimes}_K N$ such that $D\hat{\otimes}_K 1_N \circ 1_M\hat{\otimes}_K D' = D\hat{\otimes}_K D'$ (resp. $\mathcal{L}_K(1_M, D')$ and $\mathcal{L}_K(D, 1_N)$ on $\mathcal{L}_K(M, N)$, such that $\mathcal{L}_K(1_M, D') \circ \mathcal{L}_K(D, 1_N) = \mathcal{L}_K(D, D')$) commute. On the other hand, $\|D\hat{\otimes}_K 1_N\|_{M\hat{\otimes}_K N,\mathrm{sp}} = \|D\|_{M,\mathrm{sp}}$, $\|1_M\hat{\otimes}_K D'\|_{M\hat{\otimes}_K N,\mathrm{sp}} = \|D'\|_{N,\mathrm{sp}}$ (resp. $\|\mathcal{L}_K(D, 1_N)\|_{\mathcal{L}_K(M,N),\mathrm{sp}} = \|D\|_{M,\mathrm{sp}}$, $\|\mathcal{L}_K(1_M, D')\|_{\mathcal{L}_K(M,N),\mathrm{sp}} = \|D'\|_{N,\mathrm{sp}}$).

The first two formulas of i) certainly hold with a $\le$ replacing equality. The opposite inequality is checked directly. In fact $\|m \otimes n\|_{M\otimes N} = \|m\|_M\|n\|_N$ and $\|(D \otimes D')(m \otimes n)\|_{M\otimes N} = \|D(m) \otimes D'(n)\|_{M\otimes N} = \|D(m)\|_M\|D(n)\|_N$, and similarly for $\mathcal{L}_K(M, N)$. The remaining statements of i) are proven using the previous remark and the arguments of [vR, p. 222–223].

ii) (resp. iii)) follows from formula 1.2.1 (resp. 1.2.2). □

Lemma 1.5 *Let*

$$(1.5.1)\qquad (P, \|\ \|_P) = (M, \|\ \|_M) \oplus (N, \|\ \|_N)$$

be an orthogonal direct sum of K-Banach spaces and let $F \in \mathcal{L}_K(P)$ be an operator such that $F(M) \subset M$. Let $D = F_{|M} \in \mathcal{L}_K(M)$ and $D' \in \mathcal{L}_K(N)$ be the operators induced by F. Then

$$||F||_{P,\mathrm{sp}} = \max(||D||_{M,\mathrm{sp}}, ||D'||_{N,\mathrm{sp}}) .$$

Proof: Let us consider the operator $L' = D \oplus D' \in \mathcal{L}_K(P)$. Obviously, $||L'||_{P,\mathrm{sp}} = \max(||D'||_{N,\mathrm{sp}}, ||D||_{M,\mathrm{sp}})$. On the other hand, $L = L' + H$, where H is an operator that kills M and sends P to M. Such operators form a bilateral ideal $\mathcal{J}$, with $\mathcal{J}^2 = 0$, in the sub-K-Banach algebra A of $\mathcal{L}_K(P, P)$ of operators preserving M. Since $\mathcal{J} \subset \mathcal{R}_A$, we may apply Lemma 1.3.3 to deduce $||L||_{P,\mathrm{sp}} = ||L'||_{P,\mathrm{sp}}$. □

Lemma 1.6 i) *Let $(M, || \ ||_M) = (M, || \ ||_{M,a,\underline{e}})$ and $\hat{\mathcal{G}} \subset \hat{\mathcal{F}}$ be as in* (1.1.5). *Then, for any $L \in \mathcal{L}_K(M) = \mathcal{L}_K({}_{\hat{\mathcal{G}}}M)$*

$$||L||_{M,\mathrm{sp}} = ||L||^{e(v/w)}_{{}_{\hat{\mathcal{G}}}M,\mathrm{sp}} .$$

ii) *Under the assumptions of* i), *let $(M, || \ ||_M) = (\hat{\mathcal{F}}, | \ |_a)$ and let $(D_1, \ldots, D_d)$ be an adapted R_v-basis of $Der_v(\hat{\mathcal{F}}/K)$* (see (I.2.1.2)). *Then*

$$||D_i||_{\hat{\mathcal{F}},\mathrm{sp}} = ||D_i||_{{}_{\hat{\mathcal{G}}}\hat{\mathcal{F}},\mathrm{sp}} = 1 \quad for\ every \quad i = 1, \ldots, d .$$

Proof: Left to the reader. □

1.7 We prefer a valuative version of the previous notions. Let M be a finite $\hat{\mathcal{F}}$-vector space and M_0 be a *lattice* in M, *i.e.* an R_v-submodule spanned by a $\hat{\mathcal{F}}$-basis $\underline{e}$ of M. It is clear from formula 1.1.3, that the Banach norm $|| \ ||_M = || \ ||_{M,a,\underline{e}}$ on M is determined by M_0 and a. So, we define $T_{v,M_0} = T_{v,\underline{e}} : M \longrightarrow \mathbf{Z} \cup \{\infty\}$ (resp. $T_{v,M_0} = T_{v,\underline{e}} : \mathcal{L}_K(M) \longrightarrow \mathbf{Z} \cup \{\infty\}$, resp. $v : M_{\nu\times\nu'}(\hat{\mathcal{F}}) \longrightarrow \mathbf{Z} \cup \{\infty\}$) by $T_{v,\underline{e}}(m) = -\log_a ||m||_M$ (resp. $T_{v,\underline{e}}(L) = -\log_a ||L||_{\mathcal{L}_K(M)}$, resp. $v(A) = -\log_a ||A||_a = \inf v(a_{i,j})$, for $A = (a_{i,j})$). We call T_{v,M_0} a *valuative (Banach) norm* on M (resp. $\mathcal{L}_K(M)$). Whenever it makes sense,

(1.7.1)
$$v(A + B) \geq \min(v(A), v(B)),$$
$$v(AB) \geq v(A) + v(B),$$

and

(1.7.2)
$$v_{\mathrm{sp}}(L) = v_{M,\mathrm{sp}}(L) := -\log_a ||L||_{a,\mathrm{sp}} = \lim_{n\to\infty} \frac{T_{v,M_0}(L^n)}{n} = \sup_{n\geq 1} \frac{T_{v,M_0}(L^n)}{n}$$

is independent of the chosen lattice M_0 of M. We call $v_{M,\mathrm{sp}}(L)$, the *spectral valuation* of $L \in \mathcal{L}_K(M)$.

Lemma 1.8 *Let $a_{i,n}$, for $i = 1, \dots, d$ and $n \geq 0$, be real numbers such that, for any i, $\lim_{n\to\infty} a_{i,n}/n := \rho_i$ exists in $\mathbf{R}$. Then*

$$(*)_d \qquad \lim_{n\to\infty} \frac{1}{n} \sup_{\sum_i n_i = n} \sum_{i=1}^{d} a_{i,n_i} = \sup(\rho_1, \dots, \rho_d),$$

and

$$(**)_d \qquad \lim_{n\to\infty} \frac{1}{n} \sup_{\sum_i n_i \leq n} \sum_{i=1}^{d} a_{i,n_i} = \sup(0, \rho_1, \dots, \rho_d) .$$

Proof: (C. Mariconda) Clearly the conjunction of $(**)_1$ and $(*)_d$ implies $(**)_d$. On the other hand, the inductive step $(*)_d \Rightarrow (*)_{d+1}$ follows from $(*)_2$. Since $(*)_1$ is trivial, we are reduced to proving $(*)_2$ and $(**)_1$.

We first prove $(*)_2$. We set $a_{1,n} = a_n$, $a_{2,n} = b_n$. Certainly, for all $n > 0$,

$$\frac{1}{n} \sup(a_0 + b_n, \dots, a_n + b_0) \geq \sup(a_0/n + b_n/n, a_n/n + b_0/n) ,$$

so that the inequality "$\geq$" holds.

Let us fix $\epsilon > 0$. There exists n_ϵ such that

$$(\rho_1 - \epsilon)k \leq a_k \leq (\rho_1 + \epsilon)k , \ (\rho_2 - \epsilon)k \leq b_k \leq (\rho_2 + \epsilon)k \ \forall k \geq n_\epsilon .$$

For $n \geq 2n_\epsilon$ we then have

$$\sup(a_{n_\epsilon} + b_{n-n_\epsilon}, \dots, a_{n-n_\epsilon} + b_{n_\epsilon}) \leq (\max(\rho_1, \rho_2) + \epsilon)n.$$

From this we deduce, for $n \geq 2n_\epsilon$,

$$\frac{\sup(a_0 + b_n, \dots, a_n + b_0)}{n} \leq \sup\Bigg(\frac{\sup(a_0 + b_n, \dots, a_{n_\epsilon - 1} + b_{n-n_\epsilon+1})}{n},$$

$$\frac{\sup(a_n + b_0, \dots, a_{n-n_\epsilon+1} + b_{n_\epsilon-1})}{n}, \frac{\sup(a_{n_\epsilon} + b_{n-n_\epsilon}, \dots, a_{n-n_\epsilon} + b_{n_\epsilon})}{n}\Bigg)$$

$$\leq \sup\Bigg(\frac{\sup(a_0, \dots, a_{n_\epsilon-1})}{n} + \frac{\sup(b_n, \dots, b_{n-n_\epsilon+1})}{n}, \frac{\sup(a_n, \dots, a_{n-n_\epsilon+1})}{n}$$

$$+ \frac{\sup(b_0, \dots, b_{n_\epsilon-1})}{n}, \frac{\sup(a_{n_\epsilon} + b_{n-n_\epsilon}, \dots, a_{n-n_\epsilon} + b_{n_\epsilon})}{n}\Bigg)$$

$$\leq \sup\Bigg(\frac{\sup(a_0, \dots, a_{n_\epsilon-1})}{n} + \rho_2 + \epsilon, \frac{\sup(b_0, \dots, b_{n_\epsilon-1})}{n} + \rho_1 + \epsilon,$$

$$\sup(\rho_1, \rho_2) + \epsilon\Bigg).$$

Since, for the fixed ϵ,

$$\lim_{n\to\infty} \frac{\sup(a_0, \ldots, a_{n_\epsilon - 1})}{n} = \lim_{n\to\infty} \frac{\sup(b_0, \ldots, b_{n_\epsilon - 1})}{n} = 0 ,$$

formula $(*)_2$ follows.

We now prove $(**)_1$. We set $a_{1,n} = a_n$, and assume $\lim_{n\to\infty} a_n/n = \rho \in \mathbf{R}$. We prove that

$$(**)_1 \qquad \lim_{n\to\infty} \frac{1}{n} \sup(a_0, \ldots, a_n) = \sup(0, \rho).$$

Certainly, for all $n > 0$, $\frac{1}{n}\sup(a_0, \ldots, a_n) \geq \sup(a_0/n, a_n/n)$, so that the inequality "$\geq$" holds. Let us fix $\epsilon > 0$. There exists n_ϵ such that

$$(\rho - \epsilon)k \leq a_k \leq (\rho + \epsilon)k \quad \forall k \geq n_\epsilon .$$

For $n \geq n_\epsilon$ we then have $\sup(a_{n_\epsilon}, \ldots, a_n) \leq (\rho + \epsilon)n$. From this we deduce, for $n \geq n_\epsilon$,

$$\begin{aligned} \frac{\sup(a_0, \ldots, a_n)}{n} &= \frac{\sup(\sup(a_0, \ldots, a_{n_\epsilon - 1}), \sup(a_{n_\epsilon}, \ldots, a_n))}{n} \\ &= \sup\left(\frac{\sup(a_0, \ldots, a_{n_\epsilon - 1})}{n}, \frac{\sup(a_{n_\epsilon}, \ldots, a_n)}{n}\right). \end{aligned}$$

Since, for the fixed ϵ, $\lim_{n\to\infty} \frac{\sup(a_0,\ldots,a_{n_\epsilon - 1})}{n} = 0$, $(**)_1$ follows. □

Remarks 1.9 .

1.9.1 The lemma also holds (with essentially the same proof) if one replaces "$\lim_{n\to\infty}$" by "$\limsup_{n\to\infty}$" everywhere in the statement.

1.9.2 If in the previous lemma we assume that the sequences $n \longmapsto a_{i,n} - \rho_i\, n$, for $i = 1, \ldots, d$, are bounded from above, the same is obviously true for the sequence $n \longmapsto (\sup_{\sum_i n_i = n} \sum_{i=1}^d a_{i,n_i}) - \sup(\rho_1, \ldots, \rho_d)\, n$ (resp. $n \longmapsto (\sup_{\sum_i n_i \leq n} \sum_{i=1}^d a_{i,n_i}) - \sup(0, \rho_1, \ldots, \rho_d)\, n$).

§2 The generalized Poincaré-Katz rank of irregularity

2.1 Let $(\mathcal{F}, v)$ be a divisorially valued function field over K, with completion $(\hat{\mathcal{F}}, v)$ (I.1.1). So, v is a discrete valuation of $\mathcal{F}$, trivial over K, normalized by the condition that its value group is *exactly* $\mathbf{Z}$. Let (M, ∇) be an $\hat{\mathcal{F}}/K$-differential module, where $\hat{\nabla}$ is viewed as an $\hat{\mathcal{F}}$-linear map $\mathcal{D}\mathit{iff}_{cont}(\hat{\mathcal{F}}/K) \longrightarrow \mathit{End}_K(M)$ (*cf.* (I.3.1.1)).

Definition 2.1.1 We define the *Poincaré-Katz rank* of (M, ∇) (at v) as

$$\rho_v(M) = \rho_v((M, \nabla)) = -\inf\ v_{\mathrm{sp}}(\nabla(D)) \ \ (\geq 0\,), \tag{2.1.1.1}$$

where the infimum is taken over all $D \in Der_v(\hat{\mathcal{F}}/K)$. For an $\mathcal{F}/K$-differential module (M, ∇), we set $\rho_v(M) = \rho_v(M, \nabla) := \rho_v(\hat{M}, \hat{\nabla})$ and call it again the *Poincaré-Katz rank* of (M, ∇) at v.

Let $\underline{e} = (e_1, \ldots, e_\mu)$ be an $\hat{\mathcal{F}}$-basis of M. Let $(D_1, D_2, \ldots, D_d)$ be an adapted R_v-basis of $Der_v(\hat{\mathcal{F}}/K)$ and for any $\underline{\alpha} \in \mathbf{N}^d$

$$\nabla(\underline{D}^{\underline{\alpha}})\underline{e} = \underline{e}G_{\underline{\alpha}}\ .$$

Lemma 2.1.2

$$T_{v,\underline{e}}(\nabla(\underline{D}^{\underline{\alpha}})) = \inf\{v(G_{\underline{\beta}})\ :\ \beta_i \leq \alpha_i\ \forall i\}\ (\leq 0\,).$$

In particular

$$v(G_{\underline{\alpha}}) \geq \sum_{i=1}^{d} \alpha_i \min(0, v(G_{\underline{1}_i})),$$

and $v_{\mathrm{sp}}(\nabla(D)) \leq 0$ *for all* $D \in Der_v(\hat{\mathcal{F}}/K)$.

Proof: The first assertion may be checked directly. As for the second, it follows from the fact that $||\ ||_{\mathcal{L}_K(M)}$ is a K-algebra norm, so that

$$T_{v,\underline{e}}(\nabla(\underline{D}^{\underline{\alpha}})) \geq \sum_{i=1}^{d} \alpha_i T_{v,\underline{e}}(\nabla(D_i)).$$

□

Proposition 2.1.3

$$\rho_v(\hat{M}) = \lim_{n\to\infty} \frac{1}{n} \sup_{L \in Diff_v^n(\hat{\mathcal{F}}/K)} -T_{v,\underline{e}}(\nabla(L)) = \sup_i -v_{\mathrm{sp}}(\nabla(D_i)).$$

Proof: Since $||\ ||_{\mathcal{L}_K(M)}$ is a K-algebra norm

$$\inf_{L \in Diff_v^n(\hat{\mathcal{F}}/K)} T_{v,\underline{e}}(\nabla(L)) = \inf_{\sum_i n_i \leq n} \sum_i T_{v,\underline{e}}(\nabla(D_i^{n_i})).$$

It follows from Lemma 1.8 that $\lim_{n\to\infty} \frac{1}{n} \inf_{\sum_i n_i \leq n} \sum_i T_{v,\underline{e}}(\nabla(D_i^{n_i}))$ exists and equals

$$\inf_i \lim_{n\to\infty} \frac{1}{n} T_{v,\underline{e}}(\nabla(D_i^n)) = \inf_i v_{\mathrm{sp}}(\nabla(D_i)).$$

□

Corollary 2.1.3.1 *We have*

$$\rho_v(\hat{M}) = \lim_{n\to\infty} h_{v,\underline{e}}(n)/n,$$

with

$$h_{v,\underline{e}}(n) = \sup_{\sum_{i=1}^d \alpha_i \le n} -v(G_{\underline{\alpha}}).$$

Proof: From Lemma 2.1.2. □

Remark 2.1.3.2 By Remark 1.9.1 we may also say that

$$\rho_v(M) = \sup\left(0, \limsup_{\sum_{i=1}^d \alpha_i \to\infty} -v(G_{\underline{\alpha}}) \Big/ \sum_{i=1}^d \alpha_i\right).$$

Having shown that the Poincaré-Katz rank may be calculated from the spectral norms of the elements of any adapted basis of $Der_v(\mathcal{F}/K)$, we digress in the two following subsections to give an explicit procedure for the calculation of the Poincaré-Katz rank in cyclic cases.

2.2 The classical one-variable case (trivially valued constant field)

2.2.1 Here (M, ∇) is a $K((x))/K$-differential module. Let

$$\underline{e} = \left(e, \nabla\left(\frac{d}{dx}\right)e \dots, \nabla\left(\frac{d}{dx}\right)^{\mu-1} e\right)$$

be a cyclic basis of (M, ∇) and

$$\Lambda = \left(\frac{d}{dx}\right)^{\mu} - \sum_{i=0}^{\mu-1} \gamma_i(x)\left(\frac{d}{dx}\right)^i, \tag{2.2.1.1}$$

be the monic differential operator such that $\Lambda e = 0$. So, as a left $K((x))[\frac{d}{dx}]$-module, $(M, \nabla) \cong K((x))[\frac{d}{dx}]/K((x))[\frac{d}{dx}]\Lambda$, $\nabla(\frac{d}{dx})^j e \leftrightarrow (\frac{d}{dx})^j$. For any $m \ge \mu$, let Λ_m be defined by

$$\Lambda_m = \left(\frac{d}{dx}\right)^m - \sum_{j=0}^{\mu-1} \gamma_{m,j}(x)\left(\frac{d}{dx}\right)^j \in K((x))\left[\frac{d}{dx}\right]\Lambda. \tag{2.2.1.2}$$

We write

$$\begin{aligned} \gamma_j &= x^{-\delta_j}\lambda_j + \text{higher order terms}, \lambda_j \in K, \\ \sigma &:= \max\left(1, \max_{j=0,\dots,\mu-1} \frac{\delta_j}{\mu - j}\right), \\ \gamma_{m,j} &= x^{-\sigma(m-j)}\lambda_{m,j} + \text{higher order terms} \\ (\lambda_{m,j} &= 0 \text{ if } \sigma(m-j) \notin \mathbf{Z}). \end{aligned} \tag{2.2.1.3}$$

Lemma 2.2.2 **([BS, Lemma 1]).** *Assume $\sigma > 1$ and let*

(2.2.2.1) $$\tilde{\Lambda} = \left(\frac{d}{dx}\right)^{\mu} - \sum_{\substack{0 \le j \le \mu-1 \\ \delta_j = \sigma(\mu-j)}} \lambda_j \left(\frac{d}{dx}\right)^{j}.$$

A basis of solutions of $\tilde{\Lambda}$ in $K[[x]]$, is given by

(2.2.2.2) $$v_k(x) = \frac{x^k}{k!} + \frac{\lambda_{\mu,k}}{\mu!} x^{\mu} + \frac{\lambda_{\mu+1,k}}{(\mu+1)!} x^{\mu+1} + \cdots,$$

for $k = 0, \dots, \mu - 1$.

Proof: For $\sigma > 1$, one easily finds

(2.2.2.3) $$\lambda_{m+1,j} = \lambda_{m,j-1} + \lambda_{\mu,j}\lambda_{m,\mu-1},$$

for $m \ge \mu$ and $j = 0, \dots, \mu - 1$. On the other hand,

(2.2.2.4) $$\tilde{\Lambda} v_k = \sum_{i=0}^{\infty} \left(\lambda_{\mu+i,k} - \lambda_{\mu,k-i} - \sum_j \lambda_{\mu,\mu-j}\lambda_{\mu+i-j,k} \right) \frac{x^i}{i!}.$$

It is easy to check that the two inductive formulas for $\lambda_{\mu+i,k}$ in terms of $\lambda_{\mu+h,l}$, for $h < i$ and $l = 0, \dots, \mu - 1$:

$$\lambda_{\mu+i,k} = \lambda_{\mu,k-i} + \sum_j \lambda_{\mu,\mu-j}\lambda_{\mu+i-j,k}$$

and

$$\lambda_{\mu+i,k} = \lambda_{\mu+i-1,k-1} + \lambda_{\mu,k}\lambda_{\mu+i-1,\mu-1}$$

are compatible. □

2.2.3 We have from (2.2.1.2)

(2.2.3.1) $$\left(\frac{d}{dx}\right)^{m} \underline{e} = \underline{e}\Gamma_m = \underline{e} \begin{pmatrix} \gamma_{m,0} & \gamma_{m+1,0} & \cdots & \gamma_{m+\mu-1,0} \\ \gamma_{m,1} & \gamma_{m+1,1} & \cdots & \gamma_{m+\mu-1,1} \\ \vdots & & \cdots & \vdots \\ \gamma_{m,\mu-1} & \gamma_{m+1,\mu-1} & \cdots & \gamma_{m+\mu-1,\mu-1} \end{pmatrix},$$

and from the previous lemma we deduce that, at least if $\sigma > 1$, *for each column of Γ_m, one of its entries $\gamma_{m+i,j}$ has minimal value $-\sigma(m+i-j)$*. Therefore, if $\sigma > 1$,

$$-\sigma m - \sigma(\mu - 1) \le T_{v,\underline{e}}\left(\left(\frac{d}{dx}\right)^{m}\right) \le -\sigma m,$$

that is, using (I.2.1.2.1,2),

$$-(\sigma-1)m-\sigma(\mu-1)\le T_{v,\underline{e}}\Big(\Big(x\frac{d}{dx}\Big)^m\Big)\le-(\sigma-1)m, \tag{2.2.3.2}$$

and, now for *any* $\sigma\ge 1$, and for any $n\in\mathbf{N}$,

$$-(\sigma-1)n-\sigma(\mu-1)\le \inf_{L\in Diff_v^n(\hat{\mathcal{F}}/K)} T_{v,\underline{e}}(\nabla(L))\le-(\sigma-1)n. \tag{2.2.3.3}$$

We conclude (*cf.* [De, II, Theorem 1.9 and Proposition 1.10]):

Proposition 2.2.4 *Let v be the x-adic valuation on $K((x))$. For the left $K((x))[\frac{d}{dx}]$-module $(M,\nabla)\cong K((x))[\frac{d}{dx}]/K((x))[\frac{d}{dx}]\Lambda$, with Λ given by* (2.2.1.1), *we have*

$$\rho_v(M)=-v_{\mathrm{sp}}\Big(\nabla\Big(x\frac{d}{dx}\Big)\Big)=\max\Big(0,\max_{j=0,\dots,\mu-1}\Big(-\frac{v(\gamma_j)}{\mu-j}-1\Big)\Big).$$

In particular, $\rho_v(M)$ is a non-negative rational number of denominator at most μ. Let $\underline{e}$ be any $K((x))$-basis of M. The function

$$n\longmapsto |T_{v,\underline{e}}\Big(\nabla\Big(x\frac{d}{dx}\Big)^n\Big)+\rho_v(M)\,n|, \tag{2.2.4.1}$$

is bounded.

Parallel results hold for function fields $\mathcal{F}$ of one variable over K.

Corollary 2.2.5 *Let $\Lambda\in K((x))[\frac{d}{dx}]$ and (M,∇) be as in* (2.2.4). *Write*

$$x^\mu\Lambda=\Big(x\frac{d}{dx}\Big)^\mu-\sum_{i=0}^{\mu-1}\theta_i(x)\Big(x\frac{d}{dx}\Big)^i. \tag{2.2.5.1}$$

Then

$$\rho_v(M)=\max\Big(0,\max_{j=0,\dots,\mu-1}\Big(-\frac{v(\theta_j)}{\mu-j}\Big)\Big). \tag{2.2.5.2}$$

Proof: We have in $Diff\,\hat{\mathcal{F}}/K$ relations of the type (*cf.* (I.2.1.2.1,2))

$$\begin{aligned}\Big(x\frac{d}{dx}\Big)^i&=x^i\Big(\frac{d}{dx}\Big)^i+\sum_{j<i}a_{ij}x^j\Big(\frac{d}{dx}\Big)^j,\\ x^i\Big(\frac{d}{dx}\Big)^i&=\Big(x\frac{d}{dx}\Big)^i+\sum_{j<i}b_{ij}\Big(x\frac{d}{dx}\Big)^j,\end{aligned} \tag{2.2.5.3}$$

with $a_{ij}, b_{ij} \in \mathbf{Z}$. We rewrite (2.2.5.1) as

(2.2.5.4)
$$\begin{aligned} x^{\mu}\Lambda &= \left(x\frac{d}{dx}\right)^{\mu} - \sum_{i=0}^{\mu-1} \theta_i(x)\left(x\frac{d}{dx}\right)^i \\ &= x^{\mu}\left(\frac{d}{dx}\right)^{\mu} - \sum_{i=0}^{\mu-1} \tilde{\gamma}_i(x)x^i\left(\frac{d}{dx}\right)^i, \end{aligned}$$

with $\tilde{\gamma}_i(x) = x^{\mu-i}\gamma_i(x)$. It follows from (2.2.5.3) that $\forall i = 0, \ldots, \mu - 1$

$$\begin{aligned} \theta_i &= -b_{\mu i} + \tilde{\gamma}_i - \sum_{j>i} b_{ji}\tilde{\gamma}_j, \\ \tilde{\gamma}_i &= -a_{\mu i} + \theta_i - \sum_{j>i} a_{ji}\theta_j. \end{aligned}$$

Hence, $\forall i = 0, \ldots, \mu - 1$,

$$\begin{aligned} -v(\theta_i) &\le \max\left(0, \sup_{j\ge i} -v(\tilde{\gamma}_j)\right), \\ -v(\tilde{\gamma}_i) &\le \max\left(0, \sup_{j\ge i} -v(\theta_j)\right). \end{aligned}$$

A fortiori, $\forall i = 0, \ldots, \mu - 1$,

$$\begin{aligned} -\frac{v(\theta_i)}{\mu-1} &\le \max\left(0, \sup_{j\ge i} -\frac{v(\tilde{\gamma}_j)}{\mu-j}\right), \\ -\frac{v(\tilde{\gamma}_i)}{\mu-1} &\le \max\left(0, \sup_{j\ge i} -\frac{v(\theta_j)}{\mu-j}\right). \end{aligned}$$

Hence, $\forall i = 0, \ldots, \mu - 1$,

(2.2.5.5)
$$\begin{aligned} \max\left(0, \sup_{j\ge i} -\frac{v(\theta_j)}{\mu-j}\right) &= \max\left(0, \sup_{j\ge i} -\frac{v(\tilde{\gamma}_j)}{\mu-j}\right) \\ &= \max\left(0, \sup_{j\ge i} -\frac{v(\gamma_j)}{\mu-j} - 1\right). \end{aligned}$$

Formula 2.2.5.2 follows from Proposition 2.2.4 and (2.2.5.5) with $i = 0$. □

2.2.6 The results of this subsection will apply in particular to the calculation of $v_{\mathrm{sp}}(D_1)$ in the case of a divisorially valued function field $(\mathcal{F}, v)$ over K (with an adapted R_v-basis $(D_1, \ldots, D_d)$ of $Der_v(\mathcal{F}/K)$). For that application, one completes v-adically, takes $x\frac{d}{dx} = D_1$, $x = x_1$, and replaces K by the field of constants $\mathcal{F}^D$.

2.3 One-variable case with non-trivially valued field of constants

2.3.1 Let $(\mathcal{G}, v)$ be a divisorially valued function field over K, and let $(\hat{\mathcal{G}}, v)$ be its completion. We now consider the case when $\hat{\mathcal{F}}$ is a finite extension of the completion of $\hat{\mathcal{G}}(x)$ with respect to the v-adic Gauss norm ($v(\sum_i a_i x^i) = \min_i v(a_i)$, if $\sum_i a_i x^i \in \hat{\mathcal{G}}[x]$), and D is the $\hat{\mathcal{G}}$-linear derivation of $\hat{\mathcal{F}}$ such that $Dx = 1$. This situation arises for instance in the case of the completion of a divisorially valued function field $(\mathcal{F}, v)$ over K (with an adapted R_v-basis $(D_1, \ldots, D_d)$ of $Der_v(\mathcal{F}/K)$), if we take $D = D_i$, $i > 1$, $\hat{\mathcal{G}} = \hat{\mathcal{F}}^D$ (with the induced valuation v), and $x = x_i$.

We observe that in the present case for an $\hat{\mathcal{F}}/\hat{\mathcal{G}}$-differential module M, $-v_{M,\mathrm{sp}}(D)$ is (in valuative form) the v-adic "generic" radius of convergence of solutions of (M, ∇) [ABa]. This may be explained as follows. We consider a second copy $(\hat{\mathcal{F}}^{(1)}, v^{(1)})$ of $(\hat{\mathcal{F}}, v)$, isomorphic via $f \longmapsto f^{(1)}$, and the Taylor map $T_D : f \longmapsto \sum_i (D^i f)^{(1)}(T - x^{(1)})^i/i!$ of $\hat{\mathcal{F}}$ into $\hat{\mathcal{F}}^{(1)}[[T - x^{(1)}]]$. Let $(\Omega, v^{(1)})$ be the completion of the algebraic closure of $(\hat{\mathcal{F}}^{(1)}, v^{(1)})$. Let $\mathcal{B}_\Omega(\Delta)$ denote the ring of bounded analytic Ω-valued functions on the "generic" disk $\Delta = \{a \in \Omega | v^{(1)}(a - x^{(1)}) > 0\}$. The valuation $v^{(1)}$ extends to a valuation on $\mathcal{B}_\Omega(\Delta)$, on setting $v^{(1)}(f) = \inf_{a \in \Delta} v^{(1)}(f(a))$.

It is clear that T_D takes values in $\mathcal{B}_\Omega(\Delta)$ and that it is an isometric embedding of $(\hat{\mathcal{F}}, v)$ into $(\mathcal{B}_\Omega(\Delta), v^{(1)})$, with $T = T_D(x)$.

Then, $-v_{M,\mathrm{sp}}(D)$ is related to the radius of the maximum common disk of convergence of power series solutions of $M \otimes_{\hat{\mathcal{F}}} \mathcal{B}_\Omega(\Delta)$ in $\Omega[[x - x^{(1)}]]$. In fact, if $\underline{e}$ is an $\hat{\mathcal{F}}$-basis of M and $D^i \underline{e} = \underline{e} G_i$, with $G_i \in M_{\mu \times \mu}(\hat{\mathcal{F}})$, a fundamental matrix solution $\sum_i G_i^{(1)}(x - x^{(1)})^i/i!$, converges "precisely" for $v(x - x^{(1)})$ bigger than

$$- \liminf_{i \to \infty} v^{(1)}(G_i^{(1)})/i = - \liminf_{i \to \infty} v(G_i)/i = \limsup_{i \to \infty} -v(G_i)/i \,.$$

By Remark 2.1.3.2,

$$-v_{M,\mathrm{sp}}(D) = \max\left(0, - \liminf_{i \to \infty} v^{(1)}(G_i^{(1)})/i\right).$$

Let

$$\Lambda = D^\mu - \sum_{i=0}^{\mu-1} \gamma_i D^i \in \hat{\mathcal{F}}[D]. \tag{2.3.1.1}$$

For any $m \geq \mu$, let Λ_m be defined, as in (2.2.2), by

$$\Lambda_m = D^m - \sum_{j=0}^{\mu-1} \gamma_{m,j} D^j \in \hat{\mathcal{F}}[D]\Lambda. \tag{2.3.1.2}$$

We set, as in (2.2.3),

$$\sigma := \max\left(0, \max_{j=0,\ldots,\mu-1} -\frac{v(\gamma_j)}{\mu - j}\right). \tag{2.3.1.3}$$

Proposition 2.3.2[(1)]. *We have*

$$v(\gamma_{m,j}) \geq -\sigma(m-j).$$

Therefore, the solutions of Λ *at the generic point converge for* $v(x) > -\sigma$. *If* $\sigma > 0$ *the generic radius of convergence of solutions of* Λ *is precisely* $-\sigma$.

Proof: The first part of the statement follows recursively from the calculation

$$\gamma_{m+1,j} = \gamma_{m,j-1} + \gamma_{m,\mu-1}\gamma_j + D(\gamma_{m,j}) .$$

Let us now follow the reasoning of [DGS, *loc.cit.*], assuming $\sigma > 0$. For a cyclic $\hat{\mathcal{F}}$-basis $\underline{e}$ of the left *Diff*$(\hat{\mathcal{F}}/\hat{\mathcal{G}})$-module $M = \mathit{Diff}(\hat{\mathcal{F}}/\hat{\mathcal{G}})/\mathit{Diff}(\hat{\mathcal{F}}/\hat{\mathcal{G}})\Lambda$, we have $D\underline{e} = \underline{e}G$, with

$$G = \begin{pmatrix} 0 & 0 & 0 & \dots & \gamma_0 \\ 1 & 0 & 0 & \dots & \gamma_1 \\ & \dots & & \dots & \\ 0 & & & \dots & \gamma_{\mu-2} \\ 0 & 0 & \dots & 1 & \gamma_{\mu-1} \end{pmatrix}.$$

Let $\gamma \in \hat{\mathcal{G}}$ be such that $v(\gamma) = \sigma$ and let

$$H = \begin{pmatrix} 1 & & & 0 \\ & 1/\gamma & & \\ & & \ddots & \\ 0 & & & 1/\gamma^{\mu-1} \end{pmatrix}.$$

We replace the basis $\underline{e}$ by $\underline{e}' = \underline{e}H$. Then

$$D^i\underline{e}' = \underline{e}'T_i,$$

with

$$T_1 =: \gamma W = \gamma \begin{pmatrix} 0 & 0 & 0 & \dots & \gamma_0/\gamma^{\mu} \\ 1 & 0 & 0 & \dots & \gamma_1/\gamma^{\mu-1} \\ & \dots & & \dots & \dots \\ 0 & & & 0 & \gamma_{\mu-2}/\gamma^2 \\ 0 & 0 & \dots & 1 & \gamma_{\mu-1}/\gamma \end{pmatrix}.$$

(1) Proposition 2.3.2 may be regarded as an analogue of Young's theorem [DGS, VI, Lemma 2.1] for an equicharacteristic valuation.

We check that

$$T_i \equiv \gamma^i W^i \mod \gamma^{i-1},$$

for every $i \geq 1$. The eigenvalues of W are v-integral and at least one of them, say λ, is a unit. Let $\vec{u} \in \hat{\mathcal{F}}^\mu$ be an eigenvector of eigenvalue λ such that $v(\vec{u}) = 0$. We then have

$$T_i\vec{u} \equiv \gamma^i W^i \vec{u} = \gamma^i \lambda^i \vec{u} \bmod \gamma^{i-1},$$

so that

$$v(T_i\vec{u}) = i\, v(\gamma)$$

which implies that the solution

$$\sum_i \frac{T_i\vec{u}}{i!} x^i$$

can only converge for $v(x) > -\sigma$. □

Corollary 2.3.3 *Let M be the $\hat{\mathcal{F}}/\hat{\mathcal{G}}$-differential module defined by M = Diff($\hat{\mathcal{F}}/\hat{\mathcal{G}}$)/Diff($\hat{\mathcal{F}}/\hat{\mathcal{G}}$)$\Lambda$, with Λ given by* (2.3.1.1). *Then*

$$\rho_v(M) = -v_{\mathrm{sp}}(\nabla(D)) = \max\left(0, \max_{j=0,\dots,\mu-1} -\frac{v(\gamma_j)}{\mu - j}\right).$$

In particular, $\rho_v(M)$ is a non-negative rational number of denominator at most μ. Let $\underline{e}$ be any $\hat{\mathcal{F}}$-basis of M. The function

$$n \longmapsto |T_{v,\underline{e}}(\nabla(D^n)) + \rho_v(M)\, n|,$$

is bounded.

2.4 Several variables again

We are ready to synthesize the estimates of the previous sections.

Proposition 2.4.1 *Let $(\mathcal{F}, v)$ be a divisorially valued function field (or a field of formal functions) over K and M, M_1, M_2 be $\mathcal{F}/K$-differential modules. Then*

i) $\rho_v(\mathcal{F}, d_{\mathcal{F}/K}) = 0$;
ii) $\rho_v(M^\vee) = \rho_v(M)$;
iii) $\rho_v(M_1 \otimes_{\mathcal{F}} M_2) \leq \max(\rho_v(M_1), \rho_v(M_2))$;
iv) $\rho_v(Hom_{\mathcal{F}}(M_1, M_2)) \leq \max(\rho_v(M_1), \rho_v(M_2))$;
v) *if*

$$0 \longrightarrow M_1 \longrightarrow M \longrightarrow M_2 \longrightarrow 0$$

is an exact sequence of differential modules, then

$$\rho_v(M) = \max(\rho_v(M_1), \rho_v(M_2)) .$$

vi) *If* $\rho_v(M_1) \neq \rho_v(M_2)$ *we have equality in* iii) *and* iv).

If M *is of dimension* μ *as an* $\mathcal{F}$*-vector space,* $\rho_v(M)$ *is a non-negative rational number of denominator at most* μ*. Let* $\underline{e}$ *be any* $\mathcal{F}$*-basis of* M *and* $(D_1, \ldots, D_d)$ *be an adapted basis of* $Der_v(\mathcal{F}/K)$*. The function*

$$n \longmapsto |\rho_v(M)\, n + \min_{\sum \alpha_i = n} T_{v,\underline{e}}(\nabla(\underline{D}^{\underline{\alpha}}))|, \tag{2.4.1.1}$$

is bounded.

Proof: The statements $i)$ to $vi)$ follow directly from Lemmata 1.3–1.6. The last part of the lemma is a consequence of (2.2.4), (2.3.3) and of Remark 1.9.2. □

Remark 2.4.1.2 It will be proven in Corollary 3.2 below that the function (2.4.1.1) coincides with

$$n \longmapsto |\rho_v(M)\, n + \min_n T_{v,\underline{e}}(\nabla(D_1^n))|.$$

Corollary 2.4.2 *M is* regular *at* v *if and only if* $\rho_v(M) = 0$.

Proof: This is clear from the characterization (I.3.3.4) of regularity. □

Proposition 2.4.3 *Let* $(\mathcal{F}, v)/(\mathcal{G}, w)$ *be an extension of divisorially valued function fields (or fields of formal functions), of relative ramification* $e = e(v/w)$*. Let* N *be a* $\mathcal{G}/K$*-differential module. The Poincaré-Katz rank of the* $\mathcal{F}/K$*-differential module* $N_{\mathcal{F}}$ *is given by* $\rho_v(N_{\mathcal{F}}) = e\rho_w(N)$.

Proof: By (I.2.4.ii), we may choose an adapted basis $(D_1, \ldots, D_d)$ of $Der_v(\mathcal{F}/K)$, with $(D_{1|\mathcal{G}}, \ldots, D_{r|\mathcal{G}})$ an adapted basis of $Der_w(\mathcal{G}/K)$, while $D_{i|\mathcal{G}} = 0$, for $i = r+1, \ldots, d$. We write ${}_{\mathcal{G}}N_{\mathcal{F}}$ for ${}_{\mathcal{G}}(N_{\mathcal{F}})$; then we have $v_{N_{\mathcal{F}},\mathrm{sp}}(\nabla_{\mathcal{F}}(D_i)) = e\ \ v_{{}_{\mathcal{G}}N_{\mathcal{F}},\mathrm{sp}}(\nabla_{\mathcal{F}}(D_i))$. Since N is a submodule of ${}_{\mathcal{G}}N_{\mathcal{F}}$, $v_{{}_{\mathcal{G}}N_{\mathcal{F}},\mathrm{sp}}(\nabla_{\mathcal{F}}(D_i)) \leq v_{N,\mathrm{sp}}(\nabla(D_{i|\mathcal{G}}))$. On the other hand, $\nabla_{\mathcal{F}}(D_i)$ is obtained from $\nabla(D_{i|\mathcal{G}}) \otimes_K 1_{\mathcal{F}} + 1_N \otimes_K D_i$ as in ii) of Lemma 1.4, where the field $\mathcal{G}$ plays now the role played there by $\mathcal{F}$. Then $v_{{}_{\mathcal{G}}N_{\mathcal{F}},\mathrm{sp}}(\nabla_{\mathcal{F}}(D_i)) \geq \min(v_{N,\mathrm{sp}}(\nabla(D_{i|\mathcal{G}})), v_{{}_{\mathcal{G}}\mathcal{F},\mathrm{sp}}(D_i))$. We note that $v_{{}_{\mathcal{G}}\mathcal{F},\mathrm{sp}}(D_i) = 0$. Therefore, for $i \leq r$, $v_{N,\mathrm{sp}}(\nabla(D_{i|\mathcal{G}})) \leq 0$ implies that

$$v_{N_{\mathcal{F}},\mathrm{sp}}(\nabla_{\mathcal{F}}(D_i)) = e v_{N_{\mathcal{F}},\mathrm{sp}}(\nabla(D_{i|\mathcal{G}})),$$

while for $i > r$, $v_{N,\mathrm{sp}}(\nabla(D_{i|\mathcal{G}})) = v_{N,\mathrm{sp}}(0) = +\infty$ implies

$$v_{{}_{\mathcal{G}}N_{\mathcal{F}},\mathrm{sp}}(\nabla_{\mathcal{F}}(D_i)) = v_{N_{\mathcal{F}},\mathrm{sp}}(\nabla_{\mathcal{F}}(D_i)) = 0\ .$$

Since

$$\rho_v(N_{\mathcal{F}}) = \max_{i=1,\ldots,d} -v_{N_{\mathcal{F}},\mathrm{sp}}(\nabla_{\mathcal{F}}(D_i))$$

and

$$\rho_w(N) = \max_{i=1,\ldots,r} -v_{N,\mathrm{sp}}(\nabla(D_{i|\mathcal{G}})),$$

this concludes the proof. □

Proposition 2.4.4 *Assume that $[\mathcal{F} : \mathcal{G}]$ is finite, and let M be an $\mathcal{F}/K$-differential module. The Poincaré-Katz rank of the $\mathcal{G}/K$-differential module ${}_{\mathcal{G}}M$ is given by $\rho_w({}_{\mathcal{G}}M) = \max_{v|w} e(v/w)^{-1}\rho_v(M)$.*

Proof: This follows from (I.3.1.3). □

2.5 The Poincaré-Katz rank of a connection at a divisor

Let (X, Z) be a model of a divisorially valued function field $(\mathcal{F}, v)$. Let E be an $\mathcal{F}/K$-differential module, and let $(\mathcal{E}, \nabla)$ be a model of E on $X \setminus Z$, *i.e.* a $\mathcal{O}_{X\setminus Z}$-coherent module with integrable connection whose generic fiber gives rise to E.

Definition 2.5.1 We define the *Poincaré-Katz rank* of $(\mathcal{E}, \nabla)$ at Z as the number $\rho_Z(\nabla) := \rho_v(E)$.

Proposition 2.5.2 *Assume that $(x_1, \dots, x_d)$ is an adapted system of coordinates for (X, Z). Then $\rho_Z(\nabla) = \rho$ if and only if there exists a coherent extension $\tilde{\mathcal{E}} \subset j_*\mathcal{E}$ of $\mathcal{E}$ on X and an integer c such that for any n, $Diff^n_{X/K,Z}\,\tilde{\mathcal{E}} \subset x_1^{-[n\rho]-c}\,\tilde{\mathcal{E}}$.*

Proof: From (2.4.1.1), the proof is parallel to that of (I.3.4.2). We do not repeat it. □

Proposition 2.5.3 *Let (X, Z) be a model and $(\mathcal{E}, \nabla)$, $(\mathcal{E}_1, \nabla_1)$, $(\mathcal{E}_2, \nabla_2)$ be coherent sheaves with integrable connections on $X \setminus Z$.*

i) *If $\mathcal{E}$ sits in a horizontal exact sequence $0 \longrightarrow \mathcal{E}_1 \longrightarrow \mathcal{E} \longrightarrow \mathcal{E}_2 \longrightarrow 0$, then $\rho_Z(\nabla) = \max(\rho_Z(\nabla_1), \rho_Z(\nabla_2))$;*
ii) *$\rho_Z((\mathcal{E}_1, \nabla_1) \otimes (\mathcal{E}_2, \nabla_2)) \le \max(\rho_Z(\nabla_1), \rho_Z(\nabla_2))$ and $\rho_Z(\underline{Hom}((\mathcal{E}_1, \nabla_1), (\mathcal{E}_2, \nabla_2))) \le \max(\rho_Z(\nabla_1), \rho_Z(\nabla_2))$ with equality if $\rho_Z(\nabla_1) \neq \rho_Z(\nabla_2)$.*

Proof: This follows from (2.4.1). □

Proposition 2.5.4 *Let $(X, Z) \xrightarrow{f} (Y, W)$ be a morphism of models such that $f(Z) \subset W$. Let $(\mathcal{E}, \nabla)$ be a coherent sheaf with integrable connection on $Y \setminus W$. Then $\rho_Z(f^*\nabla) \le e_f \rho_W(\nabla)$, with equality if f is log-dominant.*

Proof: Starting from (2.5.2), the proof of the inequality is parallel to that of (I.3.4.4). We do not repeat it. The equality for a log-dominant morphism is a translation of (2.4.3). □

Corollary 2.5.5 *The Poincaré-Katz rank at Z is local at η_Z for the étale topology on X.*

N.B. 2.5.5.1 In order to obtain a notion local at η_Z for the *fppf* topology on X, we should *normalize* the Poincaré-Katz rank at Z on any *fppf* neighborhood $f : U \longrightarrow X$ of η_Z, by dividing it by e_f.

Proposition 2.5.6 *Let $X \xrightarrow{f} Y$ be a finite morphism, and let W be a smooth connected divisor of Y. Let us assume that $f^{-1}W = \sum e_i Z_i$, where Z_i are smooth* *connected divisors of X, and that f is étale on $X \setminus \bigcup Z_i$. Let $(\mathcal{E}, \nabla)$ be a coherent sheaf with integrable connection on $X \setminus \bigcup Z_i$. Then $\rho_W(f_*\nabla) \leq \max e_i^{-1} \rho_{Z_i}(\nabla)$.*

Proof: Let w be the valuation of $\kappa(Y)$ attached to W. Since f is finite, each valuation v of $\kappa(X)$ above w is attached to some Z_i. The result then follows from (2.4.4). □

2.5.7 Let X be a smooth connected variety over K, $\mathcal{F} = \kappa(X)$. Let $j : X \longrightarrow X'$ be a dominant open immersion of X in a normal K-variety X', and let Z' be a subvariety of $\partial X = X' \setminus X$. If $(\mathcal{E}, \nabla)$ is a coherent sheaf with integrable connection on X, we define the rank of $(\mathcal{E}, \nabla)$ at Z' as the number $\rho_{Z'}(\nabla) := \max(\rho_v(E_{\eta_X}))$, where v runs among the divisorial valuations of $\mathcal{F}$ defined by the irreducible components of Z' of codimension 1 in X'. If $\operatorname{codim}_{X'} Z' \geq 2$ (in particular, if $Z' = \emptyset$) $\rho_{Z'}(\nabla) = 0$.

The following result generalizes (I.5.4) in the irregular case, as well as (2.5.4) in the confluent situation.

Theorem 2.5.8 *Let $\overline{C} \overset{h}{\longrightarrow} X'$ be a morphism from a smooth curve $\overline{C}$ with $h(\overline{C}) \not\subset \partial X$. Let Q be a closed point of $\overline{C}$ such that $h(Q) = P \in \partial X$. Let $Z'_1, \ldots, Z'_r$ be the components of ∂X of codimension one in X' passing through P; for $i = 1, \ldots, r$, we denote by e_i the valuation of the image in $\mathcal{O}_{\overline{C},Q}$ of a local equation of Z'_i in $\mathcal{O}_{X',P}$. Then $\rho_Q(h^*\nabla) \leq \sum_{i=1}^r e_i \, \rho_{Z'_i}(\nabla)$.*

Proof: We may assume that every irreducible component of ∂X contains P and that $C = h^{-1}(X)$ is the complement of $\{Q\}$ in $\overline{C}$. We may replace X' by any affine neighborhood of P. Let $\{x_1, x_2, \ldots\}$ be a finite set of generators of the ideal of ∂X in $\mathcal{O}(X')$ such that $x_i \neq 0$ on $h(C)$. We may and shall replace ∂X by the reduced divisor defined by the ideal $\sqrt{(\Pi x_i)}$. There is a closed subset $T \subset \partial X$ such that $\tilde{Z} = \partial X \setminus T$ is a disjoint union of smooth divisors Z_i in $X' \setminus T$ defined by the equation $x_i = 0$. We denote by j' (resp. $\tilde{j}$, resp. k) the open immersion of $\tilde{X} = X' \setminus T$ into X' (resp. X into $\tilde{X}$, resp. C into $\overline{C}$), so that $j = j' \circ \tilde{j}$.

If $P \notin T$, the result follows from (2.5.4). In order to settle the case $P \in T$, we follow the method of remark I.5.6, using jets with logarithmic poles (*cf.* Appendix B.6). We have a map of functoriality

$$h^* j'_* \mathcal{P}^n_{\tilde{X}/K,\tilde{Z}} \longrightarrow \mathcal{P}^n_{\overline{C}/K,Q}. \tag{2.5.8.1}$$

Let $\tilde{\mathcal{E}} \subset k_*\mathcal{E}$ be a coherent extension of $\mathcal{E}$ on $\tilde{X}$ as in (2.5.2). Then for all n, the mapping $\tilde{\nabla}$ gives rise to a composite mapping

$$h^* j'_* \tilde{\nabla} : h^* j'_* \tilde{\mathcal{E}} \longrightarrow \left(\Pi x_i^{-[n\rho_{Z'_i}(\nabla)]-c}\right) h^* j'_* \tilde{\mathcal{E}} \otimes_{\mathcal{O}_{\overline{C}}} h^* j'_* \mathcal{P}^n_{\tilde{X}/K,\tilde{Z}}$$
$$\longrightarrow \left(x^{-[n\sum_i e_i \rho_{Z'_i}(\nabla)]-c'}\right) h^* j'_* \tilde{\mathcal{E}} \otimes_{\mathcal{O}_{\overline{C}}} \mathcal{P}^n_{\overline{C}/K,Q}$$

for some c' independent of n. We conclude by (2.5.2) again. □

Remark 2.5.9 This theorem shows in particular that in the *integrable* case, no confluence occurs at the crossing of two singular divisors, if the connection is regular along each one of them.

§3 Some consequences of the Turrittin-Levelt-Hukuhara theorem

Let us recall the well-known Turrittin-Levelt-Hukuhara theorem, in the form given in [B1] (see also [Le], [R]). We start with a field of formal functions $(\hat{\mathcal{G}}, w)$ of d variables over K, and consider an adapted R_w-basis $(D_1, \dots, D_d)$ of $Der_w(\hat{\mathcal{G}}/K)$ (*cf.* (I.1.1.1) and (I.2.1.2)). If $(\hat{\mathcal{F}}, v)$ is a finite extension of $(\hat{\mathcal{G}}, w)$, we still denote by D_j the unique extension of $D_j \in Der_w(\hat{\mathcal{G}}/K)$ to a derivation in $Der_v(\hat{\mathcal{F}}/K)$).

Theorem 3.1 *Let M be a $\hat{\mathcal{G}}/\hat{\mathcal{G}}^{D_1}$-differential module of rank μ over $\hat{\mathcal{G}}$. There exists a finite extension $(\hat{\mathcal{F}}, v)$ of $(\hat{\mathcal{G}}, w)$, with ramification index $e = e(v/w)$ dividing $\mu!$, and a well-determined finite subset $\{\overline{P}_1, \dots, \overline{P}_r\}$ of $\hat{\mathcal{F}}/(\mathfrak{m}_v + \mathbf{Z})$, such that for every choice of representatives $P_i \in \hat{\mathcal{F}}$ of $\overline{P}_i$, we have the following decomposition of left $\hat{\mathcal{F}}[D_1]$-modules,*

$$M_{\hat{\mathcal{F}}} = \oplus_{i=1}^r \hat{\mathcal{F}} \otimes_{\hat{\mathcal{F}}^{D_1}} \mathrm{Ker}_{M_{\hat{\mathcal{F}}}}(D_1 - P_i)^\mu . \tag{3.1.0}$$

The direct sum decomposition of $M_{\hat{\mathcal{F}}}$ into the $\hat{\mathcal{F}}[D_1]$-submodules

$$W_\sigma(M_{\hat{\mathcal{F}}}) = \begin{cases} \oplus_{v(P_i)\geq 0} \hat{\mathcal{F}} \otimes_{\hat{\mathcal{F}}^{D_1}} \mathrm{Ker}_{M_{\hat{\mathcal{F}}}}(D_1 - P_i)^\mu & \text{if } \sigma = 0, \\ \oplus_{v(P_i)=-\sigma} \hat{\mathcal{F}} \otimes_{\hat{\mathcal{F}}^{D_1}} \mathrm{Ker}_{M_{\hat{\mathcal{F}}}}(D_1 - P_i)^\mu & \text{if } \sigma > 0, \end{cases} \tag{3.1.1}$$

for $\sigma \in \mathbf{Q}_{\geq 0}$, descends to a decomposition (the slope decomposition*) $W_\bullet(M)$ of M by $\hat{\mathcal{G}}/\hat{\mathcal{G}}^{D_1}$-differential submodules, such that $(W_{\frac{\sigma}{e}}(M))_{\hat{\mathcal{F}}} = W_\sigma(M_{\hat{\mathcal{F}}})$.*

Let $F_\sigma(M) := \oplus_{\sigma' \leq \sigma} W_{\sigma'}(M)$ denote the associated increasing filtration of M. For any $\sigma \in \mathbf{Q}_{\geq 0}$, $F_\sigma(M)$ consists of those $m \in M$ for which the monic operator of minimal order $\Gamma_m = D_1^n - \sum_{j=0}^{n-1} \gamma_j D_1^j$, with $\gamma_j \in \hat{\mathcal{G}}$ such that $\Gamma_m m = 0$ satisfies

$$\frac{w(\gamma_j)}{n-j} \geq -\sigma. \tag{3.1.2}$$

For the proof, we refer to [B1].

Remark 3.1.3 In an estimate like (3.1.2), the number n indicates the degree of the monic polynomial involved. Note that the estimates (3.1.2) just mean that the Gauss valuation

$$\inf_{v(y)=-\sigma} v\left(y^n - \sum_{k=0}^{n-1} \gamma_k y^k\right)$$

on the commutative associated graded ring $\hat{\mathcal{F}}[y]$ of $\hat{\mathcal{F}}[D_1]$ (filtered by the order of differential operators in D_1) is $= -n\sigma$. So, if two monic polynomials in $\hat{\mathcal{F}}[D_1]$ satisfy that estimate, so does their product. Also, by the Gauss lemma, if that estimate holds for an operator Λ, it also holds for any monic factor of it.

Supplement to 3.1 *If, moreover, M is a $\hat{\mathcal{G}}/K$-differential module, then the filtration $F_\sigma(M_{\hat{\mathcal{F}}})$ is a filtration as an $\hat{\mathcal{F}}/K$-differential module, which descends to a filtration $F_\bullet(M)$ of M as a $\hat{\mathcal{G}}/K$-differential module.*

Proof: We prove the stability of $F_\sigma(M)$ under any $\partial \in \{D_2, \ldots, D_d\}$ as follows. Let $m \in W_\sigma(M)$ and $\Gamma_m m = 0$ be as in (3.1.2). Then $0 = \partial\Gamma_m m = \Gamma_m \partial m + \partial(\Gamma_m)m$, where $\partial(\Gamma_m) = -\sum_{j=0}^{n-1} \partial(\gamma_j)D_1^j$ so that $\Gamma_m \partial m = -\partial(\Gamma_m)m := p \in F_\sigma(M)$. Then $\Gamma_p \Gamma_m \partial m = 0$ and $\Gamma_p\Gamma_m$ still satisfies the estimate (3.1.2). Then so does $\Gamma_{\partial m} \mid \Gamma_p\Gamma_m$, hence $\partial m \in W_\sigma(M)$.

In the notation of (3.1.0), for $\nu = 0, 1, \ldots,$ we set

$$K_i^{(\nu)} = \mathrm{Ker}_{M_{\hat{\mathcal{F}}}}(D_1 - P_i)^\nu \text{ (a finite dimensional } \kappa(Z) - \text{space)},$$

$$M_i^{(\nu)} = \hat{\mathcal{F}} \otimes_{\kappa(Z)} K_i^{(\nu)}.$$

□

Lemma 3.1.4 *Any element $m \in M_{\hat{\mathcal{F}}}$ such that $(D_1 - P_i)m \in M_i^{(\nu)}$ belongs to $M_i^{(\nu+1)}$.*

Proof: It is clear that m belongs to the factor $\hat{\mathcal{F}} \otimes_{\kappa(Z)} \mathrm{Ker}_{M_{\hat{\mathcal{F}}}}(D_1 - P_i)^\mu$ of the Turrittin decomposition. There exists an $\hat{\mathcal{F}}$-basis $\underline{e} = (\underline{e}^{(1)}, \ldots, \underline{e}^{(r)})$, $\underline{e}^{(j)} = (e_1^{(j)}, \ldots, e_{s(j)}^{(j)})$, of this factor, such that

$$D_1 e_1^{(j)} = P_i e_1^{(j)} \text{ and } D_1 e_h^{(j)} = P_i e_h^{(j)} + e_{h-1}^{(j)},$$

for every $1 < h \leq s(j)$ $(\leq \mu)$. Let us write $m = \sum_{h,j} a_{h,j} e_h^{(j)}$, with $a_{h,j} \in \kappa(Z)((t))$, $(D_1 - P_i)m = D_1(a_{s(j),j})e_{s(j)}^{(j)} + \sum_{h<s(j)}(a_{h+1,j} + D_1(a_{h,j}))e_h^{(j)}$. Then by assumption, $a_{h+1,j} + D_1(a_{h,j}) = 0$ for $h > \nu$. By descending induction, we get $a_{h,j} = 0$ for $h > \nu + 1$ and $a_{\nu+1,j} \in \kappa(Z)$: indeed $D_1(a_{h,j}) \in \kappa(Z)$ only if $D_1(a_{h,j}) = 0$ $(D_1 = \frac{t}{e}\frac{d}{dt})$. □

Corollary 3.2 *In the notation of* (3.1),

$$v_{M,\mathrm{sp}}(D_1) = e(v/w)^{-1} \inf_{i=1,\ldots,r} v(P_i)\ .$$

Proof: According to (1.6) it is enough to establish the formula

$$v_{M_{\hat{\mathcal{F}}},\mathrm{sp}}(D_1) = \inf_{i=1,\ldots,r} v(P_i)$$

or even the analogous formula with $M_{\hat{\mathcal{F}}}$ replaced by a single summand (in (3.1.0)) $M_i^{(\mu)} = \hat{\mathcal{F}} \otimes_{\hat{\mathcal{F}}^{D_1}} \mathrm{Ker}_{M_{\hat{\mathcal{F}}}}(D_1 - P_i)^\mu$. The estimate (3.1.2) holds for $(D_1 - P_i)$, hence for $(D_1 - P_i)^\mu$; our formula then follows from (2.2.4) and (2.2.5).

The next result shows that *not only the slope filtration is horizontal* (with respect to every derivation), as we saw in (3.1), but *the Turrittin-Levelt-Hukuhara decomposition itself is horizontal.* □

Proposition 3.3 *$M_i^{(\mu)}$ is stable under $D_1, D_2, \dots, D_d$.*

Proof: We first prove that for all $\nu = 0, 1, \dots$ and $\partial \in \{D_1, \dots, D_d\}$, one has

$$(\nu) \qquad \partial M_i^{(\nu)} \subseteq M_i^{(\nu+1)} .$$

This is obvious for D_1. Let $\partial \in \{D_2, \dots, D_d\}$. The statement (ν) is trivial for $\nu = 0$. We proceed by induction: assume (ν) holds for $\nu < \nu_0$ ($\nu_0 \geq 1$) and let $m \in K_i^{(\nu_0)}$. So, $(D_1 - P_i)^{\nu_0} m = 0$ and $(D_1 - P_i)m \in K_i^{(\nu_0 - 1)}$. Then $(D_1 - P_i)\partial m = \partial(D_1 - P_i)m + (\partial P_i)m \in M_i^{(\nu_0)}$, by the induction assumption. Due to (3.1.4), this implies $\partial M_i^{(\nu_0)} \subseteq M_i^{(\nu_0+1)}$, and this achieves the proof of (3.5) since $M_i^{(\nu)} = M_i^{(\nu+1)}$ for $\nu \geq \mu$. □

Corollary 3.4 *In the notation of* (3.1), *let us assume moreover that M is a $\hat{\mathcal{G}}/K$-differential module. We have $v_{M,\mathrm{sp}}(D_1) \leq v_{M,\mathrm{sp}}(D_j)$, for $j = 2, \dots, d$.*

Proof: We may extend the scalars to $\hat{\mathcal{F}}$, so as to have the decomposition (3.1.0), and to deal with a single factor $M_{\hat{\mathcal{F}}} = M_i^{(\mu)}$, with basis as in the proof of (3.1.4). We fix a derivation $\partial \in \{D_2, \dots, D_d\}$ and, for $e \in M_{\hat{\mathcal{F}}}$, we let $\Lambda_e = \partial^n + \sum_{k=0}^{n-1} \lambda_k \partial^k$, with $\lambda_k \in \hat{\mathcal{F}}$, be the monic operator of minimal order such that $\Lambda_e e = 0$. Lemma 1.5 shows that

$$\min(0, v_{M_{\hat{\mathcal{F}}},\mathrm{sp}}(\partial)) = \min\left(0, \inf_{j,h} v_{\hat{\mathcal{F}}[\partial]e_h^{(j)},\mathrm{sp}}(\partial)\right) ,$$

where $\hat{\mathcal{F}}[\partial]e_h^{(j)}$ is the cyclic $\hat{\mathcal{F}}[\partial]$-submodule of $M_{\hat{\mathcal{F}}}$ generated by $e_h^{(j)}$. According to (2.3.2) it will be enough to prove, for $k = 0, \dots, n-1$, the estimate

$$(3.4.1) \qquad \frac{v(\lambda_k)}{n-k} \geq \begin{cases} v(P_i) & \text{if } v(P_i) < 0, \\ 0 & \text{if } v(P_i) \geq 0, \end{cases}$$

for Λ_e, for any basis element $e = e_h^{(j)} \in M_i^{(\mu)}$.

Remark 3.1.3 applies also to estimates of the type (3.4.1). So, if two monic polynomials in $\hat{\mathcal{F}}[\partial]$ satisfy them, so does their product, and if the estimates hold for the operator Λ, they hold for any monic factor of it.

We start with $e_1^{(j)} \in M_i^{(1)}$ and set $\Lambda = \Lambda_{e_1^{(j)}}$.

$$D_1 \Lambda = \Lambda D_1 + \Lambda', \quad \Lambda' = \sum_{k=0}^{n-1} D_1(\lambda_k)\partial^k,$$

and

$$\Lambda P_i = P_i \Lambda + \tilde{\Lambda}, \quad \tilde{\Lambda} = n\partial(P_i)\partial^{n-1} + \cdots .$$

We have

$$0 = D_1 \Lambda e = \Lambda D_1 e + \Lambda' e = P_i \Lambda e + \tilde{\Lambda} e + \Lambda' e = (\tilde{\Lambda} + \Lambda')e \,.$$

Therefore $\Lambda' = -\tilde{\Lambda}$, that is

$$\sum_{k=0}^{n-1} D_1(\lambda_k)\partial^k = \left[P_i, \partial^n + \sum_{k=0}^{n-1} \lambda_k \partial^k \right]. \tag{3.4.2}$$

This implies (3.4.1) for Λ.

We now consider the following statements:

$(*)_h$: $\Lambda_{e_1^{(j)}} e_h^{(j)} \in M_i^{(h-1)}$, for $j = 1, \ldots, r$;

$(**)_h$: for every non-zero basis element $e \in M_i^{(h)}$ the polynomial Λ_e satisfies the estimate (3.4.1).

Note that $(*)_1$ is obvious, while $(**)_1$ was proved above. Let us assume by induction that $(*)_{<h}$ and $(**)_{<h}$, for some $h > 1$, hold. For $\Lambda = \Lambda_{e_1^{(j)}}$ and Λ', $\tilde{\Lambda}$ as before, we compute

$$\begin{aligned} D_1 \Lambda e_h^{(j)} &= \Lambda D_1 e_h^{(j)} + \Lambda' e_h^{(j)} = \Lambda(P_i e_h^{(j)} + e_{h-1}^{(j)}) + \Lambda' e_h^{(j)} \\ &= P_i \Lambda e_h^{(j)} + \tilde{\Lambda} e_h^{(j)} + \Lambda e_{h-1}^{(j)} + \Lambda' e_h^{(j)} = P_i \Lambda e_h^{(j)} + \Lambda e_{h-1}^{(j)} \,. \end{aligned}$$

So, $(D_1 - P_i)\Lambda e_h^{(j)} \in M_i^{(h-2)}$; by lemma 3.1.4, $\Lambda e_h^{(j)} \in M_i^{(h-1)}$. This establishes $(*)_h$.

To prove $(**)_h$ we observe that, since $\Lambda e_h^{(j)} \in M_i^{(h-1)}$, the estimates (3.4.1) hold for $\Lambda_{\Lambda e_h^{(j)}}$. Hence they hold for $\Lambda_{\Lambda e_h^{(j)}} \Lambda$ and for its monic factor $\Lambda_{e_h^{(j)}}$. □

Corollary 3.5 *The slope decomposition of any $\hat{\mathcal{G}}/K$-differential module M is independent of the auxiliary choice of an adapted R_w-basis of $Der_w(\hat{\mathcal{G}}/K)$: $W_\sigma(M)$ is the maximal $\hat{\mathcal{G}}/K$-differential submodule of M of Poincaré-Katz rank $= \sigma$ containing no submodule of strictly smaller Poincaré-Katz rank.*

Remark 3.6 i) In the situation of definition 2.1.1, we may say that (M, ∇) is *purely of Poincaré-Katz rank* σ if $\rho_v((N, \nabla_{|N})) = \sigma$, for any non-zero differential submodule $(N, \nabla_{|N})$ of (M, ∇). So, $W_\sigma(M)$ is the maximal $\hat{\mathcal{G}}/K$-differential submodule of M purely of Poincaré-Katz rank σ.

ii) The elements $\overline{P}_1, \ldots, \overline{P}_r$ of $\hat{\mathcal{F}}/(\mathfrak{m}_v + \mathbf{Z})$ introduced in (3.1) depend on D_1, as may already be seen with the connection in two variables with formal solution $e^{y/x}$.

iii) The elements $\overline{P}_i \in \hat{\mathcal{F}}/(\mathfrak{m}_v + \mathbf{Z})$ need not belong to $\hat{\mathcal{G}}/(\mathfrak{m}_w + \mathbf{Z})$, as is shown by the example of the direct image of the rank one connection with formal solution $e^{y/x}$ on the plane derived from the axes, under the finite étale morphism $(x, y) \mapsto (x, y^2)$; indeed, in this example $t = x$ and $\overline{P}_i \equiv \pm\sqrt{y}\,\frac{1}{t}$.

§4 Newton polygons

4.1 Classical Newton polygons

4.1.1 Let us revert to the situation of (2.2.1), and write $\gamma_i(x) = -\sum \gamma_{ij} x^j$ for $i \leq \mu$ ($\gamma_\mu(x) = -1$), so that $\Lambda = \sum \gamma_{ij}\, x^j \frac{d^i}{dx^i}$.

We recall that the *Newton polygon* $NP(\Lambda)$ of Λ is the convex hull of the quadrants $\{u \leq i,\ v \geq j - i + \mu \mid \gamma_{ij} \neq 0\} \subset \mathbf{R}^2$. The vertices of $NP(\Lambda)$ are among the points $(i,\ ord_x(\gamma_i) - i + \mu), 0 \leq i \leq \mu$. The vertical edge of $NP(\Lambda)$ is $\{u = \mu,\ v \geq 0\}$. If we write $x^\mu \Lambda = (x\frac{d}{dx})^\mu - \sum_{i=0}^{\mu-1} \theta_i(x)(x\frac{d}{dx})^i$, $\theta_\mu(x) = 1$, as in (2.2.5.1), then $NP(\Lambda)$ can also be described as the convex hull of the quadrants $\{u \leq i,\ v \geq ord_x(\theta_i)\}$, $0 \leq i \leq \mu$.

The height of $NP(\Lambda)$ (*i.e.* the maximal distance between the ordinates of the vertices) is called the *irregularity* $ir(\Lambda)$ of Λ. All finite slopes of $NP(\Lambda)$ are ≥ 0 and the maximal one is $\max(0, \max_{j=0,\ldots,\mu-1}(-\frac{v(\gamma_j)}{\mu-j} - 1))$. The points $(0, -ir(\Lambda))$ and $(\mu, 0)$ belong to the boundary of $NP(\Lambda)$ and the latter is always a vertex.

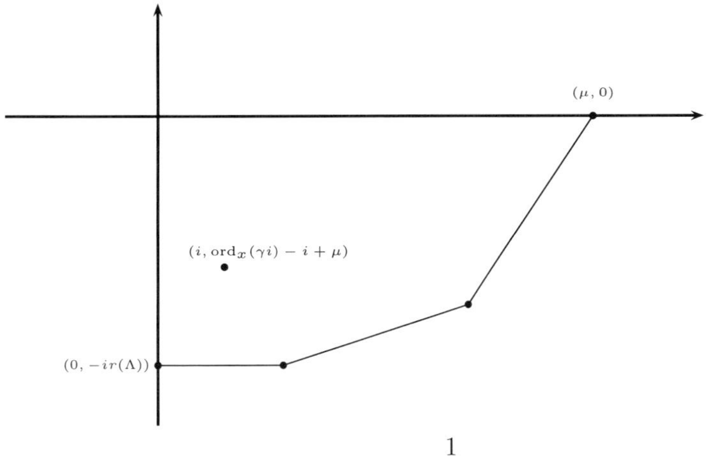

Figure 1

Proposition 4.1.2 *$NP(M) := NP(\Lambda)$ is independent of the choice of the cyclic vector e. The finite slopes of $NP(M)$ are the indices of the (Turrittin) slope decomposition* 3.1.1 (*in particular, the maximal finite slope is* $\rho_v(M)$).

Proof: See [Mal, III 1.5] or [B1, Proposition 3], and (2.2.4). □

It is clear that $NP(\Lambda)$ does not depend on the choice of the uniformizing parameter x either.

Proposition 4.1.3 i) *Let* $0 \longrightarrow M_1 \longrightarrow M \longrightarrow M_2 \longrightarrow 0$ *be an exact sequence of* $K((x))/K$*-differential modules. Then* $NP(M)$ *is the algebraic sum* $NP(M_1) + NP(M_2)$.

ii) $NP(M^\vee) = NP(M)$.

Proof: For i), see [Mal, III 1.6] or [B1, Lemma 5]. For ii), notice that if M is the differential module associated with $(x\frac{d}{dx})^\mu - \sum_{i=0}^{\mu-1} \theta_i(x)(x\frac{d}{dx})^i$, then $M^\vee$ is associated with $(x\frac{d}{dx})^\mu - \sum_{i=0}^{\mu-1}(-1)^{\mu-i}(x\frac{d}{dx})^i\, \theta_i(x)$. □

4.1.4 Here the field K is not necessarily algebraically closed. Let Q be a closed point on a smooth connected curve C defined over K. Let $(\mathcal{E}, \nabla)$ be a coherent $\mathcal{O}_{C\setminus Q}$-module with (automatically integrable) connection. We equip $\kappa(C)$ with the valuation ord_Q and set $NP_Q(\nabla) = NP(\mathcal{E}_{\eta_C} \otimes \widehat{\kappa(C)})$.

The rationality of the point Q does not matter here: if $\overline{Q}$ is any geometric point of C over Q, $\overline{C}$ is the connected component of C containing $\overline{Q}$, and $(\overline{\mathcal{E}}, \overline{\nabla})$ is the extension of $(\mathcal{E}, \nabla)$ to $\overline{C}$, we have $NP_Q(\nabla) = NP_{\overline{Q}}(\overline{\nabla})$.

4.2 Variation of the Newton polygon in a family of integral curves

4.2.1 Let X, S be K-varieties with S normal, and let $Z \subset X$ be a reduced divisor in X. We consider a diagram of the type (I.1.3.1)

$$\begin{array}{ccc} X & \overset{i}{\hookleftarrow} & Z \\ f\downarrow & \swarrow g & \\ S & & \end{array} \qquad (4.2.1.1)$$

where

i) i is the closed immersion of Z;
ii) f is smooth of pure relative dimension 1;
iii) g is quasi finite.

Let $\mathcal{E}$ be a locally free $\mathcal{O}_{X\setminus Z}$-module of finite rank μ and

$$\nabla_{(X\setminus Z)/S} : \mathcal{E} \longrightarrow \Omega^1_{(X\setminus Z)/S} \otimes \mathcal{E}$$

be a *relative* connection on $\mathcal{E}$.

Proposition 4.2.2 *In this situation, the function on* Z

$$N_f = N_f(\nabla, \bullet) : Q \longmapsto N_f(Q) = NP_Q(\nabla_{|f^{-1}(g(Q))\setminus g^{-1}(g(Q))})$$

is constructible. A fortiori, the numerical functions

$$\rho_f : Q \longmapsto \rho_f(Q) = \rho_Q(\nabla_{|f^{-1}(g(Q))\setminus g^{-1}(g(Q))})$$

on Z and

$$\rho : s \longmapsto \rho_{Z_s}(\nabla_{|(X\setminus Z)_s/\kappa(s)})$$

on S, are constructible.

Proof: To prove the first assertion, it is enough to show that, for any point Q in Z, the condition $N(P) = N(Q)$, for P in the closure Q^- of Q in Z, is *open* in Q^-. We will show, equivalently, that there exists a neighborhood U of Q in Q^- such that $N(P) = N(Q)$ for any $P \in U$.

After replacing S by the smooth part of $g(Q^-) \subset S$ and localizing around the components of Z, we are reduced to the case where X, S are smooth connected, Z is connected and étale over S, $Q = \eta_Z$ and $g(Q) = \eta_S$; then X_{η_S} is a smooth connected curve defined over $\kappa(\eta_S)$ and $\kappa(X_{\eta_S}) = \kappa(X)$. We may replace X by any affine connected neighborhood of η_Z. This allows us to assume that:

a) X, Z, S are smooth connected;
b) (*cf.* (I.1.3.1.2)) there is a transversal derivation $\partial = \partial_{t,f} \in \Gamma(X, \underline{Der}_{X/K})$ such that f is a family of integral curves of ∂; in particular, by (I.1.4.1.2), we may assume that ∂ is the image of a section $\partial_s \in \Gamma(X_s, \underline{Der}_{X_s/\kappa(s)})$ that generates $(\underline{Der}_{X_s/\kappa(s)})_P$ for any $P \in X_s$.
   We consider the restriction $(\mathcal{E}, \nabla)_{|X_{\eta_S}\setminus\{\eta_S\}} =: (\mathcal{E}', \nabla')$, a $(X_{\eta_S} \setminus \{\eta_S\})/\kappa(S)$-connection, and pick a *generic* cyclic vector $m \in \mathcal{E}'_{\eta_X}$. Then there exists a divisor $W \subset X$, $Z \not\subset W$, such that m extends to a cyclic section m of $\mathcal{E}$ on $X \setminus (Z \cup W)$ w.r.t. ∂. Since $\eta_Z \notin Z \cap W$, we may replace X by $X \setminus W$, and we may also assume
c) there is a section m of $\mathcal{E}$ over $X \setminus Z$ such that $\mathcal{E}$ is free with basis $(m, \nabla(\partial)m, \ldots, \nabla(\partial)^{\mu-1}m)$.

Let us write the associated differential equation in the form $\partial^\mu m - \sum_{j<\mu} b_j \partial^j m = 0$, with $b_j \in \mathcal{O}(X \setminus Z)$. By further shrinking X around η_Z, we may assume that the functions $t^{-\mathrm{ord}_Z b_j} b_j$ extend to invertible functions on X. It is then clear that in this situation, the Newton polygon of $\nabla_{|f^{-1}(g(P))\setminus g^{-1}(g(P))}$ is independent of $P \in Z$. The second assertion is obvious. As for the last assertion, we recall that it suffices to prove that ρ is constant on a dense open subset of S. We may then assume that $Z \xrightarrow{g} S$ is an étale covering, and we see that $\rho(s) = \sup_{g(Q)=s} \rho_f(Q)$. The last assertion follows from the fact that the supremum of a finite set of real-valued constructible functions is constructible. □

A function N with polygonal values on a topological space T is *lower semicontinuous* if, for any given polygon N_0, $\{t \in T | N(t) \subset N_0\}$ is closed in T. If T is a noetherian scheme and N is constructible, N is lower semicontinuous as soon as for any specialization $Q \longrightarrow P$ in T, $N(P) \subseteq N(Q)$.

Proposition 4.2.3 *Assume moreover that in* (4.2.1.1), $Z \xrightarrow{g} S$ *is étale. Then the function* N_f *is lower semi-continuous. In particular, the function* $\rho_f : Z \longrightarrow (\mu!)^{-1}\mathbf{Z}_{\geq 0}$

is lower semi-continuous. If g is an étale covering, the function ρ is also lower semi-continuous.

Proof: The assertion about ρ follows from the corresponding assertion for ρ_f, due to the fact that the sup of (any set of) lower semi-continuous real-valued functions is lower semi-continuous. The assertion for ρ_f is an obvious consequence of the statement about N_f, considering that for all Newton polygons under consideration, $(\mu, 0)$ is a vertex and ρ_f is the last non-infinite slope.

We already know that N_f is constructible. We must prove that if Q and P are points of Z, with P a specialization of Q, then $N_f(P) \subseteq N_f(Q)$. We may assume that the relative dimension of the specialization is 1. If the (reduced locally closed subscheme of S) $g(Q^-)$ is normal, we may replace S by $g(Q^-)$ and assume that $Q = \eta_Z$; then P is a smooth point of Z. On replacing S by an étale neighborhood of $g(P)$, we may assume that (4.2.1.1) is a strict coordinatized tubular neighborhood of Q (I.1.3.1.2). The definition (4.1.4) shows that we may extend our picture to $\kappa(g(P))$, and then to the algebraic closure of $\kappa(g(P))$. This allows us to assume that $g(P) = P$ is a closed point of the smooth curve $S = Z$ (of equation $t = 0$) over K; f is a smooth retraction $f : X \longrightarrow Z$ with geometrically connected fibers. As in the proof of (4.2.2), we let $\partial = \partial_{t,f}$ pick a cyclic section m of $\mathcal{E}$ on $X \setminus (Z \cup f^{-1}(P) \cup W)$ w.r.t. ∂, for some divisor W of $X \setminus (Z \cup f^{-1}(P))$.

Let

$$\Lambda(m) = \partial^{\mu} m - \sum_{j<\mu} b_j \partial^j m = 0$$

be the associated differential equation, with $b_j \in \mathcal{O}(X \setminus (Z \cup f^{-1}(P) \cup W))$. Let $R = \mathcal{O}_{Z,P}$, x be a uniformizer of R and let η_P be the generic point of the fiber $f^{-1}(P)$. Then $(\mathcal{E}_{\eta_X}, \nabla_{\eta_X})$ is a $\kappa(X)/\kappa(S)$-differential module and m is a cyclic vector for it. We equip $\kappa(X)$ with the valuation v with valuation ring $\mathcal{O}_{X,\eta_P}$, a discrete valuation of $\kappa(X)/K$ extending the x-adic valuation of $\kappa(S)$. Since $\nabla(\partial)(\mathcal{E}_{\eta_P}) \subset \mathcal{E}_{\eta_P}$, (2.3.2) shows that $v(b_j) \geq 0$, for $j = 0, \ldots, \mu$, so that $b_j \in \mathcal{O}(X \setminus (Z \cup W))$. We replace X by $X \setminus W$ and then by $X_P = X \times_K R$. We define a cyclic X_P/R-differential module $\mathcal{F} = \mathcal{O}_{X_P}[\partial]/\mathcal{O}_{X_P}[\partial]\Lambda$, with the natural action $\aleph(\partial)$ of ∂. By construction, $(\mathcal{E}_{\eta_X}, \nabla_{\eta_X}) \cong (\mathcal{F}_{\eta_{X_P}}, \aleph_{\eta_{X_P}})$. By Gabber's theorem [Ka 4, 2.5.2], the semisimplifications of $(\mathcal{F}_{|f^{-1}(P)}, \aleph_{|f^{-1}(P)})$ and $(\mathcal{E}_{|f^{-1}(P)}, \nabla_{|f^{-1}(P)})$ are also isomorphic. By (4.1.3), we conclude that $N_f(P) = NP((\mathcal{F}_{|f^{-1}(P)}, \aleph_{|f^{-1}(P)})) = NP(\Lambda_{x=0})$. The assertion is now clear in this particular case.

If $g(Q^-)$ is not normal, we consider the natural morphism $S' \longrightarrow S$ from the normalization S' of $g(Q^-)$ to S, and let $X' = X \times_S S'$. The consideration of the inverse image of ∇ on $X' \setminus Z'$ reduces us to the former case. □

Remarks 4.2.4

i) Example (I.3.4.8.1) shows that the rank can indeed drop.

ii) The condition that Z is *étale* over S (or, equivalently, that Z meets all the integral curves transversally) is essential in order to prevent confluence phenomena, as in the following examples.

Examples 4.2.5 We take $X =$ the (x, y)-plane with the line $x = 0$ removed.

4.2.5.1 $Z =$ the divisor $y = 0$,
$\mathcal{E} = \mathcal{O}_{X\setminus Z}$ with the integrable connection ∇ whose formal solution is $e^{1/xy}$. Let us consider the family of lines C_λ of equation $x - y = \lambda$. Then $\rho_Z((\mathcal{E}, \nabla)) = 1$, $\rho_{x=0}((\mathcal{E}, \nabla)_{|C_\lambda}) = 1$ for $\lambda \neq 0$, but $\rho_{(0,0)}((\mathcal{E}, \nabla)_{|C_0}) = 2$.

4.2.5.2 Here
$Z =$ the parabola $y = x^2$,
$\mathcal{E} = \mathcal{O}_{X\setminus Z}$ with the integrable connection ∇ whose formal solution is $e^{1/(y-x^2)}$. For the family of horizontal lines $C_\lambda : y = \lambda$, $\rho_Z((\mathcal{E}, \nabla)) = 1$, $\rho_{x=0}((\mathcal{E}, \nabla)_{|C_\lambda}) = 1$ for $\lambda \neq 0$, but $\rho_{(0,0)}((\mathcal{E}, \nabla)_{|C_0}) = 2$. (However C_0 is not transversal to Z).

4.2.5.3 We fix two constants b, c, and take
$Z =$ the hyperbola $xy = 1$,
$L_y \in \mathcal{D} := \mathit{Diff}((X \setminus Z)/S) = K[x, y, \frac{1}{x(1-xy)}, \frac{\partial}{\partial x}]$ is the differential operator

$$L_y := x(1 - xy)\left(\frac{\partial}{\partial x}\right)^2 + (c - (1/y + b + 1)xy)\frac{\partial}{\partial x} - b$$

attached to the hypergeometric function ${}_2F_1(1/y, b, c;\ xy)$,
$(\mathcal{E}, \nabla_{(X\setminus Z)/S}) =$ the free $\mathcal{O}_{(X\setminus Z)}$-module of rank 2 with relative connection associated to the $\mathcal{D}$-module $\mathcal{D}/\mathcal{D}L_y$.

The restriction of ∇ to $(X \setminus Z)_y$ is regular. But the differential operator

$$L_0 = x\left(\frac{d}{dx}\right)^2 + (c - x)\frac{d}{dx} - b$$

attached to the Kummer function ${}_1F_1(b, c; x)$ is irregular at infinity. This phenomenon could not occur for an integrable connection $(\mathcal{E}, \nabla)$ on $X \setminus Z$ (I.5.4). All this should be compared with [De, II, 1.23] which, as it stands, is false. (The mistake comes from the fact that his $\overline{X_s} \setminus X_s$ is not necessarily reduced. The example also shows that the argument given in [*loc.cit.* 2.4, case 2], is wrong).

4.2.5.4 A simpler example of confluence is the elementary formula

$$e^{-\frac{1}{x}} = \lim_{y\to 0}\left(1 - \frac{y}{x}\right)^{\frac{1}{y}} = \lim_{y\to 0}(x - y)^{\frac{1}{y}} x^{-\frac{1}{y}}.$$

In this example, $L_0 = x^2\frac{d}{dx} - 1$ appears as a specialization of the generically regular relative differential operator $L_y = x(x - y)\frac{d}{dx} - 1$.

Proposition 4.2.6 *In the situation of* (4.2.1), *assume moreover that S is smooth and g is étale (resp. an étale covering), so that (X, Z) is a model. Assume moreover that ∇ comes from an (absolute) integrable $(X \setminus Z)/K$-connection. Then the maximal value of the function ρ_f (resp. ρ) is $\rho_Z(\nabla)$.*

Proof: The fact that $\rho_Z(\nabla)$ is an upper bound comes from (2.5.8). The fact that $\rho_Z(\nabla) = \rho_{\eta_Z}(\nabla_{|f^{-1}(g(\eta_Z))\setminus g^{-1}(g(\eta_Z))})$ follows from (3.2). □

Remark 4.2.7 The difference between the integrable and non-integrable cases is more striking if one considers the irregularity ir (*i.e.* the height of the Newton polygon) instead of the Poincaré-Katz rank ρ. Let us consider a diagram as in (4.2.1.1), with X, S, Z smooth and Z *finite* over S, and let $\mathcal{E}$ be again a locally free $\mathcal{O}_{X\setminus Z}$-module of finite rank μ endowed with a *relative* connection on ∇. Following [Me 2, Section 4], let us consider the functions

$$\Psi : s \in S \longmapsto \sum_{Q \in Z_s} ir_Q(\nabla_{(X\setminus Z)_s/\kappa(s)}) \text{ and } \Phi : s \in S \longmapsto \Psi(s) + \mu \cdot \sharp Z_s \,.$$

P. Deligne proved (at least if $\dim S = 1$) that Φ is lower semicontinuous (letter to N. Katz 1976). Examples like (4.2.5.3) (or rather its variant in which $X = \mathbf{P}^1 \times \mathbf{A}^1$, $f = pr_2$, $Z = \{0, \infty\} \times \mathbf{A}^1 \cup hyperbola$) show that Ψ may not be lower semicontinuous. If ∇ comes from an (absolute) integrable X/K-connection, it is known that Φ is lower semicontinuous and that Ψ is constructible [Me 1]; moreover, B. Malgrange *conjectures* that Ψ is also lower semicontinuous (at least if $\dim S = 1$). In this direction, Z. Mebkhout proved that if Φ is constant, then Z is étale over S [Me 2, Section 4].

4.3 The case of several variables

Definition 4.3.1 Let $(\hat{\mathcal{F}}, v)$ be a field of formal functions over K, and let $\hat{M}$ be an $\hat{\mathcal{F}}/K$-differential module of rank μ. The *Newton polygon* of $\hat{M}$ is the convex polygon $NP(\hat{M}) \subset \mathbf{R}^2$ characterized by the following properties:

i) $NP(\hat{M})$ contains the quadrant $\{x \leq \mu, y \geq 0\}$, and its extreme right vertex is $(\mu, 0)$,
ii) for any $\sigma > 0$, the horizontal length of the edge of slope σ is $\dim Gr_\sigma(\hat{M})$, the dimension of the graded piece of index σ in the slope filtration of $\hat{M}$ (3.3).

In particular, the slopes of $NP(\hat{M})$ are the indices σ of the slope filtration for which $Gr_\sigma(\hat{M}) \neq 0$, and 0, ∞, and the maximal finite slope is $\rho(\hat{M})$.

4.3.2 Due to (3.3), this is a well-posed definition, independent of the choice of any adapted basis of $Der_v(\hat{\mathcal{F}}/K)$. However, if $(D_1, \dots, D_d)$ is such a basis, let us set $C = \hat{\mathcal{F}}^{D_1}$ so that $\hat{\mathcal{F}} = C((x))$, for a parameter x of $\hat{R}_v$ such that $D_1 x = x$; so $D_j C \subset C$ and $D_j x = 0$, $j = 2, \dots, d$. Then $NP(\hat{M})$ is the Newton polygon of the $C((x))/C$-differential module $M_{/C}$ in the sense of (4.1.1). In particular, the vertices of $NP(\hat{M})$ have *integral coordinates*.

Proposition 4.3.3 i) *Let* $0 \longrightarrow \hat{M}_1 \longrightarrow \hat{M} \longrightarrow \hat{M}_2 \longrightarrow 0$ *be an exact sequence of* $\hat{\mathcal{F}}/K$*-differential modules. Then* $NP(\hat{M})$ *is the algebraic sum of* $NP(\hat{M}_1)$ *and* $NP(\hat{M}_2)$.

ii) $NP(\hat{M}^\vee) = NP(\hat{M})$.

Proof: (4.3.2) reduces the statement to (4.1.3).

In the sequel, we denote by $\varphi_{a,b}$, for any $a, b \in \mathbf{Q}^\times$, the automorphism of $\mathbf{R}^2$ given by $\varphi_{a,b}(x, y) = (ax, by)$. □

Proposition 4.3.4 *Let* $(\hat{\mathcal{F}}, v)/(\hat{\mathcal{G}}, w)$ *be an extension of fields of formal functions, as in* (1.1). *Let* $\hat{N}$ *be a* $\hat{\mathcal{G}}/K$*-differential module. The Newton polygon of the* $\hat{\mathcal{F}}/K$*-differential module* $\hat{N}_{\hat{\mathcal{F}}}$ *is given by* $NP(\hat{N}_{\hat{\mathcal{F}}}) = \varphi_{1,e_v} NP(\hat{N})$.

Proof: Thanks to (3.3), this follows from the application of (2.4.3) to $F_\sigma(\hat{M})$. □

Proposition 4.3.5 *Assume moreover that* $d_v = [\hat{\mathcal{F}} : \hat{\mathcal{G}}]$ *is finite. Let* $\hat{M}$ *be an* $\hat{\mathcal{F}}/K$*-differential module. The Newton polygon of the* $\hat{\mathcal{G}}/K$*-differential module* ${}_{\hat{\mathcal{G}}}\hat{M}$ *is given by* $NP({}_{\hat{\mathcal{G}}}\hat{M}) = \varphi_{d_v, d_v/e_v} NP(\hat{M})$.

Proof: Thanks to (3.3), this follows from the application of (2.4.4) to $F_\sigma(\hat{M})$. □

4.3.6 Let $(\mathcal{F}, v)$ be a divisorially valued function field with completion $\hat{\mathcal{F}}$. We define the *Newton polygon* of any $\mathcal{F}/K$-differential module M at v to be $NP_v(M) = NP(M \otimes_{\mathcal{F}} \hat{\mathcal{F}})$.

Proposition 4.3.7 *Let* $\mathcal{G} \hookrightarrow \mathcal{F}$ *be a finite extension of function fields, and let* w *be a divisorial valuation of* $\mathcal{G}$. *For any valuation* v *of* $\mathcal{F}$, *we denote by* $d_v = [\hat{\mathcal{F}} : \hat{\mathcal{G}}]$ *(resp.* e_v*) the local degree (resp. the ramification index). Let* M *be an* $\hat{\mathcal{F}}/K$*-differential module. Then* $NP_w({}_{\mathcal{G}}M)$ *is the algebraic sum of the polygons* $\varphi_{d_v, d_v/e_v} NP_v(M)$, $v \mid w$.

Proof: Thanks to the decomposition $({}_{\mathcal{G}}M)\hat{} \cong \oplus_{v|w}\, {}_{\hat{\mathcal{G}}}(M^{\wedge(v)})$ (I.3.1.3), this follows from (4.3.4) and (4.3.3.i). □

§5 Stratification of the singular locus by Newton polygons

5.1 Newton polygon of an integrable connection at a singular divisor

5.1.1 Let (X, Z) be a model of a divisorially valued function field $(\mathcal{F}, v)$ (*cf.* (I.1.1.)), and let $(\mathcal{E}, \nabla)$ be a coherent sheaf with integrable connection on $X \setminus Z$. We define the *(principal) Newton polygon* of ∇ at Z to be the Newton polygon of the generic fiber E at v: $NP_Z(\nabla) = NP_v(E)$. The vertices of $NP_Z(\nabla)$ have integral coordinates, and the maximal slope is $\rho_Z(\nabla)$ (*cf.* (4.3.1), (4.3.2)).

The behaviour of this invariant under extensions, duality, log-dominant inverse images and finite étale direct images of connections is given by the results (4.3.3.i,ii), (4.3.4) and (4.3.7) respectively (compare also with (2.5)). In particular $NP_Z(\nabla)$ does not change if one replaces ∇ by its inverse image on any étale neighborhood of Z on X.

Theorem 5.1.2 *Let $(X, Z) \xrightarrow{h} (Y, W)$ be a morphism of models such that $h(Z) \subset W$. Let $(\mathcal{E}, \nabla)$ be a coherent sheaf with integrable connection on $Y \setminus W$. Then $NP_Z(h^*\nabla) \subset \varphi_{1,e_h} NP_W(\nabla)$, with equality if h is log-dominant* (with the notation of (4.3.4)).

Proof: The equality for a log-dominant morphism is a translation of (4.3.4), as was said above. After possibly shrinking X around η_Z (I.1.3.2), we may assume that X is a coordinatized tubular neighborhood of P as in (I.1.3.1.2). By (4.3.2), $NP_Z(h^*\nabla) = NP_{\eta_Z}((f^*\nabla)_{|(h^{-1}(\eta_S)\setminus\eta_Z)})$ is the Newton polygon at η_Z of the restriction of $h^*\nabla$ to the generic fiber of h. By (4.2.2) this coincides with $NP_Q((h^*\nabla)_{|f^{-1}(f(Q))})$, for Q in an open dense subset of Z. This shows that we may assume that X is a curve. On the other hand, we may replace X by any étale neighborhood of Z. Factorizing h as in (I.1.2.8) then leaves us with the case when h is a closed embedding (with X a curve). By (I.1.3.3), we may assume that X is a fiber in an affine coordinatized tubular neighborhood $h : Y \longrightarrow S$ of W in Y (in particular, X is an integral curve for a derivation D_1 with simple zero at W). We conclude by (4.2.3). □

5.2 Gauss maps

5.2.1 Let X be a smooth connected K-variety. The scheme $\mathrm{Inf}^1 X$ of infinitely near points of first order on X is the projective bundle of tangent directions, *i.e.* $\mathbf{P}(\Omega^1_{X/K})$. Let $C \subset X$ be a locally closed smooth curve. The *Gauss map* $\gamma_C : C \longrightarrow \mathrm{Inf}^1 X$ attaches to any point $Q \in C$ its tangent direction in $\mathcal{T}_{X,Q}$. It is a locally closed immersion.

One can interpret this map as follows: $\mathrm{Inf}^1 X$ is the exceptional divisor in the blow-up of the diagonal in X^2. The strict transform of $C \times X$ meets $\mathrm{Inf}^1 X$ transversally and their intersection is canonically isomorphic via the projection on the first factor to C: γ_C is the inverse of that isomorphism.

5.2.2 Let C' be another curve in X such that $\gamma_C(C)$ and $\gamma_{C'}(C')$ meet at some point Q^1 in $\mathrm{Inf}^1 X$. Let Q be the image of Q^1 in X. Then the strict transforms of C and C' meet at some point Q_1 of the exceptional divisor in the blow-up of Q in X, as is easily seen in local coordinates.

5.2.3 We set $\mathrm{Inf}^0 X = X$. For any integer $n \geq 1$, we construct the scheme $\mathrm{Inf}^n X$ of infinitely near points of order n on X inductively, as being $\mathrm{Inf}^1(\mathrm{Inf}^{n-1} X)$. This iterative process gives us a higher Gauss map $\gamma_C^n : C \longrightarrow \mathrm{Inf}^n X$. For $n \geq m$, we denote by $\pi^{n,m} : \mathrm{Inf}^n X \longrightarrow \mathrm{Inf}^m X$ the natural projection. For any locally closed subset $Z \subset X$, $(\mathrm{Inf}^n X)_{|Z}$ sits atop a tower of projective bundles over Z

$$\pi_Z^n : (\mathrm{Inf}^n X)_{|Z} \longrightarrow (\mathrm{Inf}^{n-1} X)_{|Z} \longrightarrow \cdots \longrightarrow Z .$$

5.2.4 Let us assume that Z is a smooth closed subvariety of pure codimension 1, and that C meets Z transversally at some point Q. Then $\gamma_C^n(C)$ meets the divisor $(\mathrm{Inf}^n X)_{|Z}$ transversally in $\mathrm{Inf}^n X$ at the point $Q^n = \gamma_C^n(Q)$ above Q.

Proposition 5.2.5 *Let $(\mathcal{E}, \nabla)$ be a coherent sheaf with integrable connection on $X \setminus Z$, and let n be $> \rho_Z(\nabla)$. Then for any locally closed curve C which meets Z transversally at some point Q, the polygon $NP_Q(\nabla_{|(C \setminus Q)})$ depends only on the point $\gamma_C^n(Q) \in (\mathrm{Inf}^n X)_{|Z}$.*

Proof: We consider a coherent extension $\tilde{\mathcal{E}}$ of $\mathcal{E}$ in a neighborhood of Q, as in (2.5.2). Let C' be another curve in X such that $\gamma_{C'}^n(Q) = \gamma_C^n(Q)$. We set $X_0 = X,\ Z_0 = Z,\ C_0 = C,\ C'_0 = C',\ Q_0 = Q$. Using the Remark (5.2.2), we construct inductively $(X_m,\ Z_m,\ C_m,\ C'_m,\ Q_m)$, for $m \leq n$, in the following way: X_m is the blow-up of X_{m-1} with center Q_{m-1} deprived of the strict transform of Z_{m-1}, Z_m is the exceptional divisor in X_m; $C_m,\ C'_m$ are the strict transforms of $C_{m-1},\ C'_{m-1}$ respectively, and Q_m is the intersection point of C_m and C'_m in Z_m. We denote by π the natural morphism (of log-schemes, if one likes) $(X_n, Z_n) \longrightarrow (X, Z)$. We may localize the situation around Q and thus assume that (X, Z) admits a global set of adapted coordinates $(x_1, \ldots, x_d)$ centered at Q. Then (X_n, Z_n) admits a global set of adapted coordinates $(y_1, \ldots, y_d)$ centered at Q_n such that π is given by $x_1 = y_1, x_2 = y_2 y_1^n, \ldots, x_d = y_d y_1^n$. Note that π identifies $C_n \setminus Q_n$ and $C'_n \setminus Q_n$ with $C \setminus Q$ and $C' \setminus Q$ respectively, and that under this identification, $\nabla_{|(C \setminus Q)} = (\pi^* \nabla)_{|(C_n \setminus Q_n)}$, $\nabla_{|(C' \setminus Q)} = (\pi^* \nabla)_{|(C'_n \setminus Q_n)}$.

Let $\hat{E}_n$ denote the $K[[y_1, \ldots, y_d]][\frac{1}{y_1}]$-module $\tilde{\mathcal{E}}_Q \otimes_{\mathcal{O}_{X,Q}} \hat{\mathcal{O}}_{X_n,Q_n}[\frac{1}{y_1}]$. We note that $\hat{\mathcal{O}}_{X_n,Q_n}[\frac{1}{y_1}]$ is a noetherian local ring with maximal ideal $(y_2, \ldots, y_d)$ and $\hat{E}_n$ is a finitely generated $\hat{\mathcal{O}}_{X_n,Q_n}[\frac{1}{y_1}]$-module. Let $||$ be an absolute value on $K((y_1))$ trivial on K such that $0 < |y_1| < 1$. If, for $m \in \hat{E}_n$, we set $||m|| = |y_1^n|$, where $n \in \mathbf{N}$ is minimal with

$$y_1^n m \in \mathcal{E}_Q \otimes_{\mathcal{O}_{X,Q}} \hat{\mathcal{O}}_{X_n,Q_n} ,$$

then $(\hat{E}_n, ||\ ||)$ is also a Banach space over $(K((y_1)), ||)$. Due to the hypothesis $n > \rho_Z(\nabla)$, it follows from (2.4.1) that there is a well-defined $K((y_1))$-linear operator

$$\Pi = \sum_{\underline{\alpha}=(\alpha_2,\ldots,\alpha_d)} (-1)^{\underline{\alpha}} (\underline{\alpha}!)^{-1} \underline{x}^{\underline{\alpha}} \nabla\left(\frac{\partial}{\partial \underline{x}}\right)^{\underline{\alpha}} : \mathcal{E}_Q \longrightarrow \hat{E}_n .$$

Notice that Π involves only the well-defined derivations $\frac{\partial}{\partial x_2} = \frac{1}{y_1^n}\frac{\partial}{\partial y_2}, \cdots, \frac{\partial}{\partial x_d} = \frac{1}{y_1^n}\frac{\partial}{\partial y_d}$, of $\hat{\mathcal{O}}_{X_n,Q_n}[\frac{1}{y_1}]$. We have the formula

$$\Pi(hm) = (h_{|x_2=\cdots=x_d=0})\Pi(m)$$

which allows us to extend Π to an element of $\mathcal{L}_{K((y_1))}(\hat{E}_n)$. By the classical telescoping [Ka1, 8.9], $\mathrm{Im}(\Pi)$ is killed by $\nabla(\frac{\partial}{\partial x_2}), \ldots, \nabla(\frac{\partial}{\partial x_d})$. On the other hand, it follows from Nakayama's lemma and the congruence $\Pi(m) \equiv m \bmod (y_2, \ldots, y_d)$ that the $K((y_1))$-vector space $\mathrm{Im}(\Pi)$ generates $\hat{E}_n$ over $\hat{\mathcal{O}}_{X_n,Q_n}[\frac{1}{y_1}]$. The argument of [*loc. cit.*, 8.9.12–15] shows that

$$\hat{E}_n \cong \mathrm{Im}(\Pi) \otimes_{K((y_1))} \hat{\mathcal{O}}_{X_n,Q_n}\left[\frac{1}{y_1}\right].$$

In particular, the $K((y_1))$-vector space $\mathrm{Im}(\Pi)$ is of finite dimension equal to the rank of the free $\hat{\mathcal{O}}_{X_n,Q_n}[\frac{1}{y_1}]$-module $\hat{E}_n$.

Moreover, $\mathrm{Im}(\Pi)$ is canonically endowed with a structure of $K((y_1))[\frac{\partial}{\partial y_1}]$-module. Let N be its Newton polygon. The composed map

$$\begin{aligned} Der_{cont}\hat{\mathcal{O}}_{C,Q} &\to \hat{\mathcal{O}}_{C,Q} \otimes Der_{cont}\hat{\mathcal{O}}_{X_n,Q_n} \\ &\to \hat{\mathcal{O}}_{C,Q} \otimes Der_{cont}\hat{\mathcal{O}}_{X_n,Q_n}/\left(\frac{\partial}{\partial y_2}, \ldots, \frac{\partial}{\partial y_d}\right) \cong K[[y_1]]\frac{\partial}{\partial y_1} \end{aligned}$$

is an isomorphism, and via this isomorphism, the $K((y_1))[\frac{\partial}{\partial y_1}]$-module

$$(\tilde{\mathcal{E}}_{|C})_Q \otimes_{\mathcal{O}_{C,Q}} \mathrm{Frac}\, \hat{\mathcal{O}}_{C,Q} = \hat{E}_n \otimes_{\hat{\mathcal{O}}_{X_n,Q_n}[\frac{1}{y_1}]} \mathrm{Frac}\, \hat{\mathcal{O}}_{C,Q}$$

is isomorphic to $\mathrm{Im}(\Pi)$. Hence $NP_Q(\nabla_{|(C\setminus Q)}) = N$. Similarly $NP_Q(\nabla_{|(C'\setminus Q)}) = N$. □

5.3 The big cell

Theorem 5.3.1 *Let $Z \subset X$ be a smooth irreducible divisor in a smooth k-variety, and let $(\mathcal{E}, \nabla)$ be a coherent sheaf with integrable connection on $X \setminus Z$. There is an open dense subset $Z_0 \subset Z$ such that for any locally closed curve C which meets Z transversally at some point Q of Z_0, $NP_Q(\nabla_{|(C\setminus Q)}) = NP_Z(\nabla)$.*

Proof: Let n be $> \rho_Z(\nabla)$. For any point $P \in (\mathrm{Inf}^n X)_{|Z}\setminus(\pi^{n,1})^{-1}(Inf^1 Z)$, there is a family of smooth locally closed curves transversal to Z: $Y \hookrightarrow X \times S$ parametrized by a smooth K-variety S, such that $\bigcup_{s\in S} \gamma^n_{Y_s}(Y_s \cap Z)$ contains some neighborhood of P. In fact, by quasi-compactness of $(\mathrm{Inf}^n X)_{|Z} \setminus (\pi^{n,1})^{-1}(\mathrm{Inf}^1 Z)$, we may even assume that $\bigcup_{s\in S} \gamma^n_{Y_s}(Y_s \cap Z) = (\mathrm{Inf}^n X)_{|Z}\setminus(\pi^{n,1})^{-1}(\mathrm{Inf}^1 Z)$ if we allow S to have several connected components. We set $W = Y \cap (Z \times S)$, which is a divisor in Y étale over S (since every Y_s meets Z transversally). We are thus in the situation (4.2.1) of a tubular neighborhood of W, denoting by f (resp. g) the projection $Y \to S$ (resp. $W \to S$).

According to (4.2.2), the subset $W_1 = \{Q \in W \mid N_Q(\nabla_{|(Y_{g(Q)}\setminus Q)}) \neq N_Z(\nabla)\}$ is constructible, and so is its projection Z_1 in Z. By construction of Y, for any locally closed curve C which meets Z transversally at some closed point Q of Z_0, there is a point $s \in S$ such that $\gamma^n_{Y_s}(Q) = \gamma^n_C(Q)$; if $Q \notin Z_1$, we thus have, by (5.2.5),

$NP_Q(\nabla_{|(C\setminus Q)}) = NP_Z(\nabla)$. What remains to be shown is that Z_1 is not dense in Z. Since Z_1 is constructible, this means that $\eta_Z \notin Z_1$, which follows from the fact that for any $s \in S$ such that Y_s meets Z transversally at η_Z, $NP_{\eta_Z}(\nabla_{|(Y_s\setminus \eta_Z)}) = NP_Z(\nabla)$ (see 4.3.2). □

Remark 5.3.2 It already follows from (4.3.2) and (4.2.2) that for any tubular neighborhood f of Z in X, there is a (maximal) dense open subset Z_f of Z with the property that $N_Q(\nabla_{|(C\setminus Q)}) = NP_Z(\nabla)$ for the fibers C of f which meet Z in Z_f, and any locally closed curve can be considered locally as a fiber of f for a suitable tubular neighborhood f (I.1.3.3). The previous theorem can be roughly interpreted as saying that the intersection of the Z_f for all tubular neighborhoods f is non-empty.

We next complete our results of (I) concerning the regularity.

Corollary 5.3.3 *Let X be a smooth connected K-variety, Z a smooth closed subvariety of pure codimension one, $(\mathcal{E}, \nabla)$ be a coherent sheaf with integrable connection on $X\setminus Z$. Then ∇ is regular at Z if and only if there exists a dense subset of closed points $P \in Z$ and, for each P, a locally closed smooth curve C in X which meets Z transversally at P and for which $\nabla_{|(C\setminus C\cap Z)}$ is regular at P.*

Proof: This follows from (I.3.4.7) and (5.3.1) (by localizing around each component of Z and noting that the regularity means that the Newton polygon has only two edges (horizontal and vertical)). □

Remark 5.3.4 As Mebkhout pointed out to us, this criterion can be made more precise: it suffices that there exists one *non-characteristic* curve C which meets Z transversally at P and for which $\nabla_{|(C\setminus C\cap Z)}$ is regular at P.

Corollary 5.3.5 *Let X be a smooth connected K-variety, and let $j : X \longrightarrow \overline{X}$ be an open dominant immersion of X in a proper normal K-variety $\overline{X}$. Let $(\mathcal{E}, \nabla)$ be a coherent sheaf with integrable connection on X. Then ∇ is regular if and only if for every irreducible component $\overline{Z}$ of $\overline{X}\setminus j(X)$ of codimension 1, there exists a dense subset of simple closed points $P \in \overline{Z}$ and, for each P, a locally closed smooth curve $\overline{C}$ in $\overline{X}$ which meets $\overline{Z}$ transversally at P and for which $\nabla_{|\overline{C}\cap X}$ is regular.*

Proof: This is a combination of (I.5.10) and (5.3.1). □

5.4 Stratification by Newton polygons

5.4.1 Let X be a smooth connected K-variety, and Q be a closed point of X. We say some property $\mathcal{P}$ is shared by *almost every* smooth locally closed curve containing Q if there is a dense open subset $U \subset (\mathrm{Inf}^1 X)_{|Q}$ such that every smooth locally closed curve in X containing Q and such that $\gamma_C^1(Q) \in U$ satisfies $\mathcal{P}$.

Theorem 5.4.2 *Let $Z = \cup Z_{(i)} \subset X$ be a union of smooth closed subvarieties of pure codimension one, and let $(\mathcal{E}, \nabla)$ be a coherent sheaf with integrable connection on $X\setminus Z$. There is a finite set $\mathcal{N}$ of polygons with integral vertices, and a unique finite* *partition of Z into non-empty constructible subsets Z_N indexed by the elements of $\mathcal{N}$, with the following property:*

$(*)$ for every closed point $Q \in Z_N$ and almost every smooth locally closed curve C which meets each relevant component of Z transversally at Q, $NP_Q(\nabla_{|(C\setminus Q)}) = N$.

Proof: It is clear that the partition $Z = \coprod Z_N$ is unique if it exists (a constructible subset is determined by its subset of closed points). We can show its existence locally on Z. By applying (5.3.1) to $X \setminus \cup_{j \neq i} Z_j$, we obtain dense open subsets $Z_{i,0} \subset Z_i$ such that $NP_Q(\nabla_{|(C\setminus Q)}) = NP_{Z_i}(\nabla)$ for any smooth locally closed curve C which meets $Z_{i,0}$ transversally at some point Q. We then blow up $Z \setminus \cup Z_{i,0}$ (which is of codimension ≥ 2 in X). By applying (5.3.1) to the smooth part of the components of the exceptional divisor, we obtain dense open subsets $Z_{i,1} \subset (Z \setminus \cup Z_{i,0})$ such that $NP_Q(\nabla_{|(C\setminus Q)})$ is constant for every point $Q \in Z_{i,1}$ and almost every smooth locally closed curve C which meets each relevant component of Z transversally at Q. We then blow up $(Z \setminus \cup Z_{i,0}) \setminus \cup Z_{i,1}$ (which is of codimension ≥ 3 in X), and iterate the process. The construction of the partition is complete in finitely many steps: at the end, we glue together the strata corresponding to the same Newton polygon, taking into account the fact that a finite union of constructible subsets is constructible. □

Remark 5.4.3 Even if Z is smooth and connected, one cannot remove the restriction "for almost every" in $(*)$, as is shown for instance by the case of the rank 1 connection ∇ with formal solution $e^{\frac{y}{x^2}}$: the curve $y = 0$ is exceptional for the stratum $\{0\}$, since the restriction of ∇ to it is regular, while the restriction of ∇ to almost every curve passing through the origin has Poincaré-Katz rank 1.

If Z is smooth and connected, the principal Newton polygon $NP_Z(\nabla)$ is a privileged element of $\mathcal{N}$. We call the other elements the *secondary Newton polygons of* ∇ at Z.

We now state the *semicontinuity theorem*:

Theorem 5.4.4 *Assume that Z is smooth. Then $\bigcup_{N' \subset N} Z_{N'}$ is closed in Z. If moreover Z is connected, then every secondary Newton polygon lies above (i.e. is contained in) the principal Newton polygon $NP_Z(\nabla)$.*

Proof: We have to show that if $P' \in Z_{N'}$ lies in the Zariski closure of Z_N, then $N' \subset N$. We may replace X by an affine neighborhood of P'. Let C' be a smooth locally closed curve in X which meets Z transversally at P', such that $NP_{P'}(\nabla_{|(C'\setminus P')}) = N'$, and let C be a smooth locally closed curve in X which meets Z transversally at some point $P \in Z_N$, such that $NP_P(\nabla_{|(C\setminus P)}) = N$. We may and shall assume that $P' \notin C$, and $P \notin C'$. We need the following generalization of (I.1.3.3): □

Proposition 5.4.5 *Let $(C, P) \longrightarrow (X, Z)$, $(C', P') \longrightarrow (X, Z)$ be closed immersions of affine models, with $P' \notin C$, $P \notin C'$. There exist an open common neighborhood U of P and P' in X and a tubular neighborhood $(U \xrightarrow{f} S, t)$ of $Z \cap U$, such that $C \cap U = f^{-1}(f(P))$, $C' \cap U = f^{-1}(f(P'))$.*

Proof: One can find elements $\lambda_1, \dots, \lambda_{d-1}$ (resp. $\mu_1, \dots, \mu_{d-1}$) of $\mathcal{J}_C$ (resp. $\mathcal{J}_{C'}$) which are local generators at P (resp. P') and such that the λ_i (resp. μ_i) do not vanish at P' (resp. P). Then $((\lambda_i, \mu_i)_{i=1,\dots,d-1})$ defines a regular map $f' : U' \longrightarrow (\mathbf{P}^1)^{d-1}$ on some open common neighborhood U' of P and P' in X, which induces a tubular neighborhood after shrinking U'.

We apply this proposition to our previous situation. Let T be the normalization of some irreducible curve joining $f(P)$ and $f(P')$ in $f(\{Z_N\})$. Let X_T (resp. Z_T) be the pull-back of U (resp. $Z \cap U$) on T, and let ∇_{X_T} be the pull-back of ∇ on $X_T \backslash Z_T$. We have $NP_{Z_T}(\nabla_{X_T}) = N$. The result then follows from (5.1.2) applied to the closed immersion of C' in X_T.

The second assertion follows from the first, and from the fact (5.3.1) that $Z_{NP_Z(\nabla)}$ is dense. □

Remark 5.4.6 (5.4.4) is no longer true if one removes the assumption that Z is smooth, as is shown for instance by the case of the rank one connection with formal solution $e^{\frac{1}{xy}}$.

§6 Formal decomposition of an integrable connection at a singular divisor

6.1 We recall that K is an algebraically closed field of characteristic zero. Let $(\hat{\mathcal{G}}, w)$ be a field of formal functions over K, $(\hat{Y}, W)$ a smooth affine formal model of $(\hat{\mathcal{G}}, w)$ (I.1.1.6), $(x_1 = t_w, x_2, \dots, x_d)$ étale coordinates on $\hat{Y}$ adapted to W, and $(D_1, \dots, D_d)$ the corresponding adapted $\mathcal{O}(\hat{Y})$-basis of $Der_{\hat{Y}/K,W}$. According to (I.1.4.5), the transversal derivation $t_w^{-1} D_1$ determines a canonical isomorphism $\mathcal{O}(\hat{Y}) \cong \mathcal{O}(W)[[t_w]]$ and a family of (formal) integral curves $\{C_P\}_{P \in W}$ of $t_w^{-1} D_1$.

Let $(\mathcal{M}, \nabla)$ be a projective $\mathcal{O}(\hat{Y})[\frac{1}{t_w}]$-module of finite rank μ with integrable connection. Then $\mathcal{M}_{\hat{\mathcal{G}}} := \mathcal{M} \otimes \hat{\mathcal{G}}$ is naturally a $\hat{\mathcal{G}}/K$-differential module. According to (3.1), there exists a finite extension $(\hat{\mathcal{F}}, v)$ of $(\hat{\mathcal{G}}, w)$, with ramification index $e = e(v/w)$ (the *Turrittin index*) dividing $\mu!$, and a well-determined finite subset $\{\overline{P}_1, \dots, \overline{P}_r\}$ of $\hat{\mathcal{F}}/(\mathfrak{m}_v + \mathbf{Z})$, such that for every choice of representatives $P_i \in \hat{\mathcal{F}}$ of $\overline{P}_i$, we have the following decomposition of left $\hat{\mathcal{F}}[D_1]$-modules,

$$\text{(6.1.1)} \qquad \mathcal{M}_{\hat{\mathcal{F}}} = \oplus_{i=1}^{r} \hat{\mathcal{F}} \otimes_{\hat{\mathcal{F}}^{D_1}} \operatorname{Ker}_{\mathcal{M}_{\hat{\mathcal{F}}}} (D_1 - P_i)^{\mu}.$$

We set $t = (t_w)^{\frac{1}{e}}$, a generator of $\mathfrak{m}_v$ integral over $\mathcal{O}(\hat{Y})$. We denote by Z the normalization of W in $\hat{\mathcal{F}}^{D_1}$; this is an affine normal connected variety over K, finite over W. Then $\hat{\mathcal{F}} \cong \kappa(Z)((t))$, $D_1 = \frac{t}{e}\frac{d}{dt}$, $\mathcal{O}(Z)((t)) := \mathcal{O}(Z)[[t]][\frac{1}{t}] \cong \mathcal{O}(W)((t)) \otimes_{\mathcal{O}(W)} \mathcal{O}(Z)$ is finite over $\mathcal{O}(\hat{Y})[\frac{1}{t_w}] \cong \mathcal{O}(W)((t_w))$, and $\{C_P\}_{P \in W}$ lifts to a family of formal integral curves $\{\tilde{C}_Q\}_{Q \in Z}$ of $\frac{d}{dt}$.

The previous decomposition can be rewritten as

(6.1.2) $$\mathcal{M}_{\kappa(Z)((t))} = \oplus_{i=1}^{r} \kappa(Z)((t)) \otimes_{\kappa(Z)} \mathrm{Ker}_{\mathcal{M}_{\kappa(Z)((t))}} (D_1 - P_i)^{\mu} \ .$$

We denote by e_i the projection of $\mathcal{M}_{\kappa(Z)((t))}$ on the factor $\kappa(Z)((t)) \otimes_{\kappa(Z)} \mathrm{Ker}_{\mathcal{M}_{\kappa(Z)((t))}} (D_1 - P_i)^{\mu}$ in the previous decomposition, so that we may rewrite

(6.1.3) $$\mathcal{M}_{\kappa(Z)((t))} = \oplus_{i=1}^{r} e_i(\mathcal{M}_{\kappa(Z)((t))}).$$

Note that $\mathcal{M}_{\mathcal{O}(Z)((t))} \hookrightarrow \mathcal{M}_{\kappa(Z)((t))}$ because $\mathcal{M}$ is projective. The actions of $D_2, \ldots, D_d$ on $\kappa(Z)$ do not respect $\mathcal{O}(Z)$ in general. In fact, there is a natural injection $Der_K \mathcal{O}(Z) \longrightarrow \mathcal{O}(Z) \otimes_{\mathcal{O}(W)} Der_K \mathcal{O}(W)$ which becomes an isomorphism after tensoring by $\kappa(Z)$. Although Z may not be smooth, we shall say, by an abuse of notation, that the commuting actions of D_1 and $Der_K \mathcal{O}(Z)$ on $\mathcal{M}_{\mathcal{O}(Z)((t))}$ define an integrable connection ∇ (in d variables).

We have seen in (3.4) that the decomposition (6.1.2) is stable under $Der_K \mathcal{O}(Z)$. In this section, we shall study the delicate problem of descending this decomposition from $\kappa(Z)((t))$ to $\mathcal{O}(Z)((t))$.

Theorem 6.2 *We make the following technical assumption*

$(*)$ *The horizontal side of the Newton polygon $N_P(End(\mathcal{M}_{|C_P}))$ does not depend on the point $P \in W$, except possibly on a subset of codimension 2 in W.*

Then, under the canonical isomorphism $\hat{\mathcal{F}}/(\mathfrak{m}_v + \mathbf{Z}) \cong \kappa(Z)[\frac{1}{t}]/\mathbf{Z}$, the elements $\overline{P}_i$ belong to $\mathcal{O}(Z)[\frac{1}{t}]/\mathbf{Z}$. For any choice of representatives P_i of $\overline{P}_i$ in $\mathcal{O}(Z)((t))$, the decomposition (6.1.3) *descends to a decomposition of $\mathcal{O}(Z)((t))$-modules with integrable connection (in d variables)*

(6.2.1) $$\mathcal{M}_{\mathcal{O}(Z)((t))} = \oplus_{i=1}^{r} e_i(\mathcal{M}_{\mathcal{O}(Z)((t))})$$

where

(6.2.2) $$e_i(\mathcal{M}_{\mathcal{O}(Z)((t))}) = \mathcal{O}(Z)((t)) \otimes_{\mathcal{O}(Z)} \mathrm{Ker}_{\mathcal{M}_{\mathcal{O}(Z)((t))}} (D_1 - P_i)^{\mu}.$$

Moreover, for any $\nu = 0, 1, \ldots,$

(6.2.3) $$\mathrm{Ker}_{\mathcal{M}_{\kappa(Z)((t))}} (D_1 - P_i)^{\nu} = \kappa(Z) \otimes_{\mathcal{O}(Z)} \mathrm{Ker}_{\mathcal{M}_{\mathcal{O}(Z)((t))}} (D_1 - P_i)^{\nu}.$$

Proof: We set $\mathcal{E} := End(\mathcal{M}_{\mathcal{O}(Z)((t))})$, and note that the e_i form a complete set of orthogonal idempotents in $\mathcal{E}_{\kappa(Z)((t))}$ which does not depend on the choice of the representatives P_i. The existence of the decomposition (6.2.1) can be expressed as: $e_i \in \mathcal{E}^{\nabla}$. It follows from (6.2.1) that

(6.2.4) $$e_i(\mathcal{M}_{\mathcal{O}(Z)((t))}) = \mathcal{M}_{\mathcal{O}(Z)((t))} \cap [\kappa(Z)((t)) \otimes_{\kappa(Z)} \mathrm{Ker}_{\mathcal{M}_{\kappa(Z)((t))}} (D_1 - P_i)^{\mu}] \ .$$

This is much weaker than (6.2.2). However, (6.2.1) allows us to prove (6.2.2) under the further assumption that in decomposition (6.1.2) $r = 1$, *i.e.* when there is only one P_i. Assumption $(*)$ is used in the proof of (6.2.1).

We shall proceed in several steps.

i) *First step:* $e_i \in (\mathcal{E}_{\kappa(Z)((t))})^{\nabla}$.

This follows from (3.5). In fact, if we set

$$K_i^{(\nu)} = \mathrm{Ker}_{\mathcal{M}_{\kappa(Z)((t))}}(D_1 - P_i)^{\nu} \ ,$$
$$M_i^{(\nu)} = \kappa(Z)((t)) \otimes_{\kappa(Z)} K_i^{(\nu)} \ (\nu = 0, 1, \ldots),$$

(3.5) tells us that $M_i^{(\mu)}$ is stable under D_1 and $Der_K \mathcal{O}(Z)$.

6.3 Interlude In various steps of the proof of (6.2), we will have to deal with the following situation. We will be given a dense affine open subset $Z' \subset Z$, and, by further shrinking of Z' if necessary, we will be able to assume that $Z \backslash Z' = \cup T_h$ is a finite union of irreducible divisors $T_h \subset Z$ (with generic point η_h). Because Z is normal, we then have

$$\mathcal{O}(Z) = \mathcal{O}(Z') \cap \bigcap_h \mathcal{O}_{Z,\eta_h}, \tag{6.3.1}$$

the intersection taking place in $\kappa(Z)$. Hence, we also have

$$\mathcal{O}(Z)((t)) = \mathcal{O}(Z')((t)) \cap \bigcap_h \mathcal{O}_{Z,\eta_h}((t)), \tag{6.3.2}$$

the intersection taking place in $\kappa(Z)((t))$.

A component T_h of $Z \backslash Z'$ corresponds to a discrete valuation v_h of $\kappa(Z)$ with valuation ring $\mathcal{O}_{Z,\eta_h}$. Let us fix one such h and let x be a uniformizer of $\mathcal{O}_{Z,\eta_h}$, *i.e.* a local equation for T_h in Z, at η_h, and $R_{v_h} = \mathcal{O}_{Z,\eta_h}((t))$. The quotient field F_{v_h} of R_{v_h} is contained in $\kappa(Z)((t))$, and R_{v_h} is the ring of integers of a discrete valuation of F_{v_h}/K extending v_h, and still denoted by v_h, with $v_h(t) = 0$, $v_h(x) = 1$.

Let ∂ be a derivation of $\mathcal{O}(Z)$ generically transversal to T_h and such that $\mathrm{tr\,deg}\,(\kappa(Z)/\kappa(Z)^{\partial}) = 1$. On replacing Z by an open neighborhood of η_h, we may assume a family of integral curves of ∂ (I.1.4.3)

$$\begin{array}{ccc} Z & \hookleftarrow & T_h \\ f\downarrow & \swarrow g & \\ S & & \end{array}$$

The generic fiber $C_h = Z_{\eta_S}$ of f is then a smooth curve over $\kappa(\eta_S) = \kappa(S) = \kappa(Z)^{\partial}$, with a distinguished closed point $Q = \eta_h$, with residue field $\kappa(Q) = \kappa(T_h)$ finite over $\kappa(S)$. We may then assume $\partial = \partial_{x,f}$ as in (1.4.1). So, x is a local coordinate on C_h centered at Q, such that $\partial_{|C_h} = \frac{d}{dx}$. We have $\kappa(C_h) = \kappa(Z)$ and $\mathcal{O}_{C,\eta_h} = \mathcal{O}_{Z,\eta_h}$. By shrinking C_h around η_h, we may and shall assume that $Z'_{\eta_S} = C_h \backslash Q$. We then have $\mathcal{O}(Z')((t)) \subset \mathcal{O}(C_h)[\frac{1}{x}]((t))$. We sometimes replace C_h by its completion $\hat{C}_h$ at the point η_h, so that $\mathcal{O}(\hat{C}_h) = \kappa(\eta_h)[[x]]$. We notice that $\mathcal{O}_{C,\eta_h}((t))$ and $\kappa(\eta_h)[[x,t]][\frac{1}{t}]$ are discrete valuation rings.

ii) *Second step:* $e_i \in \mathcal{E}^{\nabla}$ *(under assumption* $(*)$*).*
Let $F = \mathrm{Frac}(\mathcal{O}(Z)((t)))$ and let us pick a cyclic vector of the dual of $\mathcal{E}_F$ with respect to the derivation $\frac{d}{dt}$, and let

$$\Lambda = \left(\frac{d}{dt}\right)^{\mu} - \sum_{j<\mu} b_j \left(\frac{d}{dt}\right)^{j}$$

be the associated monic differential operator. Then $(\mathcal{E}_F)^{\nabla(\frac{d}{dt})} \cong \mathrm{Ker}_F \Lambda$ (see also (III.4.4) for more detail on this standard construction). Since the dual of $\mathcal{E}_F$ is an F/K-differential module, the indicial polynomial of Λ at 0 is a product $\tau\varphi$ where $\tau \in \kappa(Z)$ and φ is a polynomial with coefficients in K (I.6.1.3). □

Lemma 6.4 $\Lambda \in \mathcal{O}(Z)((t))[\frac{d}{dt}]$ *and, under assumption* $(*)$, $\tau \in \mathcal{O}(Z)^{\times}$.

Proof: It is clear that there is a dense affine open subset $Z' \subset Z$ such that $\Lambda \in \mathcal{O}(Z')((t))[\frac{d}{dt}]$ and $\tau \in \mathcal{O}(Z')^{\times}$. The considerations of the "Interlude" apply, and we have to show that (for any chosen component T_h and $v = v_h$) $\Lambda \in R_v[\frac{d}{dt}]$ and $\tau \in \mathcal{O}^{\times}_{Z,\eta_h}$. Since $\nabla(\frac{d}{dt})\mathcal{E}_{R_v} \subset \mathcal{E}_{R_v}$, (2.3.2) shows that $v(b_j) \geq 0$ for $j = 0, \ldots, \mu - 1$, *i.e.* $b_j \in R_v$. Thus we can define a cyclic $R_v/\mathcal{O}_{Z,\eta_h}$-differential module $\mathcal{F} = R_v[\frac{d}{dt}]/R_v[\frac{d}{dt}]\Lambda$, with the natural action $\aleph(\frac{d}{dt})$ of $\frac{d}{dt}$. By construction, $(\mathcal{E}_{F_v}, \nabla_{F_v}) \cong (\mathcal{F}_{F_v}, \aleph_{F_v})$. □

Sublemma 6.4.1 *The semisimplifications of the reductions of* $(\mathcal{F}, \aleph)$ *and* $(\mathcal{E}, \nabla)$ *modulo the maximal ideal of* v *are isomorphic.*

Proof: The proof is the same as for Gabber's theorem mentioned in (4.2.3). For convenience, we sketch the argument. We regard $\mathcal{E}_{R_v}$ and $\mathcal{F}_{R_v}$ as two R_v-lattices inside the F_v-vector space $\mathcal{E}_{F_v} = \mathcal{F}_{F_v}$. We introduce $\mathcal{M}(a, b) = x^a\mathcal{E}_{R_v} + x^b\mathcal{F}_{R_v}$, a submodule of $\mathcal{E}_{F_v} = \mathcal{F}_{F_v}$ horizontal w.r.t. $\frac{d}{dt}$. It is also an R_v-lattice (it is of finite rank as a quotient of $\mathcal{E}_{R_v} \oplus \mathcal{F}_{R_v}$ and has no x-torsion).

For $n >> 0$, we have $x^n\mathcal{E}_{R_v} \subset \mathcal{F}_{R_v}$, $x^n\mathcal{F}_{R_v} \subset \mathcal{E}_{R_v}$, and therefore $\mathcal{M}(0, n) = \mathcal{E}_{R_v}$, $\mathcal{M}(n, 0) = \mathcal{F}_{R_v}$. By symmetry and iteration, it suffices to compare $\mathcal{M}(a, b)$ and $\mathcal{M}(a + 1, b)$. We have

$$\begin{aligned} x\mathcal{M}(a+1, b) \subset x\mathcal{M}(a, b) &= \mathcal{M}(a+1, b+1) \\ &\subset \mathcal{M}(a+1, b) \subset \mathcal{M}(a, b), \end{aligned}$$

and associated exact sequences of R_v/xR_v-modules with K-linear connection

$$\begin{aligned} 0 &\longrightarrow \mathcal{M}(a+1, b)/x\mathcal{M}(a, b) \longrightarrow \mathcal{M}(a, b)/x\mathcal{M}(a, b) \\ &\longrightarrow \mathcal{M}(a, b)/\mathcal{M}(a+1, b) \longrightarrow 0, \\ 0 &\longrightarrow x\mathcal{M}(a, b)/x\mathcal{M}(a+1, b) \cong \mathcal{M}(a, b)/\mathcal{M}(a+1, b) \\ &\longrightarrow \mathcal{M}(a+1, b)/x\mathcal{M}(a+1, b) \longrightarrow \mathcal{M}(a+1, b)/x\mathcal{M}(a, b) \longrightarrow 0, \end{aligned}$$

which show that $\mathcal{M}(a, b)/x\mathcal{M}(a, b)$ and $\mathcal{M}(a+1, b)/x\mathcal{M}(a+1, b)$ have the same semisimplification.

Assumption $(*)$ implies that the Newton polygons of $\mathcal{E}_{F_v}$ and of $\mathcal{E}_{R_v}/x\mathcal{E}_{R_v}$ have the same horizontal slope. From (6.4.1), we deduce that the Newton polygons of $\mathcal{F}_{F_v}$ and of $\mathcal{F}_{R_v}/x\mathcal{F}_{R_v}$ also have the same horizontal slope. This implies that $\tau \in \mathcal{O}^{\times}_{Z,\eta_h}$ which proves (6.4). □

Since $\tau \in \mathcal{O}(Z)^{\times}$, the coefficient a_n in the expansion $\sum a_n t^n$ of any element of $\mathrm{Ker}_{\kappa(Z)((t))}\Lambda$, for n big enough, is a linear combination of the terms of lower order with coefficients in $\mathcal{O}(Z)$.
Hence

$$\begin{aligned}(\mathcal{E}_{\kappa(Z)((t))})^{\nabla(\frac{d}{dt})} &\cong \mathrm{Ker}_{\kappa(Z)((t))}\Lambda \\ &\cong \mathrm{Ker}_{\mathcal{O}(Z)((t))\otimes\kappa(Z)}\Lambda \cong (\mathcal{E}_{\mathcal{O}(Z)((t))\otimes\kappa(Z)})^{\nabla(\frac{d}{dt})} \cong \mathcal{E}^{\nabla(\frac{\partial}{\partial t})} \otimes \kappa(Z).\end{aligned}$$

Moreover, $\mathcal{E}^{\nabla(\frac{\partial}{\partial t})}$ is an $\mathcal{O}(Z)$-module of finite type, equipped with an action of $Der_K\mathcal{O}(Z)$. It is torsion free, as a submodule of $\mathcal{E}$. Let $Z' \subset Z$ be an open dense subvariety such that

$$e_i \in \cap_{D\in Der_K\mathcal{O}(Z)}(\mathcal{E}^{\nabla(\frac{\partial}{\partial t})} \otimes \mathcal{O}(Z'))^{\nabla(D)}.$$

Again, we use the notation of (6.3). Using the fact that $e_i \in (\mathcal{E}^{\nabla(\frac{\partial}{\partial t})} \otimes \kappa(T_h)((x)))^{\nabla(\frac{\partial}{\partial x})}$, we see that it will be enough to show that (for any h) $e_i \in \mathcal{E}^{\nabla(\frac{\partial}{\partial t})} \otimes \hat{\mathcal{O}}_{Z,\eta_h} = \mathcal{E}^{\nabla(\frac{\partial}{\partial t})} \otimes \kappa(T_h)[[x]]$. But $\mathcal{E}^{\nabla(\frac{\partial}{\partial t})} \otimes \kappa(T_h)[[x]]$ has no x-torsion, hence it is a projective $\kappa(T_h)[[x]]$-module (of finite type) with a connection (for the $\kappa(T_h)$-linear x-adically continuous derivations of $\kappa(T_h)[[x]]$). So,

$$(\mathcal{E}^{\nabla(\frac{\partial}{\partial t})} \otimes \kappa(T_h)((x)))^{\nabla(\frac{\partial}{\partial x})} = (\mathcal{E}^{\nabla(\frac{\partial}{\partial t})} \otimes \kappa(T_h)[[x]])^{\nabla(\frac{\partial}{\partial x})}. \qquad \square$$

iii) *Third step: $\overline{P}_i$ belongs to $\mathcal{O}(Z)[\frac{1}{t}]/\mathbf{Z}$.*
By the second step, we may assume that only one $P_i = P$ appears in decomposition (6.2.1). Clearly, there is a dense affine open subset $Z' \subset Z$ such that $\overline{P}$ belongs to $\mathcal{O}(Z')[\frac{1}{t}]/\mathbf{Z}$. We may and will assume that $P \in \mathcal{O}(Z')[\frac{1}{t}]$. We apply the considerations of the "Interlude" to the open subvariety Z' of Z. Then we may assume that, for any fixed component T_h and any i, $P \in \mathcal{O}(C_h)[\frac{1}{x}, \frac{1}{t}]$: we have to prove that $P \in \mathcal{O}(C_h)[\frac{1}{t}]$. We may replace C_h by its completion $\hat{C}_h$ at the point η_h, and we will only have to prove that $P \in \kappa(T_h)[[x]][\frac{1}{t}]$. So, after replacing K by the algebraic closure of $\kappa(T_h)$, and taking into account the fact that $\kappa(\eta_h)[[x,t]][\frac{1}{t}]$ is principal, we may rephrase our problem as follows.

Proposition 6.5 *Let $(\mathcal{M}, \nabla)$ be a free $K[[x,t]][\frac{1}{t}]$-module of finite rank μ with a $K[[x]]$-linear connection (for the t-adically continuous $K[[x]]$-linear derivations of $K[[x,t]]$). Let $P \in K((x))[\frac{1}{t}]$ be such that*

$$\text{(6.5.1)} \qquad \mathcal{M}_{K((x))((t))} = K((x))((t)) \otimes_{K((x))} \mathrm{Ker}_{\mathcal{M}_{K((x))((t))}} \left(\nabla\left(t\frac{\partial}{\partial t}\right) - P\right)^{\mu}.$$

Then $P \in K[[x]][\frac{1}{t}]$.

Proof: Let ρ denote the (generic) Katz-Poincaré rank of $\mathcal{M}$ along $t = 0$. We may write $P = \sum_{j=0}^{\rho} a_j(x)t^{-j}$, with $a_j(x) = \sum_{h=n_j}^{\infty} a_{jh}x^h \in K((x))$, $a_{jn_j} \neq 0$. We regard $K((x))$ as a valued field, via some x-adic absolute value, trivial over K, $|\ |_x$, with $0 < |x|_x < 1$. We use x-adic analysis and Dwork's notion of an x-adic generic point ξ_r at distance $r \in (0, 1]$ from 0. This is a point of the Berkovich x-adic analytic space $\mathbf{A}^{1\,an}_{K((x))}$ associated to the affine line, with canonical coordinate t: it is an extension $|\ |_r$ of the absolute value $|\ |_x$ to $K((x))(t)$, such that $|t|_r = r$. We denote by K_r the completion of $K((x))(t)$ w.r.t. the absolute value $|\ |_r$, and indicate by $t(\xi_r) = t_r$ the canonical image of t in K_r. Then on the extension $\mathbf{A}^{1\,an}_{K_r}$ of $\mathbf{A}^{1\,an}_{K((x))}$ over K_r, there is a canonical point ξ_r' lying over ξ_r: ξ_r' has a fundamental system of neighborhoods in the Berkovich topology isomorphic to standard disks defined over K_r.

The (easy) x-adic analogue of a result of Clark ([CL], [B1]), implies that the elements of $\mathrm{Ker}_{\mathcal{M}_{K((x))((t))}}(\nabla(t\frac{\partial}{\partial t}) - P)^{\mu}$ are defined over the field of x-adic meromorphic functions of t on $\mathbf{A}^{1\,an}_{K((x))}$ at $t = 0$: there is a radius $0 < r_0 \leq 1$, such that those elements are defined over the ring $\mathcal{A}_{r_0}$ of x-adic meromorphic functions, defined over $K((x))$, on the disk D_{r_0} of radius r_0 centered at 0, with poles only at $t = 0$.

Let $\underline{e} = (e_1, \ldots, e_{\mu})$ be a basis of $\mathcal{M}$, and let us write a full set of solutions of $\mathcal{M}$ at ξ_r as

$$\underline{y} = \underline{e}Y_{\xi_r}(x, t)$$

where $Y_{\xi_r}(x, t)$ satifies $Y_{\xi_r}(x, t_r) = 1_{\mu}$ and the differential system

$$\frac{d}{dt}Y = GY, \quad G = G(x, t) \in M_{\mu}\left(K[[x, t]]\left[\frac{1}{t}\right]\right).$$

From the previous differential system, we compute the matrices $G^{(i)} \in M_{\mu}(K[[x, t]][\frac{1}{t}])$ such that $(\frac{d}{dt})^i Y = G^{(i)}Y$. Then

$$Y_{\xi_r}(x, t) = \sum_{i \geq 0} G^{(i)}(x, t_r)\frac{(t - t(\xi_r))^i}{i!} \in M_{\mu}(K_1[[t - t(\xi_r)]]).$$

As a consequence of Proposition 2.4.1, there is a constant C such that

$$t^{(\rho+1)i+C}G^{(i)}(x, t) \in M_{\mu}(K[[x, t]]).$$

So, $|G^{(i)}(x, t_r)|_r = |G^{(i)}(x, t)|_r \leq r^{-(\rho+1)i-C}$. In particular, the solutions of $\mathcal{M}$ at ξ_1 converge and are bounded by 1 in the open unit disk centered at ξ_1 in $\mathbf{A}^{1\,an}_{K_1}$. More generally, the (maximal common) radius of (bounded) convergence R_r of solutions of $\mathcal{M}$ at the x-adic generic point ξ_r at distance r from 0 satisfies $R_r \geq r^{\rho+1}$.

For $0 \le j \le \rho$ let us denote by $\mathcal{A}_{r_0} e^{\int a_j(x)t^{-j-1}dt}$ the projective module of rank 1 over $\mathcal{A}_{r_0}$

$$\mathcal{A}_{r_0}\left[\frac{\partial}{\partial t}\right] \Big/ \mathcal{A}_{r_0}\left[\frac{\partial}{\partial t}\right]\left(t\frac{\partial}{\partial t} - a_j(x)t^{-j}\right),$$

equipped with the natural connection (relative to $K((x))$) (for $j = 0$, we may write $\mathcal{A}_{r_0} t^{a_0(x)}$ for $\mathcal{A}_{r_0} e^{\int a_0(x)t^{-1}dt}$). An easy calculation shows that, for any $j > 0$ and $r < r_0$, the radius of convergence of $\mathcal{A}_{r_0} e^{\int a_j(x)t^{-j-1}dt}$ at ξ_r is at least $r^{j+1}|x|_x^{-n_j}$; besides, if $n_j < 0$ the previous estimate gives the *exact* radius of convergence at ξ_r.

We start by considering the term $a_\rho(x)t^{-\rho}$ of P to show that $n_\rho \ge 0$. In fact, assuming the contrary, let us pick $r < r_0$ such that all values $r^{j+1}|x|_x^{-n_j}$ are *distinct*. Then the radius of convergence at ξ_r of

$$\mathcal{A}_{r_0} e^{\int P\frac{dt}{t}} := \mathcal{A}_{r_0}\left[\frac{\partial}{\partial t}\right] \Big/ \mathcal{A}_{r_0}\left[\frac{\partial}{\partial t}\right]\left(t\frac{\partial}{\partial t} - P\right),$$

tensor product of the previous $\mathcal{A}_{r_0} e^{\int a_j(x)t^{-j-1}dt}$, is at most $r^{\rho+1}|x|_x^{-n_\rho}$. Under assumption 6.5.1, $\mathrm{Ker}_{\mathcal{M}_{\mathcal{A}_{r_0}}}(\nabla(t\frac{\partial}{\partial t}) - P) \ne 0$ and any of its non-zero elements provides a solution (*i.e.* a full set of solutions) of $\mathcal{A}_{r_0} e^{\int P\frac{dt}{t}}$ at ξ_r of radius of convergence $R_r \ge r^{\rho+1}$. This is a contradiction. So, $n_\rho \ge 0$. We may then replace $\mathcal{M}$ by its tensor product $\mathcal{N}$ with the inverse of

$$\begin{aligned} &K[[x,t]]\left[\frac{1}{t}\right] e^{\int a_\rho(x)t^{-\rho-1}dt} \\ &:= K[[x,t]]\left[\frac{1}{t}\right]\left[\frac{\partial}{\partial t}\right] / K[[x,t]]\left[\frac{1}{t}\right]\left[\frac{\partial}{\partial t}\right]\left(t\frac{\partial}{\partial t} - a_\rho(x)t^{-\rho}\right). \end{aligned}$$

Since $a_\rho(x) \in K[[x]]$, $\mathcal{N}$ has the same properties as $\mathcal{M}$, for P replaced by $P - a_\rho(x)t^{-\rho}$, but its Poincaré-Katz rank is less than ρ. We then iterate our reasoning and eliminate step by step the possibility of a pole at $x = 0$ for any $a_j(x)$, $j = \rho, \rho - 1, \ldots, 0$. □

From now on, we may and shall assume that *the representatives P_i of $\overline{P}_i$ belong to $\mathcal{O}(Z)((t))$.*

iv) *Fourth step: reduction of* (6.2.2) *to the case where Z is one-dimensional.* We may assume here again that there is only one P_i in the decomposition (6.1.2). We may also find $Z' \subset Z$, as in the "Interlude", such that

$$\mathcal{M}_{\mathcal{O}(Z')((t))} = \mathcal{O}(Z')((t)) \otimes_{\mathcal{O}(Z')} \mathrm{Ker}_{\mathcal{M}_{\mathcal{O}(Z')((t))}}(D_1 - P_i)^{\mu}. \tag{6.6.1}$$

We again use the notation of the "Interlude", assuming $Z\setminus Z' = \cup_h T_h$. From (6.3.1) and (6.3.2), and because $\mathcal{M}_{\mathcal{O}(Z)((t))}$ is flat over $\mathcal{O}(Z)$, we draw

$$\mathcal{M}_{\mathcal{O}(Z)((t))} = \mathcal{M}_{\mathcal{O}(Z')((t))} \cap \bigcap_h \mathcal{M}_{\mathcal{O}_{Z,\eta_h}((t))}, \tag{6.6.2}$$

$$\begin{aligned} &\mathrm{Ker}_{\mathcal{M}_{\mathcal{O}(Z)((t))}}(D_1 - P_i)^\mu \\ &\quad = \mathrm{Ker}_{\mathcal{M}_{\mathcal{O}(Z')((t))}}(D_1 - P_i)^\mu \cap \bigcap_h \mathrm{Ker}_{\mathcal{M}_{\mathcal{O}_{Z,\eta_h}((t))}}(D_1 - P_i)^\mu . \end{aligned}$$

In this fourth step, we *reduce* the desired equality

$$\mathcal{M}_{\mathcal{O}(Z)((t))} = \mathcal{O}(Z)((t)) \otimes_{\mathcal{O}(Z)} \mathrm{Ker}_{\mathcal{M}_{\mathcal{O}(Z)((t))}}(D_1 - P_i)^\mu \tag{6.6.3}$$

to the following one:

$$\mathcal{M}_{\mathcal{O}_{Z,\eta_h}((t))} = \mathcal{O}_{Z,\eta_h}((t)) \otimes_{\mathcal{O}_{Z,\eta_h}} \mathrm{Ker}_{\mathcal{M}_{\mathcal{O}_{Z,\eta_h}((t))}}(D_1 - P_i)^\mu \tag{6.6.4}$$

for every h. We first note that the natural map

$$\mathcal{O}_{Z,\eta_h}((t)) \otimes_{\mathcal{O}_{Z,\eta_h}} \mathrm{Ker}_{\mathcal{M}_{\mathcal{O}_{Z,\eta_h}((t))}}(D_1 - P_i)^\mu \to \mathcal{M}_{\mathcal{O}_{Z,\eta_h}((t))}$$

is injective, since $\mathcal{O}(Z)((t))$ is flat over $\mathcal{O}(Z)$ and $M_i^{(\mu)} = \kappa(Z)((t)) \otimes_{\kappa(Z)} K_i^{(\mu)}$ can be identified with its image in $\mathcal{M}_{\kappa(Z)((t))}$. From (6.6.1) and (6.6.4), we deduce

$$\begin{aligned} M_i^{(\mu)} &= \kappa(Z) \otimes_{\mathcal{O}(Z)} \mathrm{Ker}_{\mathcal{M}_{\mathcal{O}(Z')}}(D_1 - P_i)^\mu \\ &= \kappa(Z) \otimes_{\mathcal{O}(Z)} \mathrm{Ker}_{\mathcal{M}_{\mathcal{O}_{Z,\eta_h}((t))}}(D_1 - P_i)^\mu \end{aligned}$$

whence, by (6.6.2),

$$M_i^{(\mu)} = \kappa(Z) \otimes_{\mathcal{O}(Z)} \mathrm{Ker}_{\mathcal{M}_{\mathcal{O}(Z)((t))}}(D_1 - P_i)^\mu . \tag{6.6.5}$$

This implies in turn:

$$\mathrm{Ker}_{\mathcal{M}_{\mathcal{O}(Z')((t))}}(D_1 - P_i)^\mu = \mathcal{O}(Z') \otimes_{\mathcal{O}(Z)} \mathrm{Ker}_{\mathcal{M}_{\mathcal{O}(Z)((t))}}(D_1 - P_i)^\mu , \tag{6.6.6}$$

$$\mathrm{Ker}_{\mathcal{M}_{\mathcal{O}_{Z,\eta_h}((t))}}(D_1 - P_i)^\mu = \mathcal{O}_{Z,\eta_h} \otimes_{\mathcal{O}(Z)} \mathrm{Ker}_{\mathcal{M}_{\mathcal{O}(Z)((t))}}(D_1 - P_i)^\mu , \tag{6.6.7}$$

whence

$$\mathcal{M}_{\mathcal{O}(Z')((t))} = \mathcal{O}(Z')((t)) \otimes_{\mathcal{O}(Z)} \mathrm{Ker}_{\mathcal{M}_{\mathcal{O}(Z)((t))}}(D_1 - P_i)^\mu , \tag{6.6.8}$$

$$\mathcal{M}_{\mathcal{O}_{Z,\eta_h}((t))} = \mathcal{O}_{Z,\eta_h}((t)) \otimes_{\mathcal{O}(Z)} \mathrm{Ker}_{\mathcal{M}_{\mathcal{O}(Z)((t))}}(D_1 - P_i)^\mu . \tag{6.6.9}$$

In order to conclude (6.6.3), it then suffices to combine (6.6.8), (6.6.9), (6.6.2), the remark following (6.6.4), and the fact that in $\mathcal{M}_{\kappa(Z)((t))}$ the images of

$\mathcal{M}_{\mathcal{O}(Z)((t))} \otimes (\mathcal{O}(Z')((t)) \cap \bigcap_h \mathcal{O}_{Z,\eta_h}((t)))$ and of $(\mathcal{O}(Z')((t)) \otimes_{\mathcal{O}(Z')} \mathrm{Ker}_{\mathcal{M}_{\mathcal{O}(Z')((t))}} (D_1 - P_i)^\mu) \cap \bigcap_h (\mathcal{O}_{Z,\eta_h}((t)) \otimes_{\mathcal{O}_{Z,\eta_h}} \mathrm{Ker}_{\mathcal{M}_{\mathcal{O}_{Z,\eta_h}((t))}} (D_1 - P_i)^\mu)$ coincide ([A.C.I, 2.6]).

Taking into account the fact that $\mathcal{O}_{Z,\eta_h}((t))$ is principal, as we noticed before, we have thus reduced our statement (6.6.3) to the following lemma.

Lemma 6.7 *Let $(\mathcal{M}, \nabla)$ be a free $\mathcal{O}_{C,\eta_h}((t))$-module of finite rank μ with integrable connection (with respect to $\partial = \partial_{x,f} = \partial/\partial x$ and $\partial/\partial t$). Let $P \in \mathcal{O}_{C,\eta_h}((t))$ be such that $\mathcal{M}_{\kappa(C)((t))} = \kappa(C)((t)) \otimes_{\kappa(C)} \mathrm{Ker}_{\mathcal{M}_{\kappa(C)((t))}} (\nabla(t\frac{\partial}{\partial t}) - P)^\mu$. Then $\mathcal{M} = \mathcal{O}_{C,\eta_h}((t)) \otimes_{\mathcal{O}_{C,\eta_h}} \mathrm{Ker}_{\mathcal{M}} (\nabla(t\frac{\partial}{\partial t}) - P)^\mu$.*

v) *Fifth step: proof of* (6.7).

We are allowed to tensor $\mathcal{M}$ by the inverse of the projective $\mathcal{O}_{C,\eta_h}((t))$-module of rank 1

$$\mathcal{O}_{C,\eta_h}((t)) e^{\int P \frac{dt}{t}} := \mathcal{O}_{C,\eta_h}((t)) \left[\frac{\partial}{\partial t}\right] / \mathcal{O}_{C,\eta_h}((t)) \left[\frac{\partial}{\partial t}\right] \left(t\frac{\partial}{\partial t} - P\right),$$

endowed with the canonical K-linear integrable connection. This reduces us to proving the following

Lemma 6.8 *Let $(\mathcal{M}, \nabla)$ be as in Lemma* 6.7 *but assume that*

(6.8.1) $$\mathcal{M}_{\kappa(C)((t))} = \kappa(C)((t)) \otimes_{\kappa(C)} \mathrm{Ker}_{\mathcal{M}_{\kappa(C)((t))}} \left(\nabla\left(t\frac{\partial}{\partial t}\right)\right)^\mu .$$

Then for all ν, $\mathrm{Ker}_{\mathcal{M}} \nabla(t\frac{\partial}{\partial t})^\nu$ is a free $\mathcal{O}_{C,\eta_h}$-submodule of finite rank of $\mathcal{M}$ stable under $\frac{\partial}{\partial x}$. Moreover

(6.8.2) $$\mathrm{Ker}_{\mathcal{M}_{\kappa(C)((t))}} \nabla\left(t\frac{\partial}{\partial t}\right)^\nu = \kappa(C) \otimes_{\mathcal{O}_{C,\eta_h}} \mathrm{Ker}_{\mathcal{M}} \nabla\left(t\frac{\partial}{\partial t}\right)^\nu$$

and

(6.8.3) $$\mathcal{M} = \mathcal{O}_{C,\eta_h}((t)) \otimes_{\mathcal{O}_{C,\eta_h}} \mathrm{Ker}_{\mathcal{M}} \nabla\left(t\frac{\partial}{\partial t}\right)^\mu .$$

Proof: The first assertion is clear. Formula (6.8.3) is a consequence of (6.8.2). Indeed, let us provisionally indicate by $\mathcal{N}$ the r.h.s. of (6.8.3): $\mathcal{N}$ is a free $\mathcal{O}_{C,\eta_h}((t))$-submodule of finite type of $\mathcal{M}$ stable by $\frac{\partial}{\partial t}$ and $\frac{\partial}{\partial x}$. On the other hand, by (6.8.1) and (6.8.2), $\mathcal{M}/\mathcal{N}$ is killed by a power of x. The usual argument based on the simplicity of the differential ring $\mathcal{O}_{C,\eta_h}((t))$ shows that it must be zero.

As for (6.8.2), we first observe that the formula for all $\mathcal{M}$ and all ν is equivalent to the same formula for all $\mathcal{M}$ and $\nu = 1$: in fact, assuming the formula for all modules as in (6.8) satisfying (6.8.1) and for $\nu = 1$, we observe that, given any $(\mathcal{M}, \nabla)$ and ν, we may endow $\mathcal{M}^\nu = \mathcal{M} \oplus \cdots \oplus \mathcal{M}$ of the connection

$$\nabla_\nu\left(t\frac{\partial}{\partial t}\right)(m_1, \ldots, m_\nu) = \left(\nabla\left(t\frac{\partial}{\partial t}\right)(m_1), \nabla\left(t\frac{\partial}{\partial t}\right)\right.$$
$$\left.(m_2) - m_1, \ldots, \nabla\left(t\frac{\partial}{\partial t}\right)(m_\nu) - m_{\nu-1}\right)$$
$$\nabla_\nu\left(\frac{\partial}{\partial x}\right)(m_1, \ldots, m_\nu) = \left(\nabla\left(\frac{\partial}{\partial x}\right)(m_1), \nabla\left(\frac{\partial}{\partial x}\right)(m_2), \ldots, \nabla\left(\frac{\partial}{\partial x}\right)(m_\nu)\right).$$

Obviously, $m \longmapsto (\nabla(t\frac{\partial}{\partial t})^{\nu-1}(m), \nabla(t\frac{\partial}{\partial t})^{\nu-2}(m), \ldots, \nabla(t\frac{\partial}{\partial t})(m), m)$ establishes an isomorphism of $\mathcal{O}_{C,\eta_h}$-modules between $\mathrm{Ker}_{\mathcal{M}}\nabla(t\frac{\partial}{\partial t})^\nu$ and $\mathrm{Ker}_{\mathcal{M}^\nu}\nabla_\nu(t\frac{\partial}{\partial t})$.

For $\nu = 1$, formula (6.8.2) is equivalent to

$$\text{(6.8.5)} \qquad \mathrm{Ker}_{\mathcal{M}_{\kappa(C)((t))}}\nabla\left(t\frac{\partial}{\partial t}\right) = \mathrm{Ker}_{\mathcal{M}\otimes\mathrm{Frac}(\mathcal{O}_{C,\eta_h})}\nabla\left(t\frac{\partial}{\partial t}\right).$$

This is a statement on x-adic boundedness of solutions of a linear differential equation, which follows from the following general result (replacing K by the algebraic closure of $\kappa(\eta_h)$): □

Proposition 6.9 *Let $(\mathcal{M}, \nabla)$ be a free $K[[x,t]][\frac{1}{t}]$-module of finite rank μ with a $K[[x]]$-linear connection (for the t-adically continuous $K[[x]]$-linear derivations of $K[[x,t]]$). Let us assume that*

$$\mathcal{M}_{K((x))((t))} = K((x))((t)) \otimes_{K((x))} \mathrm{Ker}_{\mathcal{M}_{K((x))((t))}}\left(\nabla\left(t\frac{\partial}{\partial t}\right)\right)^\mu.$$

Then

$$\text{(6.9.1)} \qquad \mathrm{Ker}_{\mathcal{M}_{K((x))((t))}}\nabla\left(t\frac{\partial}{\partial t}\right) = \mathrm{Ker}_{\mathcal{M}_{K[[x,t]][\frac{1}{x},\frac{1}{t}]}}\nabla\left(t\frac{\partial}{\partial t}\right).$$

For the proof, we need the following result.

Lemma 6.9.2 *In the situation of (6.9), there exists a basis $\underline{e} = (e_1, \ldots, e_\mu)$ of $\mathcal{M}_{K((x,t))}$ in which $\nabla(t\frac{\partial}{\partial t})\underline{e} = \underline{e}G$ with $G \in M_\mu(K[[x,t]])$. Moreover, the eigenvalues of $G(t=0)$ are rational integers.*

Proof: Let us pick an element m of $\mathcal{M}$ which is generically cyclic with respect to $\nabla(t\frac{\partial}{\partial t})$. In the basis $\underline{e}' = (m, \ldots, \nabla(t\frac{\partial}{\partial t})^{\mu-1}(m))$ of $\mathcal{M} \otimes K((x,t))$, where $K((x,t)) = \mathrm{Frac}(K[[x,t]])$, let us write $\nabla(t\frac{\partial}{\partial t})\underline{e}' = \underline{e}'G'$. By regularity at $t = 0$, we then have $G' \in M_\mu(K((x,t)) \cap K((x))[[t]])$ (we view $K((x,t))$ as a subfield of $K((x))((t))$). Let us choose a basis $\underline{e}''$ of $\mathcal{M}$, and let $H \in GL_\mu(K((x,t))) \cap M_\mu(K[[x,t]][\frac{1}{t}])$ be the matrix transforming $\underline{e}''$ into $\underline{e}'$. On modifying $\underline{e}''$ by powers of t if necessary, we may and shall assume also that $H \in M_\mu(K[[x,t]])$. If $\det H$ does not vanish at $t=0$, it is clear that we can take $\underline{e} = \underline{e}''$. If $\operatorname{ord}_{t=0} \det H > 0$, let $(b_1, \dots, b_\mu)$ be the coefficients of a dependence relation over $K((x))$ between the columns of $H(t=0)$. We may assume that the b_i belong to $K[[x]]$ and that some $b_{i_0} = 1$. Let

$$H_1 = \begin{pmatrix} 1 & 0 & \dots & & & & 0 \\ & \ddots & & & & & \\ 0 & \dots & 1 & & & & \\ -b_1 & \dots & -b_{i_0-1}, & t, & -b_{i_0+1}, & \dots, & -b_\mu \\ 0 & \dots & & & 1 & & \\ & & & & & \ddots & \\ & & & & & & 1 \end{pmatrix}.$$

We have

$$H_1^{-1} = \begin{pmatrix} 1 & 0 & \dots & & & & 0 \\ & \ddots & & & & & \\ 0 & \dots & 1 & & & & \\ \frac{b_1}{t} & \dots & \frac{b_{i_0-1}}{t}, & \frac{1}{t}, & \frac{b_{i_0+1}}{t}, & \dots, & \frac{b_\mu}{t} \\ 0 & \dots & & & 1 & & \\ & & & & & \ddots & \\ & & & & & & 1 \end{pmatrix},$$

hence $H_1 \in GL_\mu(K[[x,t]][\frac{1}{t}])$. On the other hand, $\operatorname{ord}_{t=0} \det H_1^{-1} H < \operatorname{ord}_{t=0} \det H$. We see that after modifying the basis $\underline{e}''$ by a finite product of matrices of type H_1, we may assume that $\operatorname{ord}_{t=0} \det H = 0$. The last assertion of the lemma is clear.

Let us return to the proof of (6.9). We write $G = \sum_{j=0}^{+\infty} G^{(j)} t^j$, with $G^{(j)} \in M_\mu(K[[x]])$ and $G^{(0)} = \Delta + N$, with Δ diagonal with coefficients in $\mathbf{Z}$, N nilpotent, $\Delta N = N \Delta$. We first consider a solution $m = \underline{e} y(t)$ of $\nabla(\frac{\partial}{\partial t})(m) = 0$ of the form $y(t) = \sum_{i=-r}^{\infty} y_i t^i$, with $y_i \in C^\mu$, $y_{-r} \neq 0$. We must prove that there exists an M such that $x^M y_i \in K[[x]]$ for all $i \geq -r$. >From $(t\frac{\partial}{\partial t} + G) y(t) = 0$ we obtain

(6.9.3) $$(\Delta + i + N) y_i = - \sum_{j=-r}^{i-1} G^{(i-j)} y_j, \quad \forall\, i\,.$$

On the other hand, there exists i_0 such that the matrix $\Delta + i + N$ is invertible in $M_\mu(K[[x]])$ for $i > i_0$. We conclude that, for any x-adic absolute value $|\ |$, $|y_i| \leq \max\{|y_r|, \dots, |y_{i_0}|\}$, for all i. The result follows. This achieves the proof of (6.7) and (6.2.2).

vi) *Last step: proof of* (6.2.3).
It follows from (6.2.1) and (6.2.2) that

$$(\mathrm{Ker}_{\mathcal{M}_{\kappa(Z)((t))}}(D_1 - P_i)^{\mu}) = \kappa(Z) \otimes_{\mathcal{O}(Z)} \mathrm{Ker}_{\mathcal{M}_{\mathcal{O}(Z)((t))\otimes\kappa(Z)}}(D_1 - P_i)^{\mu},$$

and this implies that for any $\nu = 0, 1, \dots$, $(\mathrm{Ker}_{\mathcal{M}_{\kappa(Z)((t))}}(D_1 - P_i)^{\nu}) \cap \mathcal{M}_{\mathcal{O}(Z)((t))} = \mathrm{Ker}_{\mathcal{M}_{\mathcal{O}(Z)((t))\otimes\kappa(Z)}}(D_1 - P_i)^{\nu} = \kappa(Z) \otimes_{\mathcal{O}(Z)} \mathrm{Ker}_{\mathcal{M}_{\mathcal{O}(Z)((t))}}(D_1 - P_i)^{\nu}$. □

Corollary 6.10 *(Under assumption (*) of* (6.2)*). Let $W_\bullet(M)$ be the slope decomposition of $M = \mathcal{M} \otimes \hat{\mathcal{G}}$ as in* (3.1), (3.5)*. Then $W_\bullet(\mathcal{M}) := \mathcal{M} \cap W_\bullet(M)$ is a decomposition of $\mathcal{M}$ into a direct sum of horizontal $\mathcal{O}(\hat{Y})[\frac{1}{t_w}]$-submodules of $\mathcal{M}$, and $W_\bullet(\mathcal{M}) \otimes \hat{\mathcal{G}} \cong W_\bullet(M)$. Moreover, $W_\sigma(\mathcal{M})$ is the maximal $\mathcal{O}(\hat{Y})[\frac{1}{t_w}]$-submodule of $\mathcal{M}$ stable under ∇ (whose extension over $\hat{\mathcal{G}}$ is) purely of Poincaré-Katz rank σ.*

Proof: Since $\hat{\mathcal{G}} = \mathcal{O}(\hat{Y})\{\mathcal{O}(\hat{Y}) \setminus \{0\}\}$ is flat over $\mathcal{O}(\hat{Y})$ by [EGA 0, 7.6.13], hence over $\mathcal{O}(\hat{Y})[\frac{1}{t_w}]$, $W_\sigma(\mathcal{M}) \otimes \hat{\mathcal{G}} \longrightarrow \mathcal{M} \otimes \hat{\mathcal{G}} = M$ is injective so that

$$\mathrm{Im}(W_\bullet(\mathcal{M}) \otimes \hat{\mathcal{G}} \longrightarrow M) \cong W_\bullet(\mathcal{M}) \otimes \hat{\mathcal{G}}\ .$$

The fact that $W_\bullet(\mathcal{M})$ is stable under ∇ comes from (3.3). On the other hand, with the notation of (6.1), the image of $(\mathcal{M} \cap W_\bullet(M)) \otimes \mathcal{O}(Z)((t))$ in $M \otimes_{\mathcal{O}(\hat{Y})[\frac{1}{t_w}]} \mathcal{O}(Z)((t)) \cong M_{\hat{\mathcal{F}}}$ coincides with the intersection of the images of $\mathcal{M} \otimes \mathcal{O}(Z)((t))$ and $W_\bullet(M) \otimes \mathcal{O}(Z)((t))$ in $M_{\hat{\mathcal{F}}}$ respectively ([A.C.I, 2.6]), *i.e.* with $\mathcal{M}_{\mathcal{O}(Z)((t))} \cap W_{e\bullet}(M_{\hat{\mathcal{F}}})$. We get

$$\mathrm{Im}(W_\bullet(\mathcal{M}) \otimes \mathcal{O}(Z)((t)) \longrightarrow M_{\hat{\mathcal{F}}}) = \mathcal{M}_{\mathcal{O}(Z)((t))} \cap W_{e\bullet}(M_{\hat{\mathcal{F}}})\ .$$

It follows from the decomposition (6.2.1) that

$$(\mathcal{M}_{\mathcal{O}(Z)((t))} \cap W_{e\bullet}(M_{\hat{\mathcal{F}}})) \otimes_{\mathcal{O}(Z)((t))} \hat{\mathcal{F}} = W_{e\bullet}(M_{\hat{\mathcal{F}}}).$$

Hence

$$\begin{aligned}\mathrm{Im}(W_\bullet(\mathcal{M}) \otimes \mathcal{O}(Z)((t)) &\longrightarrow M \otimes_{\mathcal{O}(\hat{Y})[\frac{1}{t_w}]} \mathcal{O}(Z)((t)))\\ &\otimes_{\mathcal{O}(Z)((t))} \hat{\mathcal{F}} = W_{e\bullet}(M_{\hat{\mathcal{F}}}) = W_\bullet(M) \otimes_{\hat{\mathcal{G}}} \hat{\mathcal{F}}.\end{aligned}$$

We deduce that

$$\begin{aligned}\mathrm{Im}(W_\bullet(\mathcal{M}) \otimes \mathcal{O}(Z)((t)) \longrightarrow M_{\hat{\mathcal{F}}}) \otimes \hat{\mathcal{F}} &= \mathrm{Im}((W_\bullet(\mathcal{M}) \otimes \hat{\mathcal{F}}) \to M_{\hat{\mathcal{F}}})\\ = \mathrm{Im}((W_\bullet(\mathcal{M}) \otimes \hat{\mathcal{G}}) \to M) \otimes_{\hat{\mathcal{G}}} \hat{\mathcal{F}} &= (W_\bullet(\mathcal{M}) \otimes \hat{\mathcal{G}}) \otimes_{\hat{\mathcal{G}}} \hat{\mathcal{F}}\end{aligned}$$

coincides with $W_\bullet(M) \otimes_{\hat{\mathcal{G}}} \hat{\mathcal{F}}$. Hence $W_\bullet(\mathcal{M}) \otimes \hat{\mathcal{G}} \cong W_\bullet(M)$. The last assertion is obvious. □

Remark 6.11 A slight difficulty arises here from the fact that $\mathcal{O}(Z)((t))$ is not necessarily flat over $\mathcal{O}(\hat{Y})[\frac{1}{t_w}]$ (Z is not necessarily flat over W). In fact, as M. Chardin pointed out to us, Z is flat over W if and only if Z is Cohen-Macaulay. However, there is a closed subset $T \subset W$ of codimension ≥ 2 such that $Z_{|(W \setminus T)}$ is flat over $W \setminus T$ and smooth over K (it suffices to take T to be the image in W of the singular locus of Z).

Corollary 6.12 *If in the theorem we replace assumption (*) by the stronger condition that the Newton polygon $N_P(End(\mathcal{M}_{|C_P}))$ itself does not depend on the point $P \in W$, except possibly on a subset of codimension 2 in W. Then, the P_i being taken in $\mathcal{O}(Z)[\frac{1}{t}]$, for every pair (i, i') with $i \neq i'$, $P_i - P_{i'}$ is a unit in $\mathcal{O}(Z)((t))$ of non-positive t-order.*

This is obvious, taking into account the fact that the exponents of an integrable differential module are constant [I.6.3.2].

Remark 6.13 It is likely that assumption (*) is not necessary for (6.2). This would for instance follow from a positive answer to the following

Open question 6.14 *Let $(\mathcal{M}, \nabla)$ be a free $K[[x, t]][\frac{1}{t}]$-module of finite rank with integrable connection (for the (x, t)-adically continuous K-linear derivations of $K[[x, t]]$) and let ∇' be the $K((x))((t))/K((x))$-connection inherited by $\mathcal{M}_{K((x))((t))}$. Is it true that $(\mathcal{M}_{K((x))((t))})^{\nabla'} = K((x)) \otimes \mathcal{M}^{\nabla}$?*

§7 Cyclic vectors, indicial polynomials and tubular neighborhoods

7.1 Exponents and indicial polynomials

7.1.1 In the notation of (3.1), let $t = t_v$ be a parameter of R_v, such that $eD_1 t = t$ and $D_j t = 0$, for $j = 2, \ldots, d$. The elements $e\mathrm{Res}_{t=0} P_i \frac{dt}{t} \in k(v)$ (well determined modulo $\mathbf{Z}$) are called the *Turrittin exponents* of (E, ∇) (along w). Robba's result reproduced in (I App.A) asserts that if ∇ is integrable, the Turrittin exponents of (E, ∇) along w are constant, *i.e.* that they belong to K. Then $W_0(E, \nabla)$ is called *the fuchsian part* of (E, ∇). The Turrittin exponents along w of the fuchsian part of (E, ∇) are also called the *Fuchs exponents* of (E, ∇). The Fuchs exponents of (E, ∇) along w, can be computed modulo $\mathbf{Z}$, in terms of a cyclic vector, as the roots of the indicial polynomial of (I.6.1.1) (*cf. loc. cit.*, or [B1]).

Proposition 7.1.2 *Let $W \subset Y$ be a smooth irreducible divisor of a smooth K-variety, and let $(\mathcal{E}, \nabla)$ be a coherent sheaf with integrable connection on $Y \setminus W$. Let ∂ be any derivation transversal to W admitting a system of integral curves*

$$\begin{array}{ccc} U & \overset{i}{\hookleftarrow} & W \cap U \\ f \downarrow & \swarrow g & \\ S & & \end{array}$$

(with U open dense in Y, cf. (I.1.4.1)). *Let us assume that the horizontal edge of $NP_Q(\nabla_{|f^{-1}(f(Q))})$ does not depend on the point $Q \in W \cap U$. Then for any cyclic vector (if any) of $\mathcal{E}(U)$ with respect to $\nabla(\partial)$, with associated (monic) differential polynomial Λ, the leading coefficient of the indicial polynomial of Λ at $W \cap U$ is a unit in $\mathcal{O}(W \cap U)$.*

Proof: Let us shrink U in such a way that $\hat{U}_{W\cap U} \cong \mathrm{Spf}(\mathcal{O}(W \cap U)[[t]])$ with $\partial = \frac{d}{dt}$ as in (I.1.4.5). Then the coefficients of Λ are in $\mathcal{O}(W \cap U)[[t]]$, and they specialize at each $Q \in W \cap U$, to the coefficients of the analogous polynomial for $(\mathcal{E}, \nabla)_{|f^{-1}(f(Q))}$ at Q. Since the connection is integrable, the roots of the indicial polynomial of Λ are in K (I.6.1.3), hence the vanishing of the leading coefficient at some point Q of $W \cap U$ implies the vanishing of the indicial polynomial itself at Q. This implies in turn that the horizontal edge of the Newton polygon $N_f(\Lambda, \bullet)$ is not constant, contrary to our assumption. □

7.2 Cyclic vectors

Up to now, we have used the lemma of the cyclic vector in its coarsest, birational form. We shall need refined variants in the sequel. Let us start with the existence of cyclic vectors in the neighborhood of a given *ordinary* point.

Lemma 7.2.1 (Katz' lemma of the cyclic vector). *Let R be a $\mathbf{Q}$-algebra, t an element of R, ∂ a derivation of R such that $\partial t = 1$, R^∂ the $\mathbf{Q}$-subalgebra of ∂-constants. Let N be an $R[\partial]$-module, and let us assume that N is a free R-module of finite rank μ, with fixed basis of vectors $(e_0, \ldots, e_{\mu-1})$. Let $\alpha_0, \ldots, \alpha_{\mu(\mu-1)}$ be elements in R^∂ such that $\alpha_i - \alpha_j$ is a unit whenever $i \neq j$. Then locally on* Spec *R, one of the vectors*

$$(7.2.1.1) \qquad v_{\alpha_i} = \sum_{j=0}^{\mu-1} \frac{(t-\alpha_i)^j}{j!} \sum_{k=0}^{j} (-1)^k \binom{j}{k} \partial^k(e_{j-k}), \quad 0 \le i \le \mu(\mu-1),$$

is a cyclic vector with respect to ∂.

Proof: See [Ka3, Theorem 2]. □

Corollary 7.2.2 *Let L be a subfield of R^∂ and let $P \in$* Spec*R. Then v_α is a cyclic vector at P for all $\alpha \in L$ except possibly $\mu(\mu-1)$ of them.*

Corollary 7.2.3 *Let C be a smooth affine curve over K, and let $(\mathcal{N}, \nabla)$ be a coherent $\mathcal{O}_C$-module with connection. For any proper closed subset $T \subset C$, there is a closed subset $T' \subset C$ disjoint from T and a global section v of $\mathcal{N}$ which is a cyclic vector over $C \setminus T'$.*

Proof: Let μ be the rank of the projective $\mathcal{O}_C$-module $\mathcal{N}$. By Nakayama's lemma, one can find global sections $e_0, \ldots, e_{\mu-1}$ of $\mathcal{N}$ which are linearly independent at the points of T. One can then apply (7.2.2) to the valuation rings of C at these points. □

7.2.4 We now turn to the more delicate existence problem of cyclic vectors in the neighborhood of a *singularity*. The natural question in this context is:

Question 7.2.5 *Assume that f is a coordinatized tubular neighborhood, i.e. factors through an étale map $\phi : X \longrightarrow \mathbf{A}^1_S$ (with coordinate x on $\mathbf{A}^1$), and that $\mathcal{E}$ extends to a locally free $\mathcal{O}_X$-module. For any point $s \in S$, does there exist an affine neighborhood U of the finite subset $g^{-1}(s)$ of X and a* cyclic vector *$v \in \Gamma(U \backslash (Z \cap U), \mathcal{E})$ with respect to $\nabla(\frac{\partial}{\partial x})$?*

The answer is *no* in general, even in the integrable case. Indeed, let X be the (x, y)-plane, $Z =$ the y-axis, $\mathcal{E} = \mathcal{O}_{X \backslash Z} e_1 \oplus \mathcal{O}_{X \backslash Z} e_2$, where the first (resp. second) factor is endowed with the trivial connection (resp. the connection with formal solution $e^{-\frac{y}{x}}$). If $v = a e_1 + b e_2$, $\frac{\partial}{\partial x} v = \frac{\partial a}{\partial x} e_1 + (\frac{\partial b}{\partial x} - \frac{by}{x^2}) e_2$; by looking at the lowest x-order, we see that for any affine neighborhood U of the origin, the determinant $a\frac{\partial b}{\partial x} - b\frac{\partial a}{\partial x} - \frac{aby}{x^2}$ cannot be a unit in $\mathcal{O}(U \backslash (Z \cap U))$. However we shall give a qualified positive answer to the previous question.

7.2.6 We first consider the formal situation, as in (6.1): $(\hat{\mathcal{G}}, w)$ is a field of formal functions over K, $(\hat{Y}, W)$ a smooth affine formal model of $(\hat{\mathcal{G}}, w)$, $(D_1, \ldots, D_d)$ an adapted $\mathcal{O}(\hat{Y})$-basis of $Der_{\hat{Y}/K,W}$ and $(\mathcal{M}, \nabla)$ is a projective $\mathcal{O}(\hat{Y})[\frac{1}{t_w}]$-module of finite rank μ with integrable connection. Let $e|\mu!$, Z, and $P_i \in \kappa(Z)[\frac{1}{t}]$ be as in (6.1).

Lemma 7.2.7 *Assume that the Newton polygon of the $\kappa(P)((t_w))[D_1]$-module $End(\mathcal{M}_{|C_P})$ does not depend on the point $P \in W$, except possibly on a closed subset of codimension 2 in W. Then any point $Q \in Z$ outside some closed subset $T \subset Z$ of codimension ≥ 2 has an affine smooth neighborhood Z_Q such that $\mathcal{M}_{\mathcal{O}(Z)((t))\otimes\mathcal{O}(Z_Q)}$ is cyclic with respect to D_1.*

Proof: Since we allow exclusion of a closed subset of codimension ≥ 2 from Z, we may restrict to the part of W where the Newton polygon of $\underline{End}(\mathcal{M}_P)$ is constant, and over which Z is étale. Also, by quasi-compactness of Z, it is enough to consider points Q of codimension 1. The assumption that the Newton polygon of $End(\mathcal{M}_{|C_P})$ is constant on W implies that the Newton polygon of $(End(\mathcal{M}_{\mathcal{O}(Z)((t))}))_{|\tilde{C}_Q}$ is constant on Z, hence that $P_i - P_{i'}$ is either zero or a unit in $\mathcal{O}(Z)((t))$ of non-positive t-order (*cf.* (6.2), (6.12)).

We saw in (6.2.3) that

$$\mathrm{Ker}_{\mathcal{M}_{\kappa(Z)((t))}}(D_1 - P_i)^{\nu} = \kappa(Z) \otimes_{\mathcal{O}(Z)} \mathrm{Ker}_{\mathcal{M}_{\mathcal{O}(Z)((t))}}(D_1 - P_i)^{\nu}.$$

Moreover $\mathrm{Ker}_{\mathcal{M}_{\mathcal{O}(Z)((t))}}(D_1 - P_i)^{\nu}/\mathrm{Ker}_{\mathcal{M}_{\mathcal{O}(Z)((t))}}(D_1 - P_i)^{\nu-1}$ has no $\mathcal{O}(Z)$-torsion and is finitely generated (as a subquotient of the noetherian module $\mathrm{Ker}_{\mathcal{M}_{\mathcal{O}(Z)((t))}}(D_1 - P_i)^{\mu}$). Hence by shrinking Z_Q, we may assume that there exist elements $e_{\nu,i,1},\ldots,e_{\nu,i,s_{\nu,i}}$ in $\mathrm{Ker}_{\mathcal{M}_{\mathcal{O}(Z)((t))\otimes\mathcal{O}(Z_Q)}}(D_1-P_i)^\nu$ whose classes form a basis in the quotient

$$\mathrm{Ker}_{\mathcal{M}_{\mathcal{O}(Z)((t))\otimes\mathcal{O}(Z_Q)}}(D_1-P_i)^\nu/\mathrm{Ker}_{\mathcal{M}_{\mathcal{O}(Z)((t))\otimes\mathcal{O}(Z_Q)}}(D_1-P_i)^{\nu-1}.$$

Altogether, the elements $e_{\nu,i,k}$ then form a $\mathcal{O}(Z)((t))\otimes\mathcal{O}(Z_Q)$-basis of $\mathcal{M}_{\mathcal{O}(Z)((t))\otimes\mathcal{O}(Z_Q)}$. We reorder them in lexicographical order, and rename them $(m_0,\ldots,m_{\mu-1})$. Let us remark that, as a D_1-module, $\mathcal{M}_{\mathcal{O}(Z)((t))\otimes\mathcal{O}(Z_Q)}$ is an iterated extension of rank one differential modules: $D_1m_j=Q_jm_j+\sum_{k>j}a_{kj}m_k$, where Q_j is one of the P_l, and $a_{kj}\in\mathcal{O}(Z_Q)$. We set $-s=\min ord_t\ P_i$, and choose a large integer n (e.g. $n>s\mu$).

We claim that $m:=m_0+t^{-en}m_1+\cdots+t^{-en(\mu-1)}m_{\mu-1}$ *is a cyclic vector* of $\mathcal{M}_{\mathcal{O}(Z)((t))\otimes\mathcal{O}(Z_Q)}$.

Let us set $m_j^0=t^{-enj}m_j$, $m^0:=m=m_0+\sum_{j=1}^{\mu-1}m_j^0$, $Q_j^0=Q_j-nj$, so that $D_1m_j^0=Q_j^0m_j^0+\sum_{k>j}a_{kj}^0m_k^0$, with $ord_t\ a_{kj}^0>0$ and, for $j\neq k$, $Q_j^0-Q_k^0$ is a unit in $\mathcal{O}(Z)((t))\otimes\mathcal{O}(Z_Q)$, $ord_t\ (Q_j^0-Q_k^0)\leq 0$. We have $m^1:=(Q_1^0-Q_0^0+a_{1,0}^0)^{-1}(D_1-Q_0^0)m=m_1^0+\sum_{k>1}u_{k,1}m_k^0$, where $u_{k,1}$ is a unit in $\mathcal{O}(Z)((t))\otimes\mathcal{O}(Z_Q)$, $-s\leq ord_tu_{k,1}\leq s$. If we set $m_j^1=u_{j,1}m_j^0$, $m^1=m_1^0+\sum_{j=2}^{\mu-1}m_j^1$, $Q_j^1=Q_j^0+\frac{D_1(u_{j,1})}{u_{j,1}}$ for $j>1$, we get $D_1m_j^1=Q_j^1m_j^1+\sum_{k>j}a_{kj}^1m_k^1$, with $ord_t\ a_{kj}^1>0$ and, for $j\neq k$, $Q_j^1-Q_k^1$ is a unit in $\mathcal{O}(Z)((t))\otimes\mathcal{O}(Z_Q)$, $ord_t\ (Q_j^1-Q_k^1)\leq 0$. We then construct $m^2:=(Q_2^1-Q_1^1+a_{2,1}^1)^{-1}(D_1-Q_1^1)m^1=m_2^1+\sum_{k>2}u_{k,2}m_k^1$, and iterate $\mu-1$ times. Iteration step ν shows that $\sum_{j\leq\nu}\mathcal{O}(Z)((t))\otimes\mathcal{O}(Z_Q)D_1^jm$ projects onto $\mathcal{M}_{\mathcal{O}(Z)((t))\otimes\mathcal{O}(Z_Q)}/\sum_{k>\nu}\mathcal{O}(Z)((t))\otimes\mathcal{O}(Z_Q)m_k$. Hence m is cyclic. □

7.2.8 We now consider a diagram of smooth K-varieties

$$\begin{array}{ccccc} Y & \hookrightarrow & \overline{Y} & \hookleftarrow & W=\coprod_{i=0}^r W_i \\ & \overset{f}{\searrow} & \downarrow\overline{f} & \swarrow^{g} & \\ & & S & & \end{array} \qquad (7.2.8.1)$$

where

i) Y is the complement of W in $\overline{Y}$,
ii) Y and S are affine, and each W_i is a divisor defined by one equation $t_i=0$,
iii) $\overline{f}$ is smooth surjective with geometric fibers irreducible of dimension one,
iv) each W_i is an étale covering of S.

Let ∂ be a global section of $\mathcal{D}er_{\overline{Y}/S}$ having a simple zero at each W_i.

Proposition 7.2.9 *Let $(\mathcal{E},\nabla)$ be an $\mathcal{O}_Y$-module with integrable connection. We assume that:*

i) *$\mathcal{E}$ is the restriction to Y of a locally free $\mathcal{O}_{\overline{Y}}$-module $\overline{\mathcal{E}}$ of rank μ;*

ii) *the Turrittin index e_i of $(\mathcal{E}, \nabla)$ at each singular divisor W_i is equal to* 1 *(this is stronger than requiring that the slopes are integral, cf. [Ba1, Remark 9]);*
iii) *the Newton polygon of the $\mathcal{O}(Y_{g(P)})[\partial]$-module $End(\mathcal{E}_{|Y_{g(P)}})$ at $P \in W_i$ does not depend on the point P (nor on i).*

Then there exists a smooth K-variety S' and a flat morphism $h : S' \to S$ such that

i) *S' is finite over $h(S')$,*
ii) *$S \backslash h(S')$ has codimension ≥ 2 in S,*
iii) *for any $Q \in S'$, there is an affine neighborhood U of the finite subset $g_{S'}^{-1}(Q)$ of $\overline{Y} \times_S S'$ such that the inverse image of $\mathcal{E}$ on $U \cap (Y \times_S S')$ is* cyclic *with respect to $\nabla(\partial)$.*

Roughly speaking, this result establishes a close link between the set of points $s \in S$ for which question (7.2.5) has a negative answer, and the set of points s where the Newton polygon of $End(\mathcal{E}_{|f^{-1}(s)})$ at some $Q \in g^{-1}(s)$ is strictly above the generic one.

Proof: We first notice that we may replace S by any finite étale covering, hence assume that each W_i is the image of a section σ_i of $\overline{f}$. Then the completion of $\overline{Y}$ at W_i is isomorphic to the affine t_i-adic formal scheme $\mathrm{Spf}\,\mathcal{O}(S)[[t_i]]$.

We shall apply the previous lemma to $\mathcal{M}_i := \mathcal{E}(Y) \otimes_{\mathcal{O}(Y)} \mathcal{O}(S)((t_i))$ (with the induced integrable connection). We notice that the derivation D_1 considered in that lemma is a multiple of ∂ by a unit in $\mathcal{O}(S)[[t_i]]$, hence that the condition of being cyclic with respect to D_1 or to ∂ are equivalent. Also, we can arrange that the normal variety Z is independent of i, the finite morphism $Z \to W_i$ being the composite of a finite (surjective) morphism $Z \to S$ and σ_i. By assumption, the Turrittin indices are 1, so that what was denoted by $\mathcal{O}(Z)((t))$ in *loc. cit.* is now $\mathcal{O}(Z)((t_i)) \cong \mathcal{O}(S)((t_i)) \otimes_{\mathcal{O}(S)} \mathcal{O}(Z)$ (since Z is finite over S). There is a closed subset $T \subset S$ of codimension ≥ 2 such that $S' := Z_{|(S \backslash T)}$ is flat over $S \backslash T$ and smooth over K (it suffices to take T to be the image in S of the singular locus of Z).

It then follows from (7.2.7) that one can adjust T in such a way that any $Q \in S'$ has an affine neighborhood S'_Q such that the $\mathcal{O}(S)((t_i)) \otimes_{\mathcal{O}(S)} \mathcal{O}(S'_Q)[\partial]$-module $\mathcal{M}_i \otimes_{\mathcal{O}(S)} \mathcal{O}(S'_Q)$ is *cyclic*. We denote by m_i a cyclic vector in $\mathcal{M}_i \otimes_{\mathcal{O}(S)} \mathcal{O}(S'_Q)$.

We have to find a cyclic vector m for $\mathcal{E} \otimes_{\mathcal{O}_S} \mathcal{O}_{S'_Q}$ in some affine neighborhood of $g_{S'}^{-1}(h(Q))$ in $Y' := Y \times_S S'$. The question is now Zariski-local on $S \backslash T$. In particular, we may replace $S \backslash T$ by an affine open subset S_0 of $S \backslash h(S' \backslash S'_Q)$ such that $\overline{\mathcal{E}}$ is free in some affine neighborhood of $W \times_S S_0$ in $\overline{Y} \times_S S_0$, and replace S' by $S' \times_S S_0$. After shrinking S in such a way, we have the following properties:

i) in some affine neighborhood V of W in $\overline{Y}$, $\overline{\mathcal{E}}(V) = \oplus_{j=1}^{\mu} \mathcal{O}(V) e_j$,
ii) S' is finite flat over S,

iii) m_i is a cyclic vector in $\mathcal{M}_i \otimes_{\mathcal{O}(S)} \mathcal{O}(S') \cong \oplus_{j=1}^{\mu} \mathcal{O}(S)((t_i)) \otimes_{\mathcal{O}(S)} \mathcal{O}(S')\ e_j$ with respect to ∂.

We thus have $m_i \wedge \partial m_i \wedge \cdots \wedge \partial^{\mu-1} m_i = u_i\ e_1 \wedge e_2 \wedge \cdots \wedge e_\mu$, where u_i is a unit in $\mathcal{O}(S)((t_i)) \otimes_{\mathcal{O}(S)} \mathcal{O}(S')$.

On the other hand, the vectors m_i can be simultaneously approximated to arbitrarily high (t_i-)order by elements of

$$\overline{\mathcal{E}}(V \times_S S')\left[\frac{1}{t_0 \ldots t_r}\right] = \oplus_{j=1}^{\mu} \mathcal{O}(V \times_S S')\left[\frac{1}{t_0 \ldots t_r}\right] e_j.$$

For a good enough approximation m, we have

$$m \wedge \partial m \wedge \cdots \wedge \partial^{\mu-1} m = u\ e_1 \wedge e_2 \wedge \cdots \wedge e_\mu,$$

where $u \in \mathcal{O}(V \times_S S')$ is a unit in $\mathcal{O}(S)((t_i)) \otimes_{\mathcal{O}(S)} \mathcal{O}(S')[\frac{1}{t_0 \ldots t_r}]$ for every i. This implies that for some affine neighborhood U of $W \times_S S'$ in Y', u is a unit in $\mathcal{O}(U \cap Y')$, hence that m is a cyclic vector in $\mathcal{E}(S) \otimes_{\mathcal{O}(S)} \mathcal{O}(U \cap Y')$ with respect to ∂ . $\square$

3 Direct images (the Gauss-Manin connection)

Introduction

This chapter develops a new approach to the study of direct images of connections (*i.e.* De Rham cohomology with coefficients, endowed with the Gauss-Manin connection) with respect to a smooth, not necessarily proper, morphism of smooth algebraic varieties in characteristic 0.

We present elementary and purely algebraic proofs of the generic finiteness and base change theorems, as well as of the fundamental finiteness, regularity, monodromy and base change theorems for direct images of regular algebraic connections.

Our proofs use neither resolution of singularities (beyond the classical case of embedded resolution of curves in surfaces), nor the theory of holonomy. In fact, in contrast to the now standard methods which consist of trying to extend all objects to a good compactification and studying ramification at infinity, our strategy relies upon a dévissage inspired by Artin's theory of elementary fibrations. This approach allows us to reduce problems to the simple case of an ordinary differential operator in one variable (but in a relative situation), which we handle directly.

Let us now describe in more detail the content of each section.

We begin with M. Artin's technique of good neighborhoods, *i.e.* local fibrations of a smooth variety by curves. This stands beside the technique of Lefschetz pencils (global fibrations of a smooth quasi-projective variety by hypersurfaces) as a powerful tool in algebraic geometry, offering the possibility of reasoning by induction on the dimension.

We combine both techniques to unscrew smooth morphisms into towers of *coordinatized* elementary fibrations; the basic pieces of this dévissage are thus relative versions of *the projective line minus a few points*, which we call *rational elementary fibrations* (I.2,3,4). By means of such elementary surgery, we attach to any smooth morphism f of smooth varieties an open dense subset $A(f)$ of the base, the *Artin set* (1.8), which will play in this chapter the role held in P. Deligne's theory [De] by those open subsets of the base over which f admits a good compactification à la Hironaka.

Section 2 is a review, in the elementary style of [Ka1], of connections $(\mathcal{E}, \nabla)$ and their direct images $R^i_{DR} f_*(\mathcal{E}, \nabla) := \mathbf{R}^i f_* DR_{X/S}(\mathcal{E}, \nabla)$ under a smooth morphism $X \xrightarrow{f} S$ over an algebraically closed field K of characteristic 0.

Section 3 is fundamental to the whole chapter. We consider properties $\mathcal{P}$ of connections, which are local for the étale topology, and strongly exact (in the sense that $\mathcal{P}$ holds for the middle term of a short exact sequence if and only if $\mathcal{P}$ holds for the other terms). The main lemma on *dévissage* asserts that *in order to establish that for any smooth morphism of smooth K-varieties $X \xrightarrow{f} S$, for any (quasi-coherent) connection $(\mathcal{E}, \nabla)$ on X satisfying $\mathcal{P}$, and for any $i \geq 0$, the restriction of $R^i_{DR} f_*(\mathcal{E}, \nabla)$ to the Artin set $A(f) \subset S$ satisfies $\mathcal{P}$, it suffices to consider rational elementary fibrations*

f (in which case $A(f) = S$), and $i = 0, 1$. This essentially reduces the problem to the study of the kernel and cokernel of a connection in one variable.

There are several variants of this lemma, and furthermore, the arguments of dévissage extend to some of the properties we are interested in, and which cannot be expressed simply as a property $\mathcal{P}$ as above (*e.g.* base change). The idea of this dévissage has already appeared in our work on G-functions [ABa][1].

In Section 4, we apply this technique to establish *generic finiteness* of direct images $R^i_{DR} f_*(\mathcal{E}, \nabla)$, reducing the question to the case of a rational elementary fibration $X \xrightarrow{f} S$ (4.1); we are left with the task of proving that the cokernel of any non-zero differential operator $\Lambda \in \mathcal{O}(S)[x, \frac{d}{dx}]$ becomes finitely generated over $\mathcal{O}(S)$ after replacing S by some open dense subset. The point of the proof is that, as a consequence of integrability, the indicial polynomials ϕ_i at the singularities of Λ have the form

$$\phi_i = (\text{ non-zero element } \tau_i \text{ of } \mathcal{O}(S)) \times (\text{polynomial with constant coefficients});$$

in order to obtain finiteness of Coker Λ, it then suffices to remove from S the zero-locus of the τ_i.

In Section 5, we use the same reduction technique to prove *generic base change* for $R^i_{DR} f_*$ (5.1). Base change for a cokernel is automatic, while "generic" base change for Ker $f_*\nabla$ follows from "generic" flatness of Coker $f_*\nabla$ (a consequence of "generic" coherence proven in Section 4).

With the help of the extension with logarithmic poles, constructed in (I, Section 4), we again take up in Section 6 the question of finiteness of $R^1_{DR} f_*(\mathcal{E}, \nabla)$, in the case of a rational fibration f and of a regular connection ∇, without localizing on S.

In Section 7, we handle the question of the regularity and exponents of the Gauss-Manin connection on $R^1_{DR} f_*(\mathcal{E}, \nabla)$ by using a method of N. Katz [KaI2]. In this question, one can assume that the base S is a curve, so that X is a surface, and embedded resolution of singularities (upon which Katz's method relies) is elementary in this case.

One is then ready to prove in Section 8, by dévissage, the first main results of the theory: the *finiteness, regularity, monodromy and base change theorems* for $R^i_{DR} f_*(\mathcal{E}, \nabla)$ (endowed with the Gauss-Manin connection $\aleph$), when $f : X \longrightarrow S$ is any smooth morphism of smooth K-varieties with $S = A(f)$, and $(\mathcal{E}, \nabla)$ is any regular connection. (The so-called monodromy theorem, in the form given by J. Bernstein [Bn], states that the exponents of $\aleph$ belong to the rational space generated by 1 and by the exponents of ∇.)

We remark that unlike what occurs in the usual approaches, there is no simplification at all in our proof if the morphism f happens to be proper.

---

[1] It is interesting to note that Artin's technique of elementary fibrations was also used, independently, as a basic tool by A. de Jong in his now famous alteration theorem, as we learned shortly after beginning this work.

Appendix C is completely due to P. Berthelot. It is a comparison result between $\mathcal{O}_X$- *vs.* $\mathcal{D}_{X/S}$-linear duals of coherent $\mathcal{D}_{X/S}$-modules, for any smooth morphism $f : X \longrightarrow S$ of smooth algebraic varieties. This is preliminary to our definition of algebraic De Rham cohomology with compact supports in Appendix D.

The second appendix to this chapter provides an introduction to Dwork's favorite approach to differential modules, namely his *dual theory*. While Dwork's main concern is to give sharp p-adic growth estimates for the elements (power series) of the dual theory, we decided to restrict ourselves, for simplicity, to the algebraic situation. We propose a definition of relative algebraic De Rham cohomology with compact supports and relative algebraic Poincaré duality based on the duality for coherent sheaves suggested by Deligne in the appendix of [H1].

We define a functor $f_!$ for coherent $\mathcal{O}_X$-modules, and extend it to a triangulated category, intermediate between the bounded derived category of $\mathcal{O}_X$-modules and that of abelian sheaves, whose objects are complexes of $\mathcal{O}_X$-modules and the differentials are relative differential operators. This category, already considered by Herrera and Lieberman [HL], seems to be the natural ambiance where De Rham complexes live, and its objects should presumably play the role of analytically constructible coefficients in an algebraic Riemann-Hilbert correspondence. On applying $f_!$ to the relative De Rham complex, one gets the definition of relative algebraic De Rham cohomology with compact supports. Our main duality theorem D.2.17 is based on Berthelot's comparison theorem C.2, and shows how our definition is related to the standard algebraic definition of algebraic direct image with compact supports of a $\mathcal{D}$-module [Me5, I.5.2.4], [Bo, VI.4.0, eq.1].

After this general construction, we describe an explicit isomorphism of Dwork's "algebraic dual theory" to algebraic De Rham cohomology with compact supports. The proof of the comparison theorem D.2.20 was part of L. Fiorot's thesis work. In the last sections of appendix D we make explicit calculations for a rational elementary fibration and illustrate some results of Dwork's on the Gauss hypergeometric function. The dual theory offers alternative proofs of most of the results we obtain in this book and poses interesting comparison problems.

§1 Elementary fibrations

Let K be an algebraically closed field of characteristic 0. A *K-variety* will always be meant to be a reduced separated K-scheme of finite type, not necessarily irreducible.

Definition 1.1 A morphism $f : X \longrightarrow S$ of smooth algebraic K-varieties is called an *elementary fibration* if it can be embedded in a commutative diagram

(1.1.1)
$$\begin{array}{ccccc} X & \overset{j}{\hookrightarrow} & \overline{X} & \hookleftarrow & Z \\ & {\scriptstyle f}\searrow & \downarrow \overline{f} & \swarrow{\scriptstyle g} & \\ & & S. & & \end{array}$$

where

(i) Z is a closed reduced subscheme of $\overline{X}$ and j is the complementary open immersion;
(ii) $\overline{f}$ is projective and smooth with geometric fibers irreducible of dimension 1;
(iii) g is an étale covering.

Lemma 1.1.2 *An elementary fibration is a surjective affine morphism.*

Proof: We may then assume that Z is a disjoint union of divisors isomorphic to S via g. For each point $s \in S$, $Z_s = g^{-1}(s)$ is then a positive effective divisor on the projective curve $\overline{X}_s/\kappa(s)$, hence it is ample [H3, Chapter IV, Corollary 3.3]. By [EGA III, 4.7.1], Z is then a relatively ample divisor in the relative projective curve $\overline{X}/S$. We conclude by [EGA II, 5.5.7] that f is affine. □

Definition 1.2 An elementary fibration f as in (1.1) is *rational* if $\overline{X} = \mathbf{P}^1_S$, $\overline{f}$ is the projection on S and Z is a disjoint union of images of sections σ_i, $i = 1, \dots, r, \infty$, of $\overline{f}$, one of which is the section $\sigma_\infty : S \longrightarrow Z_\infty = \{\infty\} \times S \subset \mathbf{P}^1_S$. We thus have the following diagram

(1.2.1)
$$\begin{array}{ccccc} X & \hookrightarrow & \mathbf{A}^1_S & \hookleftarrow & \coprod_{i=1}^{r} \sigma_i(S) \\ & \overset{f}{\searrow} & \downarrow{\scriptstyle \mathrm{pr}_S} & \swarrow & \\ & & S. & & \end{array}$$

If $K = \mathbf{C}$, any rational elementary fibration is thus topologically locally trivial for the classical complex topology.

Definition 1.3 An elementary fibration, as in (1.1), is *coordinatized* if it can be embedded in a commutative diagram

(1.3.1)
$$\begin{array}{ccccc} X & \overset{j}{\hookrightarrow} & \overline{X} & \hookleftarrow & Z \\ \downarrow \pi & & \downarrow \overline{\pi} & & \downarrow \pi' \\ Y & \overset{j'}{\hookrightarrow} & \mathbf{P}^1_S & \hookleftarrow & Z' \\ & \overset{f'}{\searrow} & \downarrow{\scriptstyle \mathrm{pr}_S} & \overset{g'}{\swarrow} & \\ & & S & & \end{array}$$

where

(i) $\overline{\pi}$ is finite and π is an étale covering;
(ii) $f = f' \circ \pi$, $\overline{f} = \mathrm{pr}_S \circ \overline{\pi}$, $g = g' \circ \pi'$;
(iii) the lower part of the diagram is an elementary fibration with $j'(Y) \subset \mathbf{A}^1_S$.

We point out that the two squares in (1.3.1) are necessarily cartesian and that π' is an étale covering. A rational elementary fibration is *ipso facto* coordinatized. Conversely, given a coordinatized elementary fibration as in (1.3.1), there exists an étale covering $S' \longrightarrow S$ such that the lower part of the pull-back of (1.3.1) over S', is a rational elementary fibration.

Let $f : X \longrightarrow S$ be a smooth morphism of smooth K-varieties, of relative dimension d. Assume S is connected. It follows from M. Artin's theorem [SGA 4, Example XI, Proposition 3.3] on "good neighborhoods" that there exists an étale dominant morphism $\epsilon : S' \longrightarrow S$ and a finite open (affine) cover $\{U_\alpha\}$ of $X' = X \times_S S'$, such that each $U_\alpha \longrightarrow S'$ is a tower of elementary fibrations

(1.3.2) $$U_\alpha = V_{\alpha,d} \longrightarrow V_{\alpha,d-1} \longrightarrow \cdots \longrightarrow V_{\alpha,0} = S'.$$

The integer $d \geq 0$ will be the *height* of the tower (a tower of height 0 is then an isomorphism $U_\alpha \overset{\sim}{\longrightarrow} S'$). This result will play a fundamental role in the present chapter, in the following more precise form.

Proposition 1.4 *Let $f : X \longrightarrow S$ be a smooth morphism of smooth K-varieties, with S connected. Then there exists an étale dominant morphism $\epsilon : S' \longrightarrow S$ and a finite open (affine) cover $\{U_\alpha\}$ of $X' = X \times_S S'$ such that each $U_\alpha \longrightarrow S'$ is a tower of coordinatized elementary fibrations.*

Proof: We may assume that f is of pure relative dimension $d \geq 0$. Standard thickening arguments allow us to replace S by its geometric generic point $\mathrm{Spec} K(S)^{alg}$, and then to replace the ground field K by $K(S)^{alg}$. The cases $d = 0, 1$ are then trivial. By induction on d ($d \geq 2$), it suffices to find, for an arbitrary closed point $\xi \in X$, an open neighborhood U of ξ in X and a coordinatized elementary fibration $U \longrightarrow T$.

Let us first recall, for convenience, Artin's construction of an elementary fibration $U \longrightarrow T$, let alone the condition of being "coordinatized". We may assume X affine and consider a normal projective closure $\overline{X}$. We pick a very ample line bundle $\mathcal{L}$ on $\overline{X}$ and embed $\overline{X}$ in a projective space $\mathbf{P}^N$ via $\mathcal{L}^{\otimes 2}$. Then a "general" linear space $L \subset \mathbf{P}^N$ of codimension $d-1$ passing through ξ cuts $\overline{X}$ and $\overline{X} \setminus X$ transversally and avoids the singular locus of $\overline{X}$. Let $H \subset \mathbf{P}^N$ be a "general" hyperplane (not passing through ξ and avoiding $L \cap (\overline{X} \setminus X)$), and let us consider the blow-up $\tilde{\mathbf{P}}^N$ of $\mathbf{P}^N$ along $L \cap H$, and the morphism $p : \tilde{\mathbf{P}}^N \longrightarrow \mathbf{P}^{d-1}$ induced by the linear projection with center $L \cap H$. Let $\tilde{\overline{X}}$ be the strict transform of $\overline{X}$ in $\tilde{\mathbf{P}}^N$, and let $Z \subset \tilde{\overline{X}}$ be the disjoint union of $\overline{X} \setminus X$ (considered as a subscheme of $\tilde{\overline{X}}$ of codimension 1) and of the exceptional divisor $\simeq \mathbf{P}^{d-1} \times (X \cap L \cap H)$. Then $\tilde{\overline{X}} \setminus Z$ identifies with an open subscheme of X containing ξ, and the pull-back of the diagram

(1.4.1) $$\begin{array}{ccccc} \tilde{\overline{X}} \setminus Z & \hookrightarrow & \tilde{\overline{X}} & \hookleftarrow & Z \\ & \searrow & \downarrow p_{|\tilde{\overline{X}}} & \swarrow & \\ & & \mathbf{P}^{d-1} & & \end{array}$$

over a suitable (affine) neighborhood T of $p(\xi)$ in $\mathbf{P}^{d-1}$ defines an elementary fibration. Moreover the fiber of $p_{|\tilde{\overline{X}}}$ above $p(\xi)$ identifies with $\overline{X} \cap L$. The proposition is then a consequence of the following lemma. □

Lemma 1.5 *Given an elementary fibration*

$$\begin{array}{ccccc} X & \hookrightarrow & \overline{X} & \hookleftarrow & Z \\ & \overset{f}{\searrow} & \downarrow \overline{f} & \swarrow & \\ & & S & & \end{array} \tag{1.5.1}$$

and a closed point $\xi \in X$, with image $\zeta = f(\xi)$, there is an open neighborhood U of ξ (resp. T of ζ), such that f induces a coordinatized elementary fibration $U \longrightarrow T$.

Proof: We denote by $\{\xi_1, \ldots, \xi_r\}$ the finite set $\{\xi\} \cup Z_\zeta$ of closed points of $\overline{X}_\zeta$. Let us choose a closed S-embedding $\overline{X} \longrightarrow \mathbf{P}^M_S$. The fiber $\overline{X}_\zeta$ is a closed smooth curve in $\mathbf{P}^M$ to which the theory of Lefschetz pencils applies [SGA 7, II, Example XVII]. There exists an open dense subset $\mathcal{U}$ of the Grassmannian of lines in the dual projective space $\check{\mathbf{P}}^M$ such that for any $D_\zeta \in \mathcal{U}$:

(a) $_\zeta$ D_ζ is a Lefschetz pencil; in particular, the axis of D_ζ (which has codimension 2 in $\mathbf{P}^M$) does not cut $\overline{X}_\zeta$, so that D_ζ gives rise to a finite morphism $\overline{\pi}_\zeta$: $\overline{X}_\zeta \longrightarrow \mathbf{P}^1 \simeq D_\zeta$;

(b) $_\zeta$ if $i \neq j, \overline{\pi}_\zeta(\xi_i) \neq \overline{\pi}_\zeta(\xi_j)$;

(c) $_\zeta$ for each i, $\overline{\pi}_\zeta$ is étale at each point of $\overline{\pi}_\zeta^{-1}\overline{\pi}_\zeta(\xi_i)$.

In order to justify $(b)_\zeta$ and $(c)_\zeta$, it suffices to show that the generic pencil satisfies these conditions; but this follows from the fact [*loc. cit.* 3.2.8] that for any hyperplane $H_i \subset \mathbf{P}^M$ containing ξ_i, but not ξ_j for $j \neq i$, and cutting $\overline{X}_\zeta$ transversally, there exists a Lefschetz pencil of hyperplanes containing H_i.

Let us fix $D_\zeta \in \mathcal{U}$. Up to changing the coordinate in $\mathbf{P}^1 \simeq D_\zeta$, we may assume that

(d) D_ζ the hyperplane H_∞ corresponding to $\infty \in D_\zeta$ cuts $\overline{X}_\zeta$ transversally, and does not meet the points ξ_i.

Let us now extend D_ζ to a relative line D/S in the dual projective space $\check{\mathbf{P}}^M_S$ (D is defined by an S-point in a suitable Grassmannian). After replacing S by a suitable Zariski neighborhood T of ζ, we may assume that the axis of D does not cut $\overline{X}$, so that D gives rise to a finite T-morphism $\overline{\pi} : \overline{X} \longrightarrow \mathbf{P}^1_T$. We may also assume that (corresponding to the conditions above):

(b) $\overline{\pi}(Z)$ is étale over T, and does not contain $\overline{\pi}(\xi)$;

(c) $\overline{\pi}$ is étale at any point of $\overline{\pi}^{-1}\overline{\pi}(Z)$;

(d) $\overline{\pi}^{-1}(\infty \times T)$ does not meet $\overline{\pi}^{-1}\overline{\pi}(Z)$ nor $\{\xi\}$, and the restriction of $\overline{\pi}$ to $\overline{\pi}^{-1}(\infty \times T)$ is étale.

According to [*loc. cit.* 6.3], we may assume furthermore that:

(a) for any closed point $\zeta' \in T$, $D_{\zeta'}$ is a Lefschetz pencil, and the branch locus $B \subset \mathbf{P}^1_T$ of $\overline{\pi}$ is étale over T.

Then there is at most one ramification point in each fiber of $\overline{\pi}^{-1}(B)$, so that $\overline{\pi}^{-1}(B)$ is étale over T (and does not meet $\overline{\pi}^{-1}\overline{\pi}(Z) \coprod \overline{\pi}^{-1}(\infty \times T)$ nor $\{\xi\}$). It is now clear that $U = \overline{X} \setminus (\overline{\pi}^{-1}(B) \coprod \overline{\pi}^{-1}\overline{\pi}(Z) \coprod \overline{\pi}^{-1}(\infty \times T))$ is a coordinatized elementary fibration (via $\overline{\pi}$), and $\xi \in U \subset X$.

This concludes the proof of Lemma 1.5, hence of Proposition 1.4. □

Remark 1.6 In fact, Proposition 1.4 holds for a separably closed ground field K of any characteristic. The proof is the same, except that one must replace $K(S)^{alg}$ by the separable closure of $K(S)$, and that in Lemma 1.5, it may be necessary to replace the given embedding $\overline{X} \hookrightarrow \mathbf{P}_S^N$ by a Veronese multiple (in characteristic 2, one should also invoke [*loc. cit.* 6.4] and replace $\mathcal{L}^{\otimes 2}$ by $\mathcal{L}^{\otimes 3}$ in 1.4).

Notation 1.7 Let $f : X \longrightarrow S$ be a smooth morphism of K-varieties. We denote by d the maximum among the dimensions of the fibers. Let $\{U_\alpha\}_{\alpha=0,\dots,r}$ be a finite open cover of X. For $\underline{\alpha} = (\alpha_0, \dots, \alpha_p)$, $0 \leq \alpha_0 < \alpha_1 < \cdots < \alpha_p \leq r$, $|\underline{\alpha}| = p+1$, we denote by $U_{\underline{\alpha}}$ the open subset $U_{\alpha_0} \cap \cdots \cap U_{\alpha_p}$ of X, and by $f_{\underline{\alpha}}$ the restriction of f to $U_{\underline{\alpha}}$. We now introduce a chain of subsets of S, the Artin sets of f, which will be well-suited for the dévissage in Section 3 below.

Definition 1.8 The *Artin set* of f *of level* $i \geq 0$, denoted by $A_i(f)$, is the union of images $\epsilon(S') \subset S$, for all étale morphisms $\epsilon : S' \longrightarrow S$, finite over their image, such that:

1) $X_{S'}$ admits a finite open cover $\{U_\alpha\}$ such that the restriction of $f_{S'}$ to each U_α is a tower of coordinatized elementary fibrations;
2) each $U_{\underline{\alpha}}$, $|\underline{\alpha}| > 1$, admits a finite open cover $\{U_{\underline{\alpha},\beta}\}$ such that the restriction of $f_{S'}$ to each $U_{\underline{\alpha},\beta}$ is a tower of coordinatized elementary fibrations;
3) each $U_{\underline{\alpha},\underline{\beta}}$, $|\underline{\beta}| > 1$, admits a finite open cover $\{U_{\underline{\alpha},\underline{\beta},\gamma}\}$ such that the restriction of $f_{S'}$ to each $U_{\underline{\alpha},\underline{\beta},\gamma}$ is a tower of coordinatized elementary fibrations; ... and so on, for all $(\underline{\alpha}, \underline{\beta}, \underline{\gamma}, \dots)$ such that

$$(|\underline{\alpha}| - 1) + (|\underline{\beta}| - 1) + (|\underline{\gamma}| - 1) + \cdots \leq i.$$

Obviously, $A_i(f) \supset A_{i+1}(f)\ \forall\ i$, and, for any $U_{\underline{\alpha}}$ appearing in the previous definition, $A_{i-|\underline{\alpha}|+1}(f_{\underline{\alpha}}) \supset A_i(f)$.

We set $A(f) = A_{d+\dim X}(f)$ (this definition will be justified in 3.2.4). From Proposition 1.4, we deduce

Corollary 1.9 *For any* $i \geq 0$, $A_i(f)$ *is a dense open subset of the image of the (open) morphism* f *in* S.

Clearly, all Artin sets of a tower $X \longrightarrow S$ of coordinatized elementary fibrations coincide with S itself. If f is étale, they coincide with the maximal open subset of S on which f is an étale covering.

§2 Review of connections and De Rham cohomology

2.1 Let X be a smooth algebraic K-variety and $\mathcal{E}$ be an $\mathcal{O}_X$-module. We recall that an *integrable connection* on $\mathcal{E}$ is an additive map

(2.1.1) $$\mathcal{E} \longrightarrow \mathcal{E} \otimes_{\mathcal{O}_X} \Omega^1_X$$

satisfying the Leibniz rule and such that the usual composite arrow ("curvature")

(2.1.2) $$\mathcal{E} \longrightarrow \mathcal{E} \otimes_{\mathcal{O}_X} \Omega^1_X \longrightarrow \mathcal{E} \otimes_{\mathcal{O}_X} \Omega^2_X$$

is 0. This is equivalent to giving an additive morphism

$$\nabla : \underline{Der}_X \longrightarrow \underline{End}_K \mathcal{E}$$

compatible with the Lie bracket [,] and satisfying

$$\begin{aligned} \nabla(fD)(m) &= f\nabla(D)(m), \\ \nabla(D)(fm) &= f\nabla(D)(m) + D(f)m \end{aligned}$$

for local sections f, D, m of $\mathcal{O}_X$, $\underline{Der}_X$ and $\mathcal{E}$, respectively. In these circumstances, one may construct the *De Rham complex* of $(\mathcal{E}, \nabla)$

(2.1.3) $$DR(X/K, (\mathcal{E}, \nabla)) := [\mathcal{E} \longrightarrow \mathcal{E} \otimes_{\mathcal{O}_X} \Omega^1_X \longrightarrow \mathcal{E} \otimes_{\mathcal{O}_X} \Omega^2_X \longrightarrow \cdots].$$

We denote by $\mathbf{MIC}(X)$ the category of $\mathcal{O}_X$-modules with integrable connection. In this general setting (no assumptions on the underlying $\mathcal{O}_X$-module), $\mathbf{MIC}(X)$ is *isomorphic* to the category of left $\mathcal{D}_X$-modules ([BO, §2], [Kas]). The point is that any action, compatible with Lie brackets, of differential operators of order ≤ 1, extends to an action of $\mathcal{D}_X$.

The category $\mathbf{MIC}(X)$ is thus abelian and has enough injectives. It has an internal $\otimes$, and an internal $\underline{Hom}$ [Ka1, 1.1][(1)].

Given two objects $(\mathcal{E}, \nabla), (\mathcal{E}', \nabla') \in Ob\, \mathbf{MIC}(X)$, an $\mathcal{O}_X$-linear mapping $\mathcal{E} \longrightarrow \mathcal{E}'$ is called *horizontal* if it comes from a morphism in $\mathbf{MIC}(X)$.

Let $(\mathcal{E}, \nabla) \in Ob\, \mathbf{MIC}(X)$. We say that $(\mathcal{E}, \nabla)$ is *simple*, if it is a simple object of $\mathbf{MIC}(X)$. We say that it is *quasi-coherent* (resp. *coherent*, resp. *free*) if the underlying $\mathcal{O}_X$-module is. As is well-known, if $(\mathcal{E}, \nabla)$ is coherent, then $\mathcal{E}$ is automatically locally free [Ka1, 8.8]; this is then the usual (restricted) notion of connection on a vector bundle, which is actually our main object of interest.

If $f : X \longrightarrow S$ is smooth and $\underline{Der}_{X/S}$ is free of rank 1 with basis ∂ (*e.g.* $f =$ a rational elementary fibration *cf.* (1.2.1)), we say that $(\mathcal{E}, \nabla)$ is *cyclic with respect to* f (or ∂)

(1) We drop Katz's general quasi-coherence assumptions: see also [H2, §4]; otherwise, the internal $\underline{Hom}$ would only be partially defined.

if it is free of finite rank and admits a basis of the form $\{v, \nabla(\partial)v, \ldots, \nabla(\partial)^{\mu-1}v\}$. We then say that v is a *cyclic vector*. The definition is easily seen to be independent of the choice of ∂.

2.2 Inverse image Let $f : X \longrightarrow S$ be a morphism of smooth K-varieties. There is a functor

$$\begin{aligned} f^* : \mathbf{MIC}(S) &\longrightarrow \mathbf{MIC}(X), \\ (\mathcal{E}, \nabla) &\longmapsto (f^*\mathcal{E}, f^*\nabla), \end{aligned}$$

where $f^*\nabla$ is defined as follows: by sheaf-theoretic inverse image, ∇ defines a mapping

$$f^{-1}\mathcal{E} \longrightarrow f^{-1}\mathcal{E} \otimes_{f^{-1}\mathcal{O}_S} f^{-1}\Omega^1_S \tag{2.2.1}$$

sitting in a canonical diagram

$$\begin{array}{ccc} f^{-1}\mathcal{E} & \longrightarrow & f^{-1}\mathcal{E} \otimes_{f^{-1}\mathcal{O}_S} f^{-1}\Omega^1_S \\ \downarrow & & \downarrow \\ f^*\mathcal{E} & & f^*\mathcal{E} \otimes_{\mathcal{O}_X} \Omega^1_X. \end{array} \tag{2.2.2}$$

There is a unique integrable connection $f^*\nabla$ on $f^*\mathcal{E}$ such that

$$f^*\nabla : f^*\mathcal{E} \longrightarrow f^*\mathcal{E} \otimes_{\mathcal{O}_X} \Omega^1_X \tag{2.2.3}$$

completes (2.2.2) into a commutative square. It is not immediate that (2.2.1), which is not $f^{-1}\mathcal{O}_S$-linear, extends to (2.2.3)[(1)]. For this reason, $f^*\nabla$ is better described using Atiyah's linear description of connections (*cf.* [De, I, 2.3]): with the notation of [EGA IV, 16.7.1.2], ∇ corresponds to an $\mathcal{O}_S$-linear section of

$$\mathcal{P}^1_S(\mathcal{E}) \longrightarrow \mathcal{E}, \tag{2.2.4}$$

which defines an $\mathcal{O}_X$-linear section of

$$f^*\mathcal{P}^1_S(\mathcal{E}) \longrightarrow f^*\mathcal{E}; \tag{2.2.5}$$

since (2.2.5) factors through

$$\mathcal{P}^1_X(f^*\mathcal{E}) \longrightarrow f^*\mathcal{E} \tag{2.2.6}$$

[*loc. cit.*, 16.7.9], we obtain an $\mathcal{O}_X$-linear section of (2.2.6) which corresponds to $f^*\nabla$. One checks easily that this connection is integrable. Viewed as a $\mathcal{D}_X$-module, $f^*(\mathcal{E}, \nabla)$ corresponds to $\mathcal{D}_{X\to S} \otimes_{f^{-1}\mathcal{D}_S} f^{-1}\mathcal{E}$ (*cf.* [Bo, VI, 4.2]).

The functor f^* is right-exact.

---

(1) The same slight difficulty arises in defining the internal $\otimes$ and is solved in the same way.

2.3 Direct image Assume moreover that f is *smooth*. There is a functor

$$R^0_{DR}f_* : \mathbf{MIC}(X) \longrightarrow \mathbf{MIC}(S),$$
$$(\mathcal{E}, \nabla) \longmapsto (f_*\mathcal{E}^{\nabla_{|\underline{Der}X/S}}, \text{induced connection}),$$

which is left-exact.

Proposition 2.3.1 *Let $f : X \longrightarrow S$ be a smooth morphism of smooth K-varieties with connected geometric fibers. Then for any coherent $(\mathcal{E}, \nabla) \in Ob\mathbf{MIC}(X)$, the natural morphism $f^* R^0_{DR} f_*(\mathcal{E}, \nabla) \longrightarrow (\mathcal{E}, \nabla)$ is a monomorphism.*

Proof: Let x be any closed point of X. We first show that $\mathcal{O}_{X,x} \otimes_{\mathcal{O}_{S,f(x)}} \mathcal{E}^{\nabla}_x \longrightarrow \mathcal{E}_x$ is injective. Because the geometric fibers of f are connected, $Frac(\mathcal{O}_{X,x})/Frac(\mathcal{O}_{S,f(x)})$ is a regular extension and $Frac(\mathcal{O}_{X,x}) \otimes_{Frac(\mathcal{O}_{S,f(x)})} Frac(\hat{\mathcal{O}}_{S,f(x)}) \longrightarrow Frac(\hat{\mathcal{O}}_{X,x})$ is injective; hence, $\mathcal{O}_{X,x} \otimes_{\mathcal{O}_{S,f(x)}} \hat{\mathcal{O}}_{S,f(x)} \longrightarrow \hat{\mathcal{O}}_{X,x}$ is itself injective. Because $\mathcal{E}_x$ is free, it embeds into its $\mathfrak{m}_x$-adic completion $\hat{\mathcal{E}}_x$, and $\mathcal{E}^{\nabla}_x \subseteq \hat{\mathcal{E}}^{\nabla}_x$. We are thus reduced to showing that $\hat{\mathcal{O}}_{X,x} \otimes_{\hat{\mathcal{O}}_{S,f(x)}} \hat{\mathcal{E}}^{\nabla}_x \longrightarrow \hat{\mathcal{E}}_x$ is injective. This is in fact an isomorphism by the argument of Taylor series in [Ka1, 8.9]. We have shown that $\mathcal{O}_X \otimes_{f^{-1}\mathcal{O}_S} \mathcal{E}^{\nabla} \longrightarrow \mathcal{E}$ is injective, and we conclude by noticing that $f^{-1} f_*\mathcal{E}^{\nabla} \longrightarrow \mathcal{E}^{\nabla}$ is injective (because $f^{-1} f_*\mathcal{E} \longrightarrow \mathcal{E}$ is, $\mathcal{E}$ being locally free and X connected) and f is flat. □

2.4 Higher direct images The q-th derived functor of $R^0_{DR} f_*$ is denoted by

$$R^q_{DR} f_* : \mathbf{MIC}(X) \longrightarrow \mathbf{MIC}(S) .$$

For any $(\mathcal{E}, \nabla) \in Ob\ \mathbf{MIC}(X)$, the concrete description of $R^j_{DR} f_*(\mathcal{E}, \nabla)$ is given by the $\mathbf{R}^q f_*$ (for complexes of abelian sheaves or, which amounts to the same, of $f^{-1}\mathcal{O}_S$-modules on X) of the relative De Rham complex $DR(X/S, (\mathcal{E}, \nabla))$ of $(\mathcal{E}, \nabla)$, endowed with the Gauss-Manin connection [Ka1, 3.0]. (One checks that one obtains in this way an effaceable δ-functor, and applies [H3, III, 1.3A and 1.4]).

By [EGA 0_{III}, 12.4.3] applied to the morphism of ringed spaces

$$f : (X, f^{-1}\mathcal{O}_S) \longrightarrow (S, \mathcal{O}_S)$$

and $\mathcal{K}^{\cdot} = DR(X/S, (\mathcal{E}, \nabla))$, the sheaf underlying $R^q_{DR} f_*(\mathcal{E}, \nabla)$ is the sheaf associated to the presheaf

$$\mathcal{H}^q(f, \mathcal{K}^{\cdot}) : U \longmapsto \mathbf{H}^q(f^{-1}(U), \mathcal{K}^{\cdot}_{|f^{-1}(U)}) \qquad (2.4.1)$$

on S. We see in particular that the formation of $R^q_{DR} f_*$ is compatible with Zariski localization on S. If S is a point, we also write $H^q_{DR}(X, (\mathcal{E}, \nabla))$ for the K-vector space $R^q_{DR} f_*(\mathcal{E}, \nabla)$.

Let us recall one of the constructions of the Gauss-Manin connection (*cf.* [KaO]). We filter the absolute De Rham complex $DR(X/K, (\mathcal{E}, \nabla))$ by the subcomplexes

$$F^p = image\ \ of\quad f^*(\Omega^p_S) \otimes_{\mathcal{O}_X} \Omega^{\cdot -p}_X \otimes_{\mathcal{O}_X} \mathcal{E},$$

so that

$$gr^p = F^p/F^{p+1} \cong f^*(\Omega^p_S) \otimes_{\mathcal{O}_X} \Omega^{\cdot -p}_{X/S} \otimes_{\mathcal{O}_X} \mathcal{E},$$

$$\mathbf{R}^q f_*(gr^0) \cong R^q_{DR} f_*(\mathcal{E}, \nabla), \qquad \mathbf{R}^{q+1} f_*(gr^1) \cong \Omega^1_S \otimes_{\mathcal{O}_S} R^q_{DR} f_*(\mathcal{E}, \nabla).$$

The Gauss-Manin connection is the coboundary map in the long cohomology sequence of the $\mathbf{R}^q f_*$ arising from the short exact sequence

$$0 \longrightarrow gr^1 \longrightarrow F^0/F^2 \longrightarrow gr^0 \longrightarrow 0. \tag{2.4.2}$$

When f is a (smooth) morphism of pure relative dimension d, $R^q_{DR} f_*$ coincides with the notion of higher direct image $h^{q-d} f_+$ of $\mathcal{D}_X$-modules [Bo, VI, 5.0][(1)]

2.5 Open covers Let $\{U_\alpha\}$ be an open cover of X (not necessarily finite). In the following proposition, a variation of [EGA 0_{III}, 12.4.6], $\mathbf{h}^q(f, \mathcal{K}^\cdot)$ is the presheaf $U \longmapsto \mathbf{R}^q f_{|U*}(\mathcal{K}^\cdot_{|U})$, while $h^q(f, \mathcal{K}^j)$ is $U \longmapsto R^q f_{|U*}(\mathcal{K}^j_{|U})$, both presheaves on X taking values in the category of *sheaves* of $\mathcal{O}_S$-modules, and $h^q(f, \mathcal{K}^\cdot)$ is the complex of presheaves obtained from term by term application of the latter. For a presheaf $\mathcal{F}$ (resp. for a complex of presheaves $\mathcal{F}^\cdot$) on X, with values in any abelian category, the notation $C^\cdot(\{U_\alpha\}, \mathcal{F})$ (resp. $C^\cdot(\{U_\alpha\}, \mathcal{F}^\cdot)$) refers to the usual Čech complex (resp. bicomplex). If $\mathcal{F}$ is a presheaf of abelian groups on X, we also use the notation $C^i_f(\{U_\alpha\}, \mathcal{F}) = \oplus_{\alpha_0 < \alpha_1 < \cdots < \alpha_i} f_{\underline{\alpha}*}(\mathcal{F}_{|U_{\underline{\alpha}}})$, for the abelian presheaf of (alternating) Čech cochains.

Lemma 2.5.1 *Let $\mathcal{K}^\cdot$ be a complex of $f^{-1}\mathcal{O}_S$-modules on X. There exist two regular spectral functors ${}_I E^{\cdot\cdot}_\cdot$ and ${}_{II} E^{\cdot\cdot}_\cdot$ in $\mathcal{K}^\cdot$ taking values in the category of $\mathcal{O}_S$-modules*

---

(1) Although this fact plays no role in the present chapter, we sketch the following argument. As in the proof of [Bo, VI, 5.3.2], what has to be proven is that, for $d = \dim X - \dim S$, $DR(X/S, \mathcal{D}_X)[d]$ is a locally free left resolution of $\mathcal{D}_{S \leftarrow X}$ in the category of right $\mathcal{D}_X$-modules. Local freedom is obvious. For an étale map $f = j : X \longrightarrow S$, $\mathcal{D}_{S \leftarrow X} = \mathcal{D}_{X \to S} = j^*(\mathcal{D}_S) = \mathcal{D}_X = DR(X/S, \mathcal{D}_X)$, by definition ([Bo, VI, 4.2] and [Bo, VI, 5.1 (4)]). For a projection $f = p : \mathbf{A}^d_S \longrightarrow S$ the assertion is proven in [Bo, VI, 5.3.1 (i)]. Consider the case of $f = p \circ j : X \xrightarrow{j} X' = \mathbf{A}^d_S \xrightarrow{p} S$, composite of a projection and of an étale map j. Then, by [Bo, VI, 6.3 (2)], $\mathcal{D}_{S \leftarrow X} = j^{-1}(\mathcal{D}_{S \leftarrow X'}) \otimes_{j^{-1}(\mathcal{D}_{X'})} \mathcal{D}_{X' \leftarrow X} = j^{-1}(\mathcal{D}_{S \leftarrow X'}) \overset{\mathbf{L}}{\otimes}_{j^{-1}(\mathcal{D}_{X'})} \mathcal{D}_{X' \leftarrow X}$, and $\mathcal{D}_{X' \leftarrow X} = \mathcal{D}_X$. So, $\mathcal{D}_{S \leftarrow X} = j^*(\mathcal{D}_{S \leftarrow X'}) = Lj^*(\mathcal{D}_{S \leftarrow X'}) = j^*(DR(X'/S, \mathcal{D}_{X'})[d]) = DR(X/S, \mathcal{D}_X)[d]$. Since any smooth morphism $f : X \longrightarrow S$ is locally, for the Zariski topology on X, of the type $p \circ j$, the assertion follows. In fact, this argument accounts only for the coincidence of $R^q_{DR} f_*(\mathcal{E}, \nabla)$ and $h^{q-d} f_+(\mathcal{E}, \nabla)$ as $\mathcal{O}_S$-modules.

and converging to the relative hypercohomology $\mathbf{R}^{\cdot} f_*(\mathcal{K}^{\cdot})$. *The* E_1 *terms are given by*

$$
\begin{aligned}
{}_I E_1^{p,q} &= (C^{\cdot}(\{U_\alpha\}, h^q(f, \mathcal{K}^{\cdot})))^p_{tot} = \oplus_{i+j=p} C^i(\{U_\alpha\}, h^q(f, \mathcal{K}^j)) \\
&= \oplus_{i+j=p, \alpha_0<\alpha_1<\cdots<\alpha_i} R^q f_{\underline{\alpha}*}(\mathcal{K}^j_{|U_{\underline{\alpha}}}) ,
\end{aligned}
$$

$$
{}_{II} E_1^{p,q} = C^p(\{U_\alpha\}, \mathbf{h}^q(f, \mathcal{K}^{\cdot})) = \oplus_{\alpha_0<\alpha_1<\cdots<\alpha_p} \mathbf{R}^q f_{\underline{\alpha}*}(\mathcal{K}^{\cdot}_{|U_{\underline{\alpha}}}) ,
$$

respectively. If $\mathcal{K}^{\cdot}$ *is bounded from below, the previous spectral sequences are biregular.*

Proof: Following [EGA 0_{III}, 12.4.6], we consider an injective Cartan-Eilenberg resolution $\mathcal{L}^{\cdot,\cdot}$ of $\mathcal{K}^{\cdot}$ and the tricomplex of $\mathcal{O}_S$-modules

$$
\mathcal{C}^{\cdot}_f(\{U_\alpha\}, \mathcal{L}^{\cdot,\cdot}) = \oplus_{i,j,k} \mathcal{C}^i_f(\{U_\alpha\}, \mathcal{L}^{j,k}) .
$$

We first regard $\mathcal{C}^{\cdot}_f(\{U_\alpha\}, \mathcal{L}^{\cdot,\cdot})$ as a bicomplex for the degrees $(i, j+k)$. The second spectral sequence of this bicomplex is regular and degenerate, since H^q_I is the associated sheaf to the presheaf $V \longmapsto H^q(\{U_\alpha \cap f^{-1}(V)\}, \mathcal{L}^{j,k})$ on S, and this is 0 for $q > 0$, $\mathcal{L}^{j,k}_{|f^{-1}(V)}$ being flasque. So, $H^n(\mathcal{C}^{\cdot}_f(\{U_\alpha\}, \mathcal{L}^{\cdot,\cdot})_{\text{tot}}) \xrightarrow{\sim} \mathbf{R}^n f_* \mathcal{K}^{\cdot}$. We then define ${}_I E^{\cdot,\cdot}_{\cdot}$ as the first spectral sequence of $\mathcal{C}^{\cdot}_f(\{U_\alpha\}, \mathcal{L}^{\cdot,\cdot})$, viewed as a bicomplex for the degrees $(i+j, k)$ (this is the spectral sequence in the statement of [*loc. cit.*]). Here $H^q_{II} = \oplus_{i,j} H^q(\mathcal{C}^i_f(\{U_\alpha\}, \mathcal{L}^{j,\cdot})) = \oplus_{i,j} C^i(\{U_\alpha\}, h^q(f, \mathcal{K}^j))$, since $\mathcal{L}^{j,\cdot}_{|U}$ is an injective resolution of $\mathcal{K}^j_{|U}$. As ${}_{II} E^{\cdot,\cdot}_{\cdot}$ we take instead the first spectral sequence of $\mathcal{C}^{\cdot}_f(\{U_\alpha\}, \mathcal{L}^{\cdot,\cdot})$, viewed as a bicomplex for the degrees $(i, j+k)$. □

In the special case of $\mathcal{K}^{\cdot} = DR(X/S, (\mathcal{E}, \nabla))$, the spectral sequence ${}_{II} E^{\cdot,\cdot}_{\cdot}$ of the lemma is the *Zariski spectral sequence* of $\mathcal{O}_S$-modules of [Ka1, 3.5.1.0]:

(2.5.2) $$E_1^{p,q} = \oplus_{\alpha_0<\alpha_1<\cdots<\alpha_p} R^q_{DR} f_{\underline{\alpha}*}(\mathcal{E}, \nabla)_{|U_{\underline{\alpha}}} \Rightarrow R^{p+q}_{DR} f_*(\mathcal{E}, \nabla) .$$

If we replace $DR(X/S, (\mathcal{E}, \nabla))$ by the short exact sequence (2.4.2), and use the fact that hypercohomology of a direct image is a cohomological functor, we obtain a coboundary morphism of spectral sequences

$$
\begin{array}{ccc}
E_1^{p,q} \cong E_1^{p,q}(gr^0) & \Rightarrow & R^{p+q}_{DR} f_*(\mathcal{E}, \nabla), \\
\downarrow & & \downarrow \\
E_1^{p,q+1}(gr^1) \cong \Omega^1_{S/K} \otimes_{\mathcal{O}_S} E_1^{p,q}(gr^0) & \to & \mathbf{R}^{p+q+1} f_*(gr^1) \cong \Omega^1_{S/K} \otimes_{\mathcal{O}_S} R^{p+q}_{DR} f_*(\mathcal{E}, \nabla),
\end{array}
$$

which expresses the fact that (2.5.2) "is" a spectral sequence in $\mathbf{MIC}(S)$. On the other hand, the E_2^{pq} term of the spectral sequence ${}_I E^{\cdot,\cdot}_{\cdot}$ of the proposition is

$$
{}_I E_2^{p,q} = \text{total homology of degree } p \text{ of the bicomplex } C^{\cdot}(\{U_\alpha\}, h^q(f, \mathcal{K}^{\cdot})) .
$$

This implies, in particular, that if each U_α is affine, and if $\mathcal{E}$ is quasi-coherent, then $R^j_{DR} f_*(\mathcal{E}, \nabla)$ may be calculated "à la Čech", as the j-th (total) homology $\mathcal{O}_S$-module of the bicomplex of $\mathcal{O}_S$-modules $C^{\cdot}_f(\{U_\alpha\}, \Omega^{\cdot}_{X/S} \otimes \mathcal{E})$, of alternating Čech cochains on the nerve of the cover $\{U_\alpha\}$.

It turns out that the Gauss-Manin connection comes, upon passage to homology, from a (not integrable) connection on this bicomplex: if ∂ is a vector field on S, and if ∂_α is a lifting of ∂ on U_α, the value at ∂ of this connection can be expressed as a sum of the Lie derivative $\mathrm{L}(\partial_{\alpha_0})$ and the interior product $I(\partial_{\alpha_0} - \partial_{\alpha_1})$ (up to sign), see [KO, Section 3].

The proof of *loc. cit.* depends on a local calculation. We present here an alternative argument, based on functoriality.

Let us denote by d_{DR} the differential of the De Rham complex $\Omega^{\cdot}_{X/S} \otimes \mathcal{E}$ or of $C^p_f(\{U_\alpha\}, \Omega^{\cdot}_{X/S} \otimes \mathcal{E})$, and by $\check{d}$ the Čech differential on $C^p_f(\{U_\alpha\}, \Omega^{\cdot}_{X/S} \otimes \mathcal{E})$. We follow the convention of [KO] (not the one in [EGA III]) by putting the total differential $\delta = d_{DR} + (-)^q \check{d}$ on the term $C^{p,q}_f = C^p_f(\{U_\alpha\}, \Omega^q_{X/S} \otimes \mathcal{E})$ of the bicomplex. We denote by $C^{\cdot}_f = C^{\cdot}_f(\mathcal{E}, \nabla)$ the associated simple complex, so that

$$H^i(C^{\cdot}_f(\mathcal{E}, \nabla)) \cong \mathbf{R}^i f_*(\Omega^{\cdot}_{X/S} \otimes \mathcal{E}),$$

functorially in $(\mathcal{E}, \nabla)$. For each vector field ∂ on S, we define an additive endomorphism $\tilde{\aleph}(\partial)$ of C^i_f by

$$\tilde{\aleph}(\partial) = \mathrm{L}(\partial_{\alpha_0}) + (-)^q I(\partial_{\alpha_0} - \partial_{\alpha_1})$$

on $C^{i-q,q}_f$. This endomorphism satisfies the Leibniz rule, because the Lie derivative does; hence the $\tilde{\aleph}(\partial)$ define a connection $\tilde{\aleph}$ on $C^i_f(\mathcal{E}, \nabla)$, which is functorial in $(\mathcal{E}, \nabla)$ (because the Lie derivative $\mathrm{L}(\partial_{\alpha_0}) : \Omega^q_{U_{\alpha_0}/S} \otimes \mathcal{E}_{|U_{\alpha_0}} \longrightarrow \Omega^q_{U_{\alpha_0}/S} \otimes \mathcal{E}_{|U_{\alpha_0}}$ is functorial).

On the other hand, we have the formulas

i) $[\mathrm{L}(\partial_{\alpha_0}), d_{DR}] = 0$,
ii) $\mathrm{L}(\partial_{\alpha_0}) - \mathrm{L}(\partial_{\alpha_1}) = d_{DR} \circ I(\partial_{\alpha_0} - \partial_{\alpha_1}) + I(\partial_{\alpha_0} - \partial_{\alpha_1}) \circ d_{DR}$,
iii) $I(\partial_{\alpha_0} - \partial_{\alpha_1}) + I(\partial_{\alpha_1} - \partial_{\alpha_2}) + I(\partial_{\alpha_2} - \partial_{\alpha_0}) = 0$.
From ii) we deduce
iv) $[\mathrm{L}(\partial_{\alpha_0}), (-)^q \check{d}] + [(-)^q I(\partial_{\alpha_0} - \partial_{\alpha_1}), d_{DR}] = 0$, and iii) implies
v) $[(-)^q I(\partial_{\alpha_0} - \partial_{\alpha_1}), (-)^q \check{d}] = 0$.

Formulas $i)$, $iv)$, $v)$ put together show that $[\tilde{\aleph}(\partial), \delta] = 0$, *i.e.* $\tilde{\aleph}(\partial)$ is an (additive) endomorphism of the complex $C^{\cdot}_f(\mathcal{E}, \nabla)$. We thus obtain functors

$$\aleph^i : \mathbf{MIC}(X) \longrightarrow \mathbf{MC}(S),$$
$$(\mathcal{E}, \nabla) \longmapsto (H^i(C^{\cdot}_f(\mathcal{E}, \nabla)) \cong \mathbf{R}^i f_*(\Omega^{\cdot}_{X/S} \otimes \mathcal{E}), H^i(\tilde{\aleph})),$$

where $\mathbf{MC}(S)$ stands for the abelian category of $\mathcal{O}_S$-modules with (not necessarily integrable) connection.

By functoriality of $\tilde{\aleph}$ with respect to $(\mathcal{E}, \nabla)$ it is clear that the collection of the $\aleph^i$ form a δ-functor. This δ-functor is effaceable (because any $(\mathcal{E}, \nabla)$ imbeds into an injective $(\mathcal{I}, \nabla)$ of $\mathbf{MIC}(X)$ and $\mathbf{R}^i f_*(\Omega^{\cdot}_{X/S} \otimes \mathcal{E}) = 0$, for $i > 0$), hence $\aleph^i = R^i\aleph^0$. Let $\omega : \mathbf{MIC}(S) \longrightarrow \mathbf{MC}(S)$ stand for the (exact) imbedding. We have $\aleph^0 = \omega \circ R^0_{DR} f_*$, because $\tilde{\aleph}(\partial)$ is none but the Gauss-Manin connection on $H^0(C^{\cdot}_f) \cong f_*\mathcal{E}^{\nabla_{|\underline{Der}X/S}}$, being induced by the Lie derivative in the direction ∂ acting on $\mathcal{E}$. We conclude that $\aleph^i = R^i(\omega \circ R^0_{DR} f_*) = \omega \circ R^i_{DR} f_*$.

2.6 Flat base change Let

$$\begin{array}{ccc} X^\sharp & \xrightarrow{u^\sharp} & X \\ f^\sharp \downarrow & & \downarrow f \\ S^\sharp & \xrightarrow{u} & S \end{array}$$

be a fibered square, u being a morphism of smooth K-varieties. We have $u^{\sharp *}\Omega^{\cdot}_{X/S} \simeq \Omega^{\cdot}_{X^\sharp/S^\sharp}$ and $u^{\sharp *}DR(X/S, (\mathcal{E}, \nabla)) \simeq DR(X^\sharp/S^\sharp, u^{\sharp *}(\mathcal{E}, \nabla))$. (Notice that despite the non-linearity of the relative De Rham complex, the complex $u^{\sharp *}DR(X/S, (\mathcal{E}, \nabla))$ is well-defined because $u^{\sharp *}(-)$ may be interpreted as

$$f^{\sharp -1}(\mathcal{O}_{S^\sharp}) \otimes_{u^{\sharp -1}f^{-1}(\mathcal{O}_S)} u^{\sharp -1}(-)).$$

Moreover, if $\mathcal{E}$ is quasi-coherent, and if u is *flat* (*e.g.* an étale localization), we have $u^* R^q_{DR} f_*(\mathcal{E}, \nabla) \xrightarrow{\sim} R^q_{DR} f^\sharp_*(u^{\sharp *}(\mathcal{E}, \nabla))$, by flat base change for hypercohomology. One can also choose an affine open cover of X, use the first spectral sequence of Lemma 2.5.1 (where ${}_I E_1^{p,q} = 0$ for $q > 0$), and flat base change for $f_{\underline{\alpha}*}(\Omega^j_{U_{\underline{\alpha}}/S})$.

2.7 Vanishing Assume $\mathcal{E}$ is quasi-coherent and let d be the maximum among the dimensions of the fibers of f. Then $R^j_{DR} f_*(\mathcal{E}, \nabla) = 0$ for $j \notin [0, d + \dim X]$, as can be seen from the Hodge-De Rham spectral sequence $E_1^{p,q} = R^q f_*(\Omega^p_{X/S} \otimes \mathcal{E}) \Rightarrow R^{p+q}_{DR} f_*(\mathcal{E}, \nabla)$.

Assume moreover that f is affine. Then $R^j_{DR} f_*(\mathcal{E}, \nabla) = 0$ for $j \notin [0, d]$. Let us look at the first spectral sequence of hypercohomology [EGA 0_{III}, 12.4.1.1]

$$\mathcal{E}_2^{p,q} = \mathcal{H}^p(R^q f_* DR(X/S, (\mathcal{E}, \nabla))) \Rightarrow \mathbf{R}^{p+q} f_* DR(X/S, (\mathcal{E}, \nabla)).$$

Because f is affine, $R^q f_* DR(X/S, (\mathcal{E}, \nabla)) = 0$ for $q > 0$, hence the spectral sequence degenerates at $\mathcal{E}_2$, so that

$$R^j_{DR} f_*(\mathcal{E}, \nabla) \cong \mathcal{H}^j(f_* \mathcal{D}R_{X/S}(\mathcal{E}, \nabla)).$$

In particular $R^1_{DR} f_*(\mathcal{E}, \nabla) \cong \operatorname{Coker} f_*(\nabla)$, if $d = 1$,[(1)] and if S is affine (hence so is X), $R^j_{DR} f_*(\mathcal{E}, \nabla)$ may thus be computed as the $\mathcal{O}_S$-module attached to the j-th cohomology $\Gamma(S, \mathcal{O}_S)$-module of the global relative De Rham complex

$$\mathcal{E}(X) \longrightarrow \mathcal{E}(X) \otimes \Omega^1_{X/S}(X) \longrightarrow \cdots .$$

(We shall apply this computation in the case where f is an elementary fibration. See also Appendix D).

More generally, if S is affine and $\{U_\alpha\}$ is an affine open cover of X, $R^j_{DR} f_*(\mathcal{E}, \nabla)$ is the $\mathcal{O}_S$-module attached to the j-th total cohomology group of the Čech bicomplex $C^\cdot(\{U_\alpha\}, \Omega^\cdot_{X/S} \otimes \mathcal{E})$.

§3 Dévissage

3.1 In the sequel, we consider properties of $\mathcal{O}_X$-modules with integrable connection, for arbitrary smooth K-varieties X; we write $\mathcal{P}((\mathcal{E}, \nabla))$ to indicate that $(\mathcal{E}, \nabla)$ satisfies property $\mathcal{P}$.

We say that $\mathcal{P}$ is *local for the étale topology* (resp. *stable under finite étale direct image*) if for any X, any $(\mathcal{E}, \nabla) \in Ob\, \mathbf{MIC}(X)$ and any étale cover $\{V_i\}$ of X (resp. any finite étale $\pi : X \longrightarrow X'$), one has

$$\mathcal{P}((\mathcal{E}, \nabla)) \Leftrightarrow (\forall\, i,\ \ \mathcal{P}((\mathcal{E}, \nabla)_{|V_i}))$$
$$(\text{resp. } \ \mathcal{P}((\mathcal{E}, \nabla)) \Rightarrow \mathcal{P}(\pi_*(\mathcal{E}, \nabla))).$$

Here we write simply $\pi_*(\mathcal{E}, \nabla)$ instead of $R^0_{DR}\pi_*(\mathcal{E}, \nabla)$, the underlying $\mathcal{O}_{X'}$-module being $\pi_*\mathcal{E}$.

We say in general that a property $\mathcal{P}$ of objects of an abelian category $\mathcal{A}$ is *strongly exact* if $\mathcal{P}(0)$ and for any exact sequence

$$E_1 \longrightarrow E \longrightarrow E_2$$

in $\mathcal{A}$, one has

$$(\mathcal{P}(E_1) \ \text{ and } \ \mathcal{P}(E_2)\,) \Rightarrow \mathcal{P}(E)\,.$$

Strong exactness is equivalent to the conjunction of exactness in the sense of [EGA III, 3.1.1] and stability by taking subobjects.

[(1)] This should not be confused with $f_*(\operatorname{Coker}\nabla)$, which arises in the *second* spectral sequence of hypercohomology

$$\mathcal{E}_2^{p,q} = R^p f_* \mathcal{H}^q(DR(X/S, (\mathcal{E}, \nabla))) \Rightarrow \mathbf{R}^{p+q} f_* DR(X/S, (\mathcal{E}, \nabla)).$$

Lemma 3.1.1 *Let $\mathcal{P}$ be a property of modules with integrable connection on smooth K-varieties, which is strongly exact and local for the étale topology. Then $\mathcal{P}$ is stable under finite étale direct image.*

Proof: Let $(\mathcal{E}, \nabla) \in Ob\ \mathbf{MIC}(X)$ and $\pi : X \longrightarrow Z$ be an étale covering. By [SGA 1, Example V, Section 4 and Section 7] there exists $\tau : Y \longrightarrow X$ such that $\pi \circ \tau$ is a Galois covering of Z of Galois group, say, G. Then [*loc. cit.*, Lemma 3.3] τ is necessarily étale. Let $\sigma : Z' \longrightarrow Z$ be an étale covering such that $Y_{Z'} \longrightarrow Z'$ [*loc. cit.*, Proposition 2.6, (ii bis)] is the trivial Galois covering $Z' \times G \xrightarrow{\mathrm{pr}_1} Z'$. We have the commutative diagram with fibered squares

$$\begin{array}{ccccccc} & & Y & \xrightarrow{\tau} & X & \xrightarrow{\pi} & Z \\ & & \uparrow & & \uparrow \sigma' & & \uparrow \sigma \\ Z' \times G & = & Y_{Z'} & \xrightarrow{\tau'} & X_{Z'} & \xrightarrow{\pi'} & Z' \end{array}$$

where all morphisms are étale, π and σ are étale coverings, and $\pi' \circ \tau' = \mathrm{pr}_1$.

We have: $\mathcal{P}((\mathcal{E}, \nabla)) \Leftrightarrow \mathcal{P}(\sigma'^*(\mathcal{E}, \nabla))$ and, since $\pi'_*\sigma'^*(\mathcal{E}, \nabla) \cong \sigma^*\pi_*(\mathcal{E}, \nabla)$, $\mathcal{P}(\pi_*(\mathcal{E}, \nabla)) \Leftrightarrow \mathcal{P}(\pi'_*\sigma'^*(\mathcal{E}, \nabla))$. So we may assume that $(\mathcal{E}', \nabla')$ is an object of $\mathbf{MIC}(X_{Z'})$ and must prove that $\mathcal{P}((\mathcal{E}', \nabla')) \Rightarrow \mathcal{P}(\pi'_*(\mathcal{E}', \nabla'))$. Now, $\mathcal{P}((\mathcal{E}', \nabla')) \Rightarrow \mathcal{P}(\tau'^*(\mathcal{E}', \nabla'))$. Since $Z' \times G = \coprod_{\gamma \in G} Z' \times \{\gamma\}$, an object $(\mathcal{M}, \nabla)$ of $\mathbf{MIC}(Z' \times G)$, consists of local components $(\mathcal{M}, \nabla)_\gamma$ in $\mathbf{MIC}(Z' \times \{\gamma\}) = \mathbf{MIC}(Z')$, and $\mathcal{P}((\mathcal{M}, \nabla)) \Leftrightarrow (\mathcal{P}((\mathcal{M}, \nabla)_\gamma)\ ,\ \forall\ \gamma \in G)$. Since G operates on $Z' \times G$ by Z'-automorphisms, for any $\gamma \in G$

$$(\mathrm{pr}_1^* \circ \mathrm{pr}_{1*}(\mathcal{M}, \nabla))_\gamma \cong \oplus_{\delta \in G} (\mathcal{M}, \nabla)_\delta,$$

and $\mathcal{P}((\mathcal{M}, \nabla)) \Leftrightarrow \mathcal{P}(\mathrm{pr}_1^* \circ \mathrm{pr}_{1*}(\mathcal{M}, \nabla))$. In particular,

$$\mathcal{P}((\mathcal{E}', \nabla')) \Rightarrow \mathcal{P}(\mathrm{pr}_1^* \circ \mathrm{pr}_{1*} \circ \tau'^*(\mathcal{E}', \nabla')) \Leftrightarrow \mathcal{P}(\tau'^* \circ \pi'^* \circ \pi'_* \circ \tau'_* \circ \tau'^*(\mathcal{E}', \nabla'))\ .$$

But $(\mathcal{E}', \nabla') \hookrightarrow \tau'_* \circ \tau'^*(\mathcal{E}', \nabla')$ is a monomorphism, and the functor $\tau'^* \circ \pi'^* \circ \pi'_* = \mathrm{pr}_1^* \circ \pi'_*$ is left-exact, pr_1^* being exact and π'_* left-exact. The conclusion is that $\tau'^* \circ \pi'^* \circ \pi'_*(\mathcal{E}', \nabla')$ is a sub-object of $\tau'^* \circ \pi'^* \circ \pi'_* \circ \tau'_* \circ \tau'^*(\mathcal{E}', \nabla')$, so that $\mathcal{P}(\tau'^* \circ \pi'^* \circ \pi'_*(\mathcal{E}', \nabla'))$, *i.e.* $\mathcal{P}(\mathrm{pr}_1{}^*(\pi'_*(\mathcal{E}', \nabla')))$, holds. But pr_1 is an étale covering, hence $\mathcal{P}(\pi'_*(\mathcal{E}', \nabla'))$ must hold. □

Lemma 3.2 *(Lemma on dévissage, first form.) Let $\mathcal{P}$ be a property of modules with integrable connection on smooth K-varieties, which is strongly exact and local for the étale topology. Assume that for any rational elementary fibration $f' : X' \longrightarrow S'$ with S' affine, étale over some affine K-space, and for any quasi-coherent $(\mathcal{E}', \nabla') \in Ob\ \mathbf{MIC}(X')$,*

(*) $$\mathcal{P}((\mathcal{E}', \nabla')) \Rightarrow \mathcal{P}(R^j_{DR} f'_*(\mathcal{E}', \nabla'))\ ,\quad j = 0, 1\ .$$

Then for any $i \geq 0$, any smooth morphism $f : X \longrightarrow S$ of smooth K-varieties, and any quasi-coherent $(\mathcal{E}, \nabla) \in Ob\ \mathbf{MIC}(X)$,

$$\mathcal{P}((\mathcal{E}, \nabla)) \Rightarrow \mathcal{P}((R^i_{DR} f_*(\mathcal{E}, \nabla))_{|A_i(f)})\ .$$

Proof:

3.2.1 The case when f has relative dimension 0 is a simple consequence of the stability of $\mathcal{P}$ under finite étale direct image; we discard this trivial case in the sequel. Because $\mathcal{P}$ is local, we may assume that S is connected. On the other hand, if $f = f_1 \coprod f_2 : X = X_1 \coprod X_2 \longrightarrow S$, then $R^j_{DR} f_*(\mathcal{E}, \nabla) = R^j_{DR} f_{1*}((\mathcal{E}, \nabla)_{|X_1}) \oplus R^j_{DR} f_{2*}((\mathcal{E}, \nabla)_{|X_2})$; because $\mathcal{P}$ is strongly exact, we may thus assume that X is connected so that f has pure relative dimension d. We may also assume that $0 \leq j \leq d + \dim X$, otherwise $R^j_{DR} f_*(\mathcal{E}, \nabla) = 0$, hence the result is trivial.

3.2.2 We may replace at once S by $A_i(f)$, and then (because $\mathcal{P}$ is local for the étale topology and $R^j_{DR} f_*$ is compatible with étale localization for quasi-coherent $\mathcal{E}$ (2.3)) by an affine connected étale neighborhood S' (étale over some affine space if we wish) such that $X_{S'}$ admits a finite open cover $\{U_\alpha\}$ with the properties listed in Definition 1.8. Let us consider the Zariski spectral sequence in **MIC**(S) (2.2.2):

$$E_1^{p,i-p} = \oplus_{\alpha_0 < \alpha_1 < \cdots < \alpha_p} R^{i-p}_{DR} f_{\underline{\alpha}*}(\mathcal{E}, \nabla)_{|U_{\underline{\alpha}}} \Rightarrow R^i_{DR} f_*(\mathcal{E}, \nabla).$$

By induction on i, since $A_{i-|\underline{\alpha}|+1}(f_{\underline{\alpha}}) = A_i(f) = S$, $\mathcal{P}(R^{i-p}_{DR} f_{\underline{\alpha}*}(\mathcal{E}, \nabla)_{|U_{\underline{\alpha}}})$ for all $\underline{\alpha}$ such that $|\underline{\alpha}| = p + 1 > 1$. Thus, by strong exactness of $\mathcal{P}$, this spectral sequence reduces to proving $\mathcal{P}(R^i_{DR} f_{\alpha*}(\mathcal{E}, \nabla)_{|U_\alpha})$; hence we may assume that f itself is a tower

$$X = X_d \longrightarrow X_{d-1} \longrightarrow \cdots \longrightarrow X_0 = S$$

of coordinatized elementary fibrations over an affine connected base S (étale over some affine space if we wish). Such a tower gives rise to Leray spectral sequences in **MIC** (*cf.* [Ka1, 3.3]); by strong exactness of $\mathcal{P}$, we are reduced to proving $\mathcal{P}$ for each level of the tower, i.e. we may assume that f itself is a coordinatized elementary fibration.

3.2.3 In case f is a coordinatized elementary fibration,

$$(3.2.2.1) \qquad \begin{array}{ccccc} X & \overset{j}{\hookrightarrow} & \overline{X} & \hookleftarrow & Z \\ \downarrow \pi & & \downarrow \overline{\pi} & & \downarrow \\ Y & \hookrightarrow & \mathbf{P}^1_S & \hookleftarrow & Z' \\ & \underset{f'}{\searrow} & \downarrow & \swarrow & \\ & & S & & \end{array}$$

$(f = f' \circ \pi)$, we may assume that $i = 0, 1$: other cohomology sheaves are 0, since f is affine of relative dimension 1 (*cf.* 2.4). Because $\mathcal{P}$ is étale-local, we may replace S by an étale covering so that f' becomes a rational elementary fibration. Since π is an étale covering, the Leray spectral sequence for $f = f' \circ \pi$ degenerates: $R^i_{DR} f_*(\mathcal{E}, \nabla) =$

$R^i_{DR} f'_*(\pi_*(\mathcal{E},\nabla))$. Then $\mathcal{P}((\mathcal{E},\nabla)) \Rightarrow \mathcal{P}(\pi_*(\mathcal{E},\nabla)) \Rightarrow \mathcal{P}(R^i_{DR} f'_*(\pi_*(\mathcal{E},\nabla)))$, by $(*)$. This proves Lemma 3.2. □

Remarks 3.2.4

i) In Lemma 3.2, we may replace $A_i(f)$ by $A(f) = A_{d+\dim X}(f)$ since

$$R^j_{DR} f_*(\mathcal{E},\nabla) = 0 \;\text{ for }\; j > d + \dim X \;.$$

ii) If in Lemma 3.2, the assumption $(*)$ holds only for $j = 0$, the conclusion keeps holding for $i = 0$.

Lemma 3.3 **(Lemma on dévissage, second form.)** *Let $\mathcal{P}$ be a property of modules with integrable connection on smooth K-varieties, which implies coherence, is strongly exact and local for the étale topology. Assume that:*

$(*)$ *for any rational elementary fibration $f' : X' \longrightarrow S'$ with S' affine, étale over some affine K-space, and for any simple, cyclic $(\mathcal{E}',\nabla') \in Ob$ **MIC**(X'), there exists an open dense subset $U' \subset S'$, such that*

$$\mathcal{P}((\mathcal{E}',\nabla')) \Rightarrow \mathcal{P}((R^j_{DR} f'_*(\mathcal{E}',\nabla'))_{|U'}), \quad j = 0, 1.$$

*Then for any smooth morphism $f : X \longrightarrow S$ of smooth K-varieties, and any $(\mathcal{E},\nabla) \in Ob$ **MIC**(X), there exists a non-empty open subset $U \subset S$ (depending upon $(\mathcal{E},\nabla)$) such that, for any $i \geq 0$,*

$$\mathcal{P}((\mathcal{E},\nabla)) \Rightarrow \mathcal{P}((R^i_{DR} f_*(\mathcal{E},\nabla))_{|U}).$$

Proof:

3.3.1 As in 3.2.1, we reduce to the case when S is connected, f is equidimensional of relative dimension $d \geq 1$, and $0 \leq j \leq d + \dim X$. By Artin's theorem (*i.e.* Proposition 1.4, without insisting on "coordinatized" elementary fibrations), there exists an étale map $\epsilon : S' \longrightarrow S$ such that $X_{S'}$ admits a finite open affine cover $\{U_\alpha\}$ such that the restriction of $f_{S'}$ to each U_α is a tower of elementary fibrations. We may then replace S by S', and assume from the beginning that X admits a finite open affine cover $\{U_\alpha\}$ such that the restriction of f to each U_α is a tower of elementary fibrations. Let us assume $\mathcal{P}((\mathcal{E},\nabla))$, and let us consider the Zariski spectral sequence in **MIC**(S):

$$E_1^{p,i-p} = \oplus_{\alpha_0<\alpha_1<\cdots<\alpha_p} R^{i-p}_{DR} f_{\underline{\alpha}*}(\mathcal{E},\nabla)_{|U_{\underline{\alpha}}} \Rightarrow R^i_{DR} f_*(\mathcal{E},\nabla).$$

By induction on i, we know that after replacing S by a dense affine open subset, $R^{i-p}_{DR} f_{\underline{\alpha}*}((\mathcal{E},\nabla)_{|U_{\underline{\alpha}}})$ satisfies $\mathcal{P}$, for all $\underline{\alpha}$ such that $|\underline{\alpha}| = p+1 > 1$. Thus, by strong exactness of $\mathcal{P}$, this spectral sequence reduces us to proving $\mathcal{P}(R^i_{DR} f_{\alpha*}(\mathcal{E},\nabla)_{|U_\alpha})$, after replacing S by a dense affine open subset. Hence we may assume that f is a tower of elementary fibrations over an affine connected base S.

3.3.2 We argue by induction on the height d of the tower. The induction hypothesis will be:
$(*)_d$ *For any tower of elementary fibrations of height d*

$$X_d \xrightarrow{f_{d-1}} X_{d-1} \longrightarrow \cdots \xrightarrow{f_0} X_0,$$

with X_0 affine connected, and any $(\mathcal{E}, \nabla) \in Ob$ $\mathbf{MIC}(X_d)$, there exists a dense affine open subset X_0' of X_0 such that for any $(i_0, \ldots, i_{d-1}) \in \{0, 1\}^d$, $\mathcal{P}((\mathcal{E}, \nabla)) \Rightarrow \mathcal{P}((R^{i_0}_{DR} f_{0*} \ldots R^{i_{d-1}}_{DR} f_{d-1*}(\mathcal{E}, \nabla))_{|X_0'})$.

We point out that via the Leray spectral sequences for $f = f_0 \circ \cdots \circ f_{d-1}$, it follows from $(*)_d$, taking into account the strong exactness of $\mathcal{P}$, that for any $i \geq 0$, $\mathcal{P}((\mathcal{E}, \nabla)) \Rightarrow \mathcal{P}((R^i_{DR} f_*(\mathcal{E}, \nabla))_{|X_0'})$.

3.3.3 Let us consider the case $d = 1$. It is clear that an elementary fibration may be coordinatized after replacing S by an open dense subset. The argument in (3.2.3) then shows that $(*)_1$ is equivalent to

$(*)'$ *For any rational elementary fibration $f : X \longrightarrow S$ with S affine, étale over some affine K-space, and for any $(\mathcal{E}, \nabla) \in Ob$ $\mathbf{MIC}(X)$, there exists an open dense subset $U \subset S$ such that $\mathcal{P}(\mathcal{E}, \nabla) \Rightarrow \mathcal{P}((R^i_{DR} f_*(\mathcal{E}, \nabla))_{|U})$, $i = 0, 1$.*

Let us show that $(*) \Rightarrow (*)'$. Let $(\mathcal{E}, \nabla)$ be an object of $\mathbf{MIC}(X)$ satisfying $\mathcal{P}$. Because $\mathcal{P}$ implies coherence, $(\mathcal{E}, \nabla)$ has finite length. By applying strong exactness of $\mathcal{P}$ to long exact cohomology sequences, we see that it is enough to establish $(*)$ for the simple subquotients of $(\mathcal{E}, \nabla)$. Hence we may assume that $(\mathcal{E}, \nabla)$ is simple.

We remark that we are free of replacing S by some open dense subset; because $\mathcal{P}$ is étale-local we may even replace S by an affine étale neighborhood. Hence we may even assume that the restriction of $(\mathcal{E}, \nabla)$ to the geometric generic fiber of f is simple.

On the other hand, there exists a finite open affine cover $\{U_\alpha\}_{\alpha=0,\ldots,r}$ of X, such that, for every α, $\mathcal{E}_{|U_\alpha}$ is free and $(\mathcal{E}, \nabla)_{|U_\alpha}$ admits a cyclic vector v_α. In fact, according to [Ka3] (*cf.* also II.7.2), starting from any basis $e_0, \ldots, e_{\mu-1}$ of sections of $\mathcal{E}_{|U_\alpha}$, one can take $r = \mu(\mu - 1)$ and

$$v_\alpha = \sum_{j=0}^{\mu-1} \frac{(x-\alpha)^j}{j!} \sum_{k=0}^{j} (-1)^k \binom{j}{k} \nabla\left(\frac{d}{dx}\right)^k (e_{j-k}). \tag{3.3.3.1}$$

Then after replacing S by some affine étale neighborhood, we may assume that each U_α, and each $U_\alpha \cap U_\beta$, is a rational elementary fibration over S. The Čech spectral sequence for $\{U_\alpha\}$ then reduces us to the cyclic case, so that $(*)_1$ follows from $(*)$.

3.3.4 We now consider a tower of height $d \geq 2$ as in $(*)_d$, and $(\mathcal{E}, \nabla) \in Ob$ $\mathbf{MIC}(X_d)$ satisfying $\mathcal{P}$ (in particular, $\mathcal{E}$ is torsion-free over $\mathcal{O}_{X_d}$, (2.6)). We apply the induction hypothesis $(*)_{d-1}$ to $X_d \longrightarrow \cdots \longrightarrow X_1$, to the effect that there exists a dense affine open subset $X'_1 \subset X_1$ such that the $\mathcal{O}_{X'_1}$-module with integrable connection $(R^{i_1}_{DR} f_{1*} \dots R^{i_{d-1}}_{DR} f_{d-1*}(\mathcal{E}, \nabla))_{|X'_1}$ satisfies $\mathcal{P}$. By shrinking X_0, we may assume that the restriction of f_0, say $f'_0 : X'_1 \longrightarrow X_0$, is an elementary fibration and that X'_1 is the complement of a relatively smooth divisor in X_1. We then apply $(*)_1$ to the latter connections, with respect to f'_0. Upon replacing X_0 again by a dense affine open subset, we obtain

$$\mathcal{P}(R^{i_0}_{DR} f'_{0*}(R^{i_1}_{DR} f_{1*} \dots R^{i_{d-1}}_{DR} f_{d-1*}(\mathcal{E}, \nabla))_{|X'_1}),$$
$$\text{for any } (i_0, \dots, i_{d-1}) \in \{0, 1\}^d.$$

We summarize the geometric situation in the diagram

$$\begin{array}{ccccccccc}
X'_d & \xrightarrow{f'_{d-1}} & X'_{d-1} & \longrightarrow & \cdots & \xrightarrow{f'_1} & X'_1 & \xrightarrow{f'_0} & X'_0 \\
j_d \downarrow & & j_{d-1} \downarrow & & & & j_1 \downarrow & & =\downarrow \\
X_d & \xrightarrow{f_{d-1}} & X_{d-1} & \longrightarrow & \cdots & \xrightarrow{f_1} & X_1 & \xrightarrow{f_0} & X_0
\end{array}$$

in which the horizontal arrows are elementary fibrations, the vertical arrows are affine open embeddings, and all squares not involving the 0-subscript are fibered.

From these properties, it follows that for any $k \in \{1, \dots, d-1\}$ and any quasi-coherent $(\mathcal{N}, \nabla) \in Ob\, \mathbf{MIC}(X_{k+1})$,

$(**)_k$
$$j_{k*} j_k^*(R^{i_k}_{DR} f_{k*} \dots R^{i_{d-1}}_{DR} f_{d-1*}(\mathcal{E}, \nabla))$$
$$= R^{i_k}_{DR} f_{k*} \dots R^{i_{d-1}}_{DR} f_{d-1*}(j_{d*} j_d^*(\mathcal{E}, \nabla)).$$

On the other hand, because $\mathcal{E}$ is torsion-free, there is an exact sequence

$$0 \longrightarrow (\mathcal{E}, \nabla) \longrightarrow j_{d*} j_d^*(\mathcal{E}, \nabla) \longrightarrow \text{Coker} \longrightarrow 0\ .$$

We shall show, by descending induction on $k \in \{0, \dots, d-1\}$, that the sequence of natural morphisms

$(***)_k$
$$0 \longrightarrow R^{i_k}_{DR} f_{k*} \dots R^{i_{d-1}}_{DR} f_{d-1*}(\mathcal{E}, \nabla) \longrightarrow R^{i_k}_{DR} f_{k*} \dots$$
$$\dots R^{i_{d-1}}_{DR} f_{d-1*}(j_{d*} j_d^*(\mathcal{E}, \nabla)) \longrightarrow R^{i_k}_{DR} f_{k*} \dots R^{i_{d-1}}_{DR} f_{d-1*}\text{Coker} \longrightarrow 0$$

is a short *exact sequence*.

By induction, and via the long exact sequence attached to $R^{\cdot}_{DR} f_{k*}$, this follows from

$$(***)'_k \qquad R^0_{DR} f_{k*}(R^{i_{k+1}}_{DR} f_{k+1*} \dots R^{i_{d-1}}_{DR} f_{d-1*}\mathrm{Coker}) = 0.$$

By $(**)_{k+1}$ and $(***)_{>k}$,

$$\begin{aligned} &R^{i_{k+1}}_{DR} f_{k+1*} \dots R^{i_{d-1}}_{DR} f_{d-1*}\mathrm{Coker} \\ &\quad = \mathrm{Coker}(R^{i_{k+1}}_{DR} f_{k+1*} \dots R^{i_{d-1}}_{DR} f_{d-1*}(\mathcal{E}, \nabla) \\ &\quad \longrightarrow j_{k+1*} j^*_{k+1} R^{i_{k+1}}_{DR} f_{k+1*} \dots R^{i_{d-1}}_{DR} f_{d-1*}(\mathcal{E}, \nabla)) \end{aligned}$$

which is torsion, supported by $X_{k+1} \setminus X'_{k+1}$. This implies $(***)'_k$. Indeed, let $m \neq 0$, if any, be a section of $R^{i_{k+1}}_{DR} f_{k+1*} \dots R^{i_{d-1}}_{DR} f_{d-1*}\mathrm{Coker}$ supported on the component of $X_{k+1} \setminus X'_{k+1}$ of local equation t. Then $t^h m = 0$, for some $h \in \mathbf{Z}_{>0}$, which we assume to be minimal. So, for the $\mathcal{O}_{X_k}$-linear derivation $\frac{d}{dt}$,

$$(3.3.4.1) \qquad 0 = \nabla\left(\frac{d}{dt}\right)(t^h m) = h t^{h-1} m + t^h \nabla\left(\frac{d}{dt}\right)(m).$$

This shows that $\nabla(\frac{d}{dt})(m) \neq 0$, hence the assertion $(***)'_k$. We have now proved $(***)'_0$, which shows that $R^{i_0}_{DR} f_{0*} \dots R^{i_{d-1}}_{DR} f_{d-1*}(\mathcal{E}, \nabla)$ is a subobject of

$$\begin{aligned} R^{i_0}_{DR} f_{0*} \dots R^{i_{d-1}}_{DR} f_{d-1*}(j_{d*} j^*_d(\mathcal{E}, \nabla)) &= R^{i_0}_{DR} f_{0*} j_{1*} j^*_1 R^{i_1}_{DR} f_{1*} \\ \dots R^{i_{d-1}}_{DR} f_{d-1*}(\mathcal{E}, \nabla) &= R^{i_0}_{DR} f'_{0*}((R^{i_1}_{DR} f_{1*} \dots R^{i_{d-1}}_{DR} f_{d-1*}(\mathcal{E}, \nabla))_{|X'_1}), \end{aligned}$$

which satisfies $\mathcal{P}$. This achieves the proof of $(*)_d$ and of the lemma. □

Remark 3.3.5 This argument shows in particular that the natural map

$$R^{i_k}_{DR} f_{k*} \dots R^{i_{d-1}}_{DR} f_{d-1*}(\mathcal{E}, \nabla) \longrightarrow R^{i_k}_{DR} f_{k*} j_{k+1*} j^*_{k+1} \dots R^{i_{d-1}}_{DR} f_{d-1*}(\mathcal{E}, \nabla)$$

is an isomorphism if $(i_k, i_{k+1}, \dots, i_{d-1}) \neq (1, 1, \dots, 1)$.

Remark 3.3.6 One can give an *alternative proof* of Lemma 3.3, under the extra assumption that for any closed immersion $i : Y \longrightarrow X$ of smooth K-varieties, $\mathcal{P}((\mathcal{E}, \nabla)) \Rightarrow \mathcal{P}(i^*(\mathcal{E}, \nabla))$. This second proof uses the language and results of [Bo], but avoids the use of Proposition 1.4 (see [ABa] in a special case).

§4 Generic finiteness of direct images

Theorem 4.1 *For any smooth morphism $f : X \longrightarrow S$ of smooth K-varieties, and any coherent $\mathcal{O}_X$-module with integrable connection $(\mathcal{E}, \nabla)$, there exists a dense open subset $U \subset S$ such that for every $j \geq 0$, $R^j_{DR} f_*(\mathcal{E}, \nabla)_{|U}$ is locally free of finite rank.*

We may and shall assume that f is dominant.

4.2 We treat separately the case $j = 0$, for which one can in fact take $U = A_0(f)$, the Artin set of level 0 (1.8), independently of $(\mathcal{E}, \nabla)$.

Lemma 4.2.1 *Let $f : X \longrightarrow S$ be a smooth morphism with $S = A_0(f)$. Then for any coherent $(\mathcal{E}, \nabla) \in Ob\mathbf{MIC}(X)$, $R^0_{DR} f_*(\mathcal{E}, \nabla)$ is locally free of finite rank as an $\mathcal{O}_S$-module.*

Proof: It suffices to check the coherence of $R^0_{DR} f_*(\mathcal{E}, \nabla)$, and this is an étale-local condition on S. By definition of $A_0(f)$, we thus may assume that f is Zariski-locally on X a tower of elementary fibrations, hence (separating the connected components of X) that f has non-empty connected geometric fibers. The coherence of $R^0_{DR} f_*(\mathcal{E}, \nabla)$ then follows from Proposition 2.3.1 and faithfully flat descent [EGA IV, 2.5.2]. □

4.3 For $j > 0$, we use Lemma 3.3 on dévissage, with $\mathcal{P}$ = coherence (*i.e.* $(\mathcal{E}, \nabla)$ satisfies $\mathcal{P}$ if and only if $\mathcal{E}$ is $\mathcal{O}_X$-coherent). Strong exactness and étale-localness are fulfilled. Therefore we are left to show that for any *rational elementary fibration* $f : X \longrightarrow S$ with S affine, étale over some affine K-space, and for any *cyclic* $(\mathcal{E}, \nabla)$, there is a dense open subset $U \subset S$ such that $R^1_{DR} f_*(\mathcal{E}, \nabla)_{|U}$ is $\mathcal{O}_U$-coherent.

4.4 Because f and S are affine the $\mathcal{O}_S$-module underlying $R^1_{DR} f_*(\mathcal{E}, \nabla)$ is nothing but the $\mathcal{O}_S$-module attached to the cokernel of the relative De Rham complex $\Gamma(X, \mathcal{E}) \longrightarrow \Gamma(X, \mathcal{E})dx$. We have to show that this cokernel is an $\mathcal{O}(S)$-module of finite type. Let $(v, \nabla(\frac{d}{dx})v, \ldots, \nabla(\frac{d}{dx})^{\mu-1}v)$ be a cyclic base of $\Gamma(X, \mathcal{E})$. Notice that the dual connection $(\mathcal{E}^\vee, \nabla^\vee)$ is also cyclic: a cyclic vector is for example $w \in \Gamma(X, \mathcal{E}^\vee) = Hom_{\mathcal{O}(X)}(\Gamma(X, \mathcal{E}), \mathcal{O}(X))$, such that

$$\left\langle w, \nabla\left(\frac{d}{dx}\right)^j v\right\rangle = \delta_{j,\mu-1}.$$

We easily see in fact, by ascending induction on $h = 0, 1, \ldots, \mu - 1$, that

$$\left\langle \nabla^\vee\left(\frac{d}{dx}\right)^h w, \nabla\left(\frac{d}{dx}\right)^j v\right\rangle = \begin{cases} 0 & \text{if } j < \mu - h - 1 \\ (-1)^h & \text{if } j = \mu - h - 1 \end{cases}$$

so that $\det(\langle \nabla^\vee(\frac{d}{dx})^h w, \nabla(\frac{d}{dx})^j v\rangle)_{0 \le j,h \le \mu-1} = (-1)^{\frac{\mu(\mu-1)}{2}}$. In the basis

$$\left(w, \nabla^\vee\left(\frac{d}{dx}\right)w, \ldots, \nabla^\vee\left(\frac{d}{dx}\right)^{\mu-1} w\right)$$

of $\Gamma(X, \mathcal{E}^\vee)$, $\nabla^\vee(\frac{d}{dx})$ is represented by a matrix of the form

$$A = \begin{pmatrix} 0 & \ldots & 0 & -\gamma_0 \\ & & & -\gamma_1 \\ & I_{\mu-1} & & \ldots \\ & & & -\gamma_{\mu-1} \end{pmatrix},$$

with $\gamma_k \in \mathcal{O}(X)$.

Let us introduce the differential polynomial (*associated to* $(\mathcal{E}, \nabla)$ *via the cyclic vector* $w \in \Gamma(X, \mathcal{E}^\vee)$) $\Lambda = (\frac{d}{dx})^\mu + \sum_{k=0}^{\mu-1} \gamma_k (\frac{d}{dx})^k \in \mathcal{O}(X)[\frac{d}{dx}]$, so that $\nabla^\vee(\Lambda)w = 0$. Let $(v_0, \dots, v_{\mu-1})$ denote the basis of $\Gamma(X, \mathcal{E})$ dual to the basis $(w, \nabla^\vee(\frac{d}{dx})w, \dots, \nabla^\vee(\frac{d}{dx})^{\mu-1}w)$, *i.e.* such that

$$\left\langle \nabla^\vee \left(\frac{d}{dx}\right)^i w, v_j \right\rangle = \delta_{i,j} .$$

One has

$$\nabla\left(\frac{d}{dx}\right) v_0 = \gamma_0 v_{\mu-1}$$

and

$$\nabla\left(\frac{d}{dx}\right) v_j = \gamma_j v_{\mu-1} - v_{j-1}$$

for $j \geq 1$. There is a commutative diagram

$$\text{(4.4.1)} \qquad \begin{array}{ccc} \mathcal{O}(X) & \xrightarrow{\Lambda} & \mathcal{O}(X) \\ \downarrow & & \downarrow \\ \Gamma(X, \mathcal{E}) = \oplus_{k=0}^{\mu-1} \mathcal{O}(X) v_k & \xrightarrow{\nabla(\frac{d}{dx})} & \Gamma(X, \mathcal{E}) = \oplus_{k=0}^{\mu-1} \mathcal{O}(X) v_k \end{array} ,$$

where the vertical arrows are $\eta \longmapsto (\eta, \dots, (\frac{d}{dx})^{\mu-1}\eta)$ and $\eta \longmapsto (0, \dots, 0, \eta)$, respectively. Diagram 4.4.1 induces an isomorphism between the cokernel (resp. kernel) of Λ and the cokernel (resp. kernel) of $\nabla(\frac{d}{dx})$. Therefore

$$\operatorname{Coker}(\Gamma(X, \mathcal{E}) \xrightarrow{\nabla} \Gamma(X, \mathcal{E})dx) \simeq \mathcal{O}(X)/\Lambda\mathcal{O}(X) .$$

Let $P(x) = \prod_{i=1}^r (x - \theta_i)^{s_i} \in \mathcal{O}(S)[x]$ be such that $\Lambda' = P(x)\Lambda = \sum_{k=0}^{\mu} \gamma'_k (\frac{d}{dx})^k \in \mathcal{O}(S)[x, \frac{d}{dx}]$, with minimal $s_i \geq 0$. We also have

$$\operatorname{Coker}(\Gamma(X, \mathcal{E}) \xrightarrow{\nabla} \Gamma(X, \mathcal{E})dx) \simeq \mathcal{O}(X)/\Lambda'\mathcal{O}(X).$$

4.5 Let us write again the diagram pertaining to a rational elementary fibration

$$\text{(4.5.1)} \qquad \begin{array}{ccccc} X & \hookrightarrow & \mathbf{P}^1_S & \hookleftarrow & \left(\coprod_{i=1}^r \sigma_i(S)\right) \coprod \sigma_\infty(S) \\ & {\scriptstyle f}\searrow & \downarrow^{\text{pr}} & \swarrow & \\ & & S & & \end{array}$$

where $\sigma_\infty(S) = \infty \times S$, $\sigma_i : \zeta \longmapsto (\theta_i(\zeta), \zeta)$, $\theta_i \in \mathcal{O}(S)$, $\theta_i - \theta_j \in \mathcal{O}(S)^\times$. We have a Mittag-Leffler decomposition

$$\mathcal{O}(X) = \mathcal{O}(S)[x] \oplus \bigoplus_{i=1}^{r} \frac{1}{x-\theta_i}\mathcal{O}(S)\left[\frac{1}{x-\theta_i}\right]. \tag{4.5.2}$$

4.6 We recall (see (I.6.1.1)) that the indicial polynomial of Λ' at θ_i, denoted by $\phi_{\theta_i}(s)$ (resp. $\phi_\infty(s)$), is the element of $\mathcal{O}(S)[s]$ defined by the condition that for any $m \in \mathbf{Z}$,

$$\Lambda' \frac{1}{(x-\theta_i)^m} = \phi_{\theta_i}(-m)\frac{1}{(x-\theta_i)^{m+r_i}} + \text{lower order terms in } \frac{1}{x-\theta_i},$$

(resp.

$$\Lambda' x^m = \phi_\infty(-m)x^{m+r_\infty} + \text{l.o.t. in } x,$$

where $r_i = \max(k - \mathrm{ord}_{\theta_i}\gamma_k') \geq 0$ (resp. $r_\infty = \max(\deg_x \gamma_k' - k) \geq 0$). According to (I.6.1.4), there exist $\tau_i, \tau_\infty \in \mathcal{O}(S) \setminus \{0\}$, such that $\frac{\phi_{\theta_i}(s)}{\tau_i}, \frac{\phi_\infty(s)}{\tau_\infty} \in K[s]$. We claim that, after replacing S by its dense open subset $\mathrm{Spec}\mathcal{O}(S)[\frac{1}{\tau_\infty \prod_i \tau_i}]$, $\mathcal{O}(X)/\Lambda'\mathcal{O}(X)$ becomes finitely generated over $\mathcal{O}(S)$.

Let $M \in \mathbf{Z}_{\geq 0}$, $M \geq r_i\ \forall i$, be such that $\phi_\infty(-n + r_\infty)\prod_{j=1,\dots,r}\phi_{\theta_j}(-n + r_j) \neq 0$ for any $n \in \mathbf{Z}_{>M}$. Then

$$\frac{1}{(x-\theta_i)^n} \in \Lambda' \frac{\sigma_{i,n}}{(x-\theta_i)^{n-r_i}} + \frac{1}{(x-\theta_i)^{n-1}}\mathcal{O}(S)[x]\,,\ \sigma_{i,n} \in \mathcal{O}(S)^\times,$$

and

$$x^n \in \Lambda'\sigma_{\infty,n}x^{n-r_\infty} + \bigoplus_{k=0}^{n-1} x^k\mathcal{O}(S),\ \sigma_{\infty,n} \in \mathcal{O}(S)^\times.$$

It follows from induction and the Mittag-Leffler decomposition that $\mathcal{O}(X)/\Lambda'\mathcal{O}(X)$ is generated by the images of $1, x, \dots, x^M, \frac{1}{(x-\theta_1)}, \dots, \frac{1}{(x-\theta_1)^M}, \dots, \frac{1}{(x-\theta_r)}, \dots, \frac{1}{(x-\theta_r)^M}$. □

Remark 4.7

i) In particular, we have proved that the *if the leading coefficient of each indicial polynomial of Λ' is a unit, then $R^1_{DR}f_*(\mathcal{E}, \nabla)$ is finitely generated over the whole of S*. As a special case, we see that $R^1_{DR}f_*(\mathcal{O}_X, d)$ is finitely generated over $\mathcal{O}_S$.

ii) The indicial polynomials of Λ and Λ' at θ_i differ by multiplication by a unit in $\mathcal{O}(S)$.

iii) The proof does not use the full hypothesis that v is a cyclic vector on the whole of X (*i.e.* that one may take $P(x) = \prod_{i=1}^{r}(x-\theta_i)^{s_i}$). It would suffice to assume that v is a cyclic vector outside some divisor D whose Zariski closure in $\mathbf{P}^1_S$ is contained in X.

4.8 (Counter)example We show that, unlike the case of $R^0_{DR}f_*(\mathcal{E},\nabla)$, it is not possible to choose an open subset $U \subset S$ of coherence for $R^1_{DR}f_*(\mathcal{E},\nabla)$, which is independent of $(\mathcal{E},\nabla)$. Indeed, let $X = \mathbf{A}^2_K = \mathrm{Spec}K[x,y] \longrightarrow S = \mathbf{A}^1_K$ be the second projection, and let us consider, for $y_0 \in K$, the integrable $\mathcal{O}_X$-connection of rank 1 $(\mathcal{E} = \mathcal{O}_X e, \nabla_{y_0})$, defined by

$$\nabla_{y_0}\left(\frac{\partial}{\partial x}\right)e = (y - y_0)e,$$

$$\nabla_{y_0}\left(\frac{\partial}{\partial y}\right)e = xe.$$

The indicial polynomial $\phi_\infty(s)$ is the "constant" $y - y_0$, and it is easy to see that the maximal open subset of coherence of $\mathcal{E}/\nabla_{y_0}(\frac{\partial}{\partial x})\mathcal{E}$ is $S \setminus \{y_0\}$, which depends upon y_0. In fact, as a $K[y-y_0]$-module, $\Gamma(\mathcal{E}/\nabla_{y_0}(\frac{\partial}{\partial x})\mathcal{E})$ is isomorphic to $K((y-y_0))/(y-y_0)K[[y-y_0]]$ via $x^k \mapsto (y_0-y)^{-k}$.

We complement the previous analysis by the following purity statement.

Lemma 4.9 *Let $f : X \longrightarrow S$ be a rational elementary fibration, and let $(\mathcal{E},\nabla)$ be a coherent $\mathcal{O}_X$-module with integrable connection. Let us assume that there is a closed subset $T \subset S$ of codimension ≥ 2 such that $R^1_{DR}f_*(\mathcal{E},\nabla)$ is locally free of finite rank over $S \setminus T$. Then $R^1_{DR}f_*(\mathcal{E},\nabla)$ is locally free of finite rank over S.*

Proof: Let us construct the natural commutative square

$$\begin{array}{ccc} X' = X_{S\setminus T} & \overset{j'}{\hookrightarrow} & X \\ f' \downarrow & & \downarrow f \\ S \setminus T & \overset{j}{\hookrightarrow} & S. \end{array}$$

By using the long exact sequence of De Rham cohomology associated to a short exact sequence of connections, we may reduce to the case where $(\mathcal{E},\nabla)$ is *simple*. Because $T \subset S$ is of codimension ≥ 2, we have

$$(\mathcal{E},\nabla) \cong j'_* j'^*(\mathcal{E},\nabla)$$

(taking into account the fact that $\mathcal{E}$ is locally free of finite rank),

$$j_* R^0_{DR} f'_*(j'^*(\mathcal{E},\nabla)) \cong j_* j^* R^0_{DR} f_*(\mathcal{E},\nabla) \cong R^0_{DR} f_*(\mathcal{E},\nabla)$$

and

$$f^* j_* R^0_{DR} f'_*(j'^*(\mathcal{E},\nabla)) \cong j'_* f'^* R^0_{DR} f'_*(j'^*(\mathcal{E},\nabla))$$

is a subobject of $(\mathcal{E},\nabla)$ (taking into account (4.2.1)).

i) If this subobject is $(\mathcal{E}, \nabla)$ itself, then

$$R^1_{DR} f_*(\mathcal{E}, \nabla) \cong R^1_{DR} f_*(\mathcal{O}_X, d) \otimes_{\mathcal{O}_S} (j_* R^0_{DR} f'_*(j'^*(\mathcal{E}, \nabla))),$$

which is coherent by (4.7.i).

ii) Otherwise, we have

$$f^* j_* R^0_{DR} f'_*(j'^*(\mathcal{E}, \nabla)) = 0,$$

hence $R^0_{DR} f'_*(j'^*(\mathcal{E}, \nabla)) = 0$. We have Leray spectral sequences converging to

$$R^\bullet_{DR}(jf')_*(j'^*(\mathcal{E}, \nabla)) = R^\bullet_{DR}(fj')_*(j'^*(\mathcal{E}, \nabla))$$

and with E_2-terms

$$E_2^{ab} = R^a_{DR} j_* R^b_{DR} f'_*(j'^*(\mathcal{E}, \nabla)),$$

and

$$'E_2^{ba} = R^b_{DR} f_* R^a_{DR} j'_*(j'^*(\mathcal{E}, \nabla)),$$

respectively. Let $H^1 := R^1_{DR}(jf')_*(j'^*(\mathcal{E}, \nabla))$. In the present case, the first spectral sequence degenerates at the level E_2 in degree 1, so that $H^1 \cong E_2^{0,1} = j_* R^1_{DR} f'_*(j'^*(\mathcal{E}, \nabla)) = j_* j^* R^1_{DR} f_*(\mathcal{E}, \nabla)$, by flat base change. Since $T \subset S$ is of codimension ≥ 2, H^1 is a coherent $\mathcal{O}_S$-module. On the other hand, for the second spectral sequence we have the exact sequence [CE, Chapter XV, Proposition 5.5]

$$0 \longrightarrow {'E_2^{1,0}} \longrightarrow H^1 \longrightarrow {'E_2^{0,1}} \longrightarrow 0,$$

showing that $'E_2^{1,0} = R^1_{DR} f_*(\mathcal{E}, \nabla)$ is a (quasi-coherent) submodule of H^1. Since S is noetherian, $R^1_{DR} f_*(\mathcal{E}, \nabla)$ is coherent, too. □

§5 Generic base change for direct images

Theorem 5.1 *Let $f : X \longrightarrow S$ be a smooth morphism of smooth K-varieties, and let $(\mathcal{E}, \nabla)$ be a coherent $\mathcal{O}_X$-module with integrable connection. There is a dense open subset $U \subset S$ with the following property. For any smooth K-variety $S^\sharp$ and any morphism $u : S^\sharp \longrightarrow S$, let us construct the fibered diagram*

$$\begin{array}{ccccc} X^\sharp & \xrightarrow{f^\sharp} & S^\sharp & \hookleftarrow & U^\sharp \\ u^\sharp \downarrow & & \downarrow u & & \downarrow \\ X & \xrightarrow{f} & S & \hookleftarrow & U. \end{array}$$

Then, for any $i \geq 0$, the restriction to $U^\sharp$ of the base change morphism

$$\varphi^i : u^* R^i_{DR} f_*(\mathcal{E}, \nabla) \longrightarrow R^i_{DR} f^\sharp_* u^{\sharp *}(\mathcal{E}, \nabla)$$

is an isomorphism in $\mathbf{MIC}(U^\sharp)$.

Notation 5.1.1 In the sequel we set $(\mathcal{E}^\sharp, \nabla^\sharp) = u^{\sharp *}(\mathcal{E}, \nabla)$.

Remarks 5.2

i) $U^\sharp$ is an open subset of $S^\sharp$, possibly empty (see 5.7 below).
The following points are straightforward consequences of the compatibility of the formation of $R^i_{DR} f_*$ with flat base change (2.3).

ii) It would suffice to consider the case of a closed immersion u (write u as a closed immersion given by its graph, followed by a projection).

iii) The statement is equivalent to the following. For any $u : S^\sharp \longrightarrow S$ such that $\mathrm{Im}(u) \subset U$, φ^i is an isomorphism.

iv) In order to prove 5.1, one may replace S by any S' étale and dominant over S.

5.3 Replacing S by affine étale neighborhoods of its connected components, we may assume that X admits a finite open affine cover $\{U_\alpha\}$ such that the restriction f_α of f to U_α is a tower of elementary fibrations. Considering an affine cover of $S^\sharp$, or using Remark 5.2, ii, we may assume that $u^{\sharp -1}(U_\alpha)$ are affine. In the Notation 1.7, there is a natural morphism of Čech spectral sequences in $\mathbf{MIC}(S^\sharp)$

$$\begin{array}{ccc} \bigoplus_{\alpha_0 < \cdots < \alpha_p} u^* R^{i-p}_{DR} f_{\underline{\alpha}*}(\mathcal{E}, \nabla)_{|U_{\underline{\alpha}}} & \Rightarrow & u^* R^i_{DR} f_*(\mathcal{E}, \nabla) \\ \downarrow \varphi_{i-p} & & \downarrow \varphi_i \\ \bigoplus_{\alpha_0 < \cdots < \alpha_p} R^{i-p}_{DR} f^\sharp_{\underline{\alpha}*}(\mathcal{E}^\sharp, \nabla^\sharp)_{|u^{\sharp-1} U_{\underline{\alpha}}} & \Rightarrow & R^i_{DR} f^\sharp_*(\mathcal{E}^\sharp, \nabla^\sharp) \end{array}$$

Arguing by induction on i, we may assume that f itself is a tower of elementary fibrations.

5.4 We argue by induction on the height d of the tower, as in 3.3.2–4. The induction hypothesis will be:

$(*)_d$ *For any tower of elementary fibrations of height d*

$$X_d \xrightarrow{f_{d-1}} X_{d-1} \longrightarrow \cdots \xrightarrow{f_0} X_0,$$

with X_0 affine connected, and any coherent $(\mathcal{E}, \nabla) \in Ob$ $\mathbf{MIC}(X_d)$, there exists a dense affine open subset X'_0 of X_0 with the following property: for any smooth K-variety $X^\sharp_0$ and any morphism $u_0 : X^\sharp_0 \longrightarrow X_0$, let us construct the fibered diagram

$$\begin{array}{ccccccc} X^\sharp_d & \xrightarrow{f^\sharp_{d-1}} & X^\sharp_{d-1} & \longrightarrow & \cdots & \xrightarrow{f^\sharp_0} & X^\sharp_0 \\ u_d \downarrow & & u_{d-1} \downarrow & & & & u_0 \downarrow \\ X_d & \xrightarrow{f_{d-1}} & X_{d-1} & \longrightarrow & \cdots & \xrightarrow{f_0} & X_0. \end{array}$$

Then for any $(i_0, \ldots, i_{d-1}) \in \{0,1\}^d$,

$$(R^{i_0}_{DR} f_{0*} R^{i_1}_{DR} f_{1*} \ldots R^{i_{d-1}}_{DR} f_{d-1*}(\mathcal{E}, \nabla))_{|X'_0}$$

is coherent, and the restriction to $u_0^{-1}(X'_0)$ *of the base change morphism*

$$\begin{aligned}\psi : u_0^*(R^{i_0}_{DR} f_{0*} R^{i_1}_{DR} f_{1*} \ldots R^{i_{d-1}}_{DR} f_{d-1*}\mathcal{E}, \nabla)) \\ \longrightarrow R^{i_0}_{DR} f^\sharp_{0*} R^{i_1}_{DR} f^\sharp_{1*} \ldots R^{i_{d-1}}_{DR} f^\sharp_{d-1*} u_d^*(\mathcal{E}, \nabla)\end{aligned}$$

is an isomorphism in $\mathbf{MIC}(u_0^{-1}(X'_0))$.

We point out that via base change for the Leray spectral sequences of the composed morphism $f = f_0 \circ \cdots \circ f_{d-1}$, 5.1 (in the case of towers of elementary fibrations of height $\leq d$) follows from $(*)_{\leq d}$.

Remark 5.4.1 If $i_0 = i_1 = \cdots = i_{d-1} = 1$, then the base change morphism ψ is an isomorphism; this follows from the fact that, for f affine smooth of relative dimension 1, the formation of $R^1_{DR} f_*$, being a cokernel, commutes with arbitrary base change.

5.5 Let us consider the case $d = 1$, *i.e.* f is an elementary fibration $X_1 = X \longrightarrow S = X_0$. In this case, we can choose for $X'_0 = U$ any dense open subset such that $R^1_{DR} f_*(\mathcal{E}, \nabla)_{|U}$ is coherent (§4). According to 4.2, we may at once replace S by such an (affine) open subset. We note that only $i = 0$ or 1 occur. Since f is affine and flat, we have the exact sequence (*cf.* (2.4))

$$\begin{aligned}0 \longrightarrow R^0_{DR} f_*(\mathcal{E}, \nabla) = \operatorname{Ker} f_*(\nabla) \longrightarrow f_*\mathcal{E} \stackrel{f_*(\nabla)}{\longrightarrow} f_*(\mathcal{E} \otimes_{\mathcal{O}_X} \Omega^1_{X/S}) \\ \longrightarrow \operatorname{Coker} f_*(\nabla) \cong R^1_{DR} f_*(\mathcal{E}, \nabla) \longrightarrow 0\end{aligned}$$

in which all terms are flat over S; a fortiori, $\operatorname{Im} f_*(\nabla)$ is flat. On the other hand, $u^* f_*\mathcal{E} = f^\sharp_* \mathcal{E}^\sharp$, and $u^* f_*(\mathcal{E} \otimes_{\mathcal{O}_X} \Omega^1_{X/S}) = f^\sharp_*(\mathcal{E}^\sharp \otimes_{\mathcal{O}_{X^\sharp}} \Omega^1_{X^\sharp/S^\sharp})$ [EGA I, 9.3.3]. Hence the sequences

$$\begin{aligned}0 \longrightarrow u^*\operatorname{Im} f_*(\nabla) \longrightarrow u^* f_*(\mathcal{E} \otimes_{\mathcal{O}_X} \Omega^1_{X/S}) \\ = f^\sharp_*(\mathcal{E}^\sharp \otimes_{\mathcal{O}_{X^\sharp}} \Omega^1_{X^\sharp/S^\sharp}) \longrightarrow u^*\operatorname{Coker} f_*(\nabla) \longrightarrow 0,\end{aligned}$$

$$0 \longrightarrow u^*\operatorname{Ker} f_*(\nabla) \longrightarrow u^* f_*\mathcal{E} = f^\sharp_*\mathcal{E}^\sharp \stackrel{u^* f_*(\nabla) = f^\sharp_*(\nabla^\sharp)}{\longrightarrow} u^*\operatorname{Im} f_*(\nabla) \longrightarrow 0$$

are exact, from which we conclude that

$$u^*\operatorname{Im} f_*(\nabla) = \operatorname{Im} f^\sharp_*(\nabla^\sharp),$$

$$u^*\operatorname{Coker} f_*(\nabla) = \operatorname{Coker} f^\sharp_*(\nabla^\sharp) \cong R^1_{DR} f^\sharp_*(\mathcal{E}^\sharp, \nabla^\sharp),$$

$$u^*\operatorname{Ker} f_*(\nabla) = \operatorname{Ker} f^\sharp_*(\nabla^\sharp) \cong R^0_{DR} f^\sharp_*(\mathcal{E}^\sharp, \nabla^\sharp),$$

hence φ^0 and φ^1 are isomorphisms in $\mathbf{MIC}(S^\sharp)$, which establishes $(*)_1$.

5.6 We now consider a tower of height $d \geq 2$ as in $(*)_d$. We apply the induction hypothesis $(*)_{d-1}$ to $X_d \longrightarrow \cdots \longrightarrow X_1$, to the effect that there exists a dense affine open subset $X_1' \subset X_1$ such that $(R^{i_1}_{DR} f_{1*} \dots R^{i_{d-1}}_{DR} f_{d-1*}(\mathcal{E}, \nabla))_{|X_1'}$ is coherent and the natural morphism

$$\begin{aligned}\varphi_{(1)} : (u_1^*(R^{i_1}_{DR} f_{1*} \dots R^{i_{d-1}}_{DR} f_{d-1*}(\mathcal{E}, \nabla)))_{|u_1^{-1}(X_1')} \\ \longrightarrow (R^{i_1}_{DR} f^\sharp_{1*} \dots R^{i_{d-1}}_{DR} f^\sharp_{d-1*} u_d^*(\mathcal{E}, \nabla))_{|u_1^{-1}(X_1')}\end{aligned}$$

is an isomorphism in $\mathbf{MIC}(u_1^{-1}(X_1'))$.

Let us consider the fibered prism

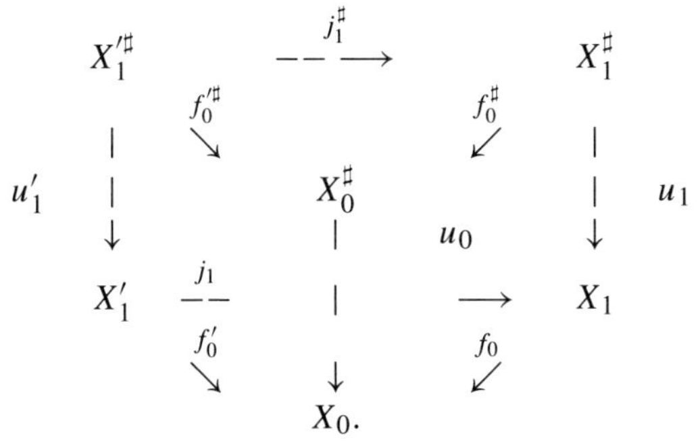

By shrinking X_0, we may assume that f_0' is an elementary fibration, and (by applying $(*)_1$ to f_0') that

(5.6.1) $R^{i_0}_{DR} f'_{0*}((R^{i_1}_{DR} f_{1*} \dots R^{i_{d-1}}_{DR} f_{d-1*}(\mathcal{E}, \nabla))_{|X_1'})$ is coherent over X_0.

(5.6.2) The natural morphism

$$\begin{aligned}&u_0^* R^{i_0}_{DR} f'_{0*}(R^{i_1}_{DR} f_{1*} \dots R^{i_{d-1}}_{DR} f_{d-1*}(\mathcal{E}, \nabla))_{|X_1'} \\ &\longrightarrow R^{i_0}_{DR} f'^\sharp_{0*} u_1'^*(R^{i_1}_{DR} f_{1*} \dots R^{i_{d-1}}_{DR} f_{d-1*}(\mathcal{E}, \nabla))_{|X_1'} \\ &\cong R^{i_0}_{DR} f'^\sharp_{0*} j_1^{\sharp*} u_1^* R^{i_1}_{DR} f_{1*} \dots R^{i_{d-1}}_{DR} f_{d-1*}(\mathcal{E}, \nabla)\end{aligned}$$

is an isomorphism in $\mathbf{MIC}(X_0^\sharp)$.

On combining with $\varphi_{(1)}$, we get a natural isomorphism in $\mathbf{MIC}(X_0^\sharp)$

$$\begin{aligned}\varphi : u_0^* R^{i_0}_{DR} f_{0*} j_{1*} j_1^* R^{i_1}_{DR} f_{1*} \dots R^{i_{d-1}}_{DR} f_{d-1*}(\mathcal{E}, \nabla) \\ \xrightarrow{\sim} R^{i_0}_{DR} f^\sharp_{0*} j^\sharp_{1*} j_1^{\sharp*} R^{i_1}_{DR} f^\sharp_{1*} \dots R^{i_{d-1}}_{DR} f^\sharp_{d-1*} u_d^*(\mathcal{E}, \nabla).\end{aligned} \tag{5.6.3}$$

According to Remarks 5.4.1 and 3.3.5, we may assume that one of the i_j is 0, to the effect that the natural maps

$$R^{i_0}_{DR} f_{0*} \dots R^{i_{d-1}}_{DR} f_{d-1*}(\mathcal{E}, \nabla)$$
$$\longrightarrow R^{i_0}_{DR} f_{0*} j_{1*} j_1^* R^{i_1}_{DR} f_{1*} \dots R^{i_{d-1}}_{DR} f_{d-1*}(\mathcal{E}, \nabla),$$
$$R^{i_0}_{DR} f^\sharp_{0*} \dots R^{i_{d-1}}_{DR} f^\sharp_{d-1*} u^*_d(\mathcal{E}, \nabla)$$
$$\longrightarrow R^{i_0}_{DR} f^\sharp_{0*} j^\sharp_{1*} j_1^{\sharp *} R^{i_1}_{DR} f^\sharp_{1*} \dots R^{i_{d-1}}_{DR} f^\sharp_{d-1*} u^*_d(\mathcal{E}, \nabla)$$

are isomorphisms. In particular, $R^{i_0}_{DR} f_{0*} \dots R^{i_{d-1}}_{DR} f_{d-1*}(\mathcal{E}, \nabla)$ is coherent (5.6.1). Moreover, these isomorphisms fit in a commutative square

$$\begin{array}{ccc} u_0^* R^{i_0}_{DR} f_{0*} \dots R^{i_{d-1}}_{DR} f_{d-1*}(\mathcal{E}, \nabla) \overset{\sim}{\longrightarrow} & u_0^* R^{i_0}_{DR} f_{0*} j_{1*} j_1^* R^{i_1}_{DR} f_{1*} \dots R^{i_{d-1}}_{DR} f_{d-1*}(\mathcal{E}, \nabla) \\ \psi \downarrow & \varphi \downarrow \sim (5.6.3) \\ R^{i_0}_{DR} f^\sharp_{0*} \dots R^{i_{d-1}}_{DR} f^\sharp_{d-1*} u^*_d(\mathcal{E}, \nabla) & \overset{\sim}{\longrightarrow} R^{i_0}_{DR} f^\sharp_{0*} j^\sharp_{1*} j_1^{\sharp *} R^{i_1}_{DR} f^\sharp_{1*} \dots R^{i_{d-1}}_{DR} f^\sharp_{d-1*} u^*_d(\mathcal{E}, \nabla). \end{array}$$

We conclude that ψ (the base change map appearing in $(*)_d$ of Theorem 5.4) is an isomorphism, and this achieves the inductive proof of Theorem 5.1. □

5.7 (Counter)example Let us consider the fibered square

$$\begin{array}{ccc} \mathrm{Spec} K[x] & \overset{u^\sharp}{\longrightarrow} & \mathrm{Spec} K[x, y] = X \\ f^\sharp \downarrow & & \downarrow f \\ \{y_0\} = \mathrm{Spec} K & \overset{u}{\longrightarrow} & \mathrm{Spec} K[y] = S \end{array}$$

in which f is the projection and u is the embedding of the point defined by $y = y_0$. Let us consider the integrable connection on X of rank 1 $(\mathcal{E}, \nabla_{y_0})$ defined in 4.8. It is clear that $R^0_{DR} f_*(\mathcal{E}, \nabla_{y_0}) = 0$, while $R^0_{DR} f^\sharp_* u^{\sharp *}(\mathcal{E}, \nabla_{y_0}) \cong K$, because $u^{\sharp *}(\mathcal{E}, \nabla_{y_0}) \cong (K[x], d)$. This shows the necessity of localizing in 5.1, when u is not flat. This example is also interesting in connection with the base change theorem of [Bo, VI, 8.4].

§6 Coherence of the cokernel of a regular connection

6.1 In this section, we establish that *for any rational elementary fibration $f : X \longrightarrow S$ as in* (4.5.1) *and any locally free $\mathcal{O}_X$-module of finite rank μ with regular integrable connection $(\mathcal{E}, \nabla)$, $R^1_{DR} f_*(\mathcal{E}, \nabla)$ is locally free of finite rank (over the whole of S).*

We may assume that S is affine and connected. The statement amounts to the finite generation of $\Gamma(X, \mathcal{E})/\nabla(\frac{d}{dx})\Gamma(X, \mathcal{E})$ over $\mathcal{O}(S)$.

6.2 Let us first remark that two special cases already follow from 4.6, namely: 1) the case when S is reduced to a point (which allows us to assume $\dim S \geq 1$ in the sequel), and 2) the case when $(\mathcal{E}, \nabla)$ is cyclic (because for regular cyclic $(\mathcal{E}, \nabla)$, the leading coefficients of the indicial polynomials are units, *cf.* Remark 4.7 (i)).

However, unlike the situation of (3.3.3), the reduction to the cyclic case is problematic: it is not clear that one can *cover* S by étale neighborhoods S' with the property that $X_{S'}$ admits a Zariski open covering $\{U_\alpha\}$ such that $U_\alpha \longrightarrow S'$ and $U_\alpha \cap U_\beta \longrightarrow S'$ are rational elementary fibrations and such that $(\mathcal{E}, \nabla)_{|U_\alpha}$ is cyclic outside some divisor whose Zariski closure in $\mathbf{P}^1_{S'}$ is contained in $|U_\alpha$, see (II, Question 7.2.5). We shall take another approach, via τ-extensions.

6.3 For notational convenience, we shall assume, by translating the "vertical" variable x, that one of the $\theta_1, \ldots, \theta_r$, say θ_1, is 0 (if $r \neq 0$). Let us consider the τ-extension $(\tilde{\mathcal{E}}, \tilde{\nabla})$ of $(\mathcal{E}, \nabla)$ on $\mathbf{P}^1_S$, as constructed in (I.4.9): $\tilde{\mathcal{E}}$ is a locally free $\mathcal{O}_{\mathbf{P}^1_S}$-module, endowed with a τ-prepared connection $\tilde{\nabla} : \mathcal{D}er_Z\mathbf{P}^1_S \longrightarrow \mathcal{E}nd_{\mathcal{O}_S}\tilde{\mathcal{E}}$ with logarithmic poles along $\mathbf{P}^1_S \setminus X = Z = \coprod_{\nu=1,\ldots r,\infty} Z_\nu$.

Lemma 6.3.1 *Let $\mathcal{F}$ be a locally free $\mathcal{O}_{\mathbf{P}^1_S}$-module of finite rank. Then there exists a locally free $\mathcal{O}_S$-module $\mathcal{M}$ such that $\mathcal{F}_{|\mathbf{A}^1_S} \cong \mathrm{pr}^*\mathcal{M}$, where* pr *stands for the projection* $\mathbf{A}^1_S \longrightarrow S$.

Proof: The graded $\mathcal{O}(S)[x_0, x_1]$-module F attached to $\mathcal{F}$ is projective of finite rank (since so are its localizations with respect to the open covering $\mathcal{U} = \{U_0, U_1\}$ of $\mathbf{P}^1_S$ with $U_0 := \mathbf{A}^1_S$, $U_1 := \mathbf{P}^1_S \setminus (\{0\} \times S)$ and

$$F \otimes_{\mathcal{O}(S)[x_0,x_1]} \mathcal{O}(S)[x_0, x_1]_{x_i} \cong \Gamma(U_i, \mathcal{F}) \otimes_{\mathcal{O}(U_i)} \mathcal{O}(S)[x_0, x_1]_{x_i},$$

for $i = 0, 1$). According to [A.C. X, p.144], there is a projective $\mathcal{O}(S)$-module M such that $F \cong M \otimes_{\mathcal{O}(S)} \mathcal{O}(S)[x_0, x_1]$. One may then take for $\mathcal{M}$ the locally free $\mathcal{O}_S$-module attached to M. □

Remark 6.3.2 Lemma 6.3.1 is a special, elementary case of a result of H. Lindel [Li] extending the work of Quillen-Suslin on the Serre conjecture, and which says that in fact, for affine smooth S/K, any locally free $\mathcal{O}_{\mathbf{A}^1_S}$-module of finite rank has the form $\mathrm{pr}^*\mathcal{M}$, for some locally free $\mathcal{O}_S$-module $\mathcal{M}$.

We apply (6.3.1) to $\mathcal{F} = \tilde{\mathcal{E}}$, to the effect that there are projective $\mathcal{O}(S)$-modules M and M' of rank μ, such that $\Gamma(\mathbf{A}^1_S, \tilde{\mathcal{E}}) = M[x]$, $\Gamma(\mathbf{P}^1_S \setminus (\{0\} \times S), \tilde{\mathcal{E}}) = M'[\frac{1}{x}]$, and $M[x, \frac{1}{x}] = M'[x, \frac{1}{x}]$. Since M and M' are finitely generated, there is a positive integer N such that

$$x^N M'[x] \subset M[x] \subset x^{-N} M'[x].$$

6.4 We have $\tilde{\nabla}(\prod_{i=1}^r (x - \theta_i)\frac{d}{dx})M[x] \subset M[x]$, and, for any $i = 1, \ldots, r$, we are given an endomorphism $\mathrm{Res}_{\theta_i}\tilde{\nabla}$ of M which enjoys the following properties (*cf.* (I.4)):

(6.4.1) the characteristic polynomial of $\mathrm{Res}_{\theta_i}\tilde{\nabla}$ has coefficients in K and roots in the image of τ (in particular, none of the roots is a non-zero integer);

(6.4.2) for any $m \in M$, identified with $m(x-\theta_i)^0 \in M[x-\theta_i] = M[x]$, one has

$$\tilde{\nabla}\left((x-\theta_i)\frac{d}{dx}\right)m - (\mathrm{Res}_{\theta_i}\tilde{\nabla})m \in (x-\theta_i)M\left[x, \frac{1}{\prod_{k\neq i}(x-\theta_k)}\right].$$

Let us set $\gamma_i = \prod_{k\neq i}(\theta_i - \theta_k)$, a unit in $\mathcal{O}(S)$. It follows from (6.4.2) that, for any $m \in M \cong M\cdot 1 \subset M[x, \frac{1}{x-\theta_i}]$,

$$(6.4.3)\qquad \begin{aligned} &\tilde{\nabla}\left(\prod_{k=1}^{r}(x-\theta_k)\frac{d}{dx}\right)(m(x-\theta_i)^{-n}) - \gamma_i \\ &\qquad (\mathrm{Res}_{\theta_i}\tilde{\nabla} - n1_M)(m)(x-\theta_i)^{-n} \in (x-\theta_i)^{-n+1}M[x]. \end{aligned}$$

On the other hand, (6.4.1) implies that $\mathrm{Res}_{\theta_i}\tilde{\nabla} - n1_M$ is an invertible endomorphism of M for any $n > 0$ (by Cayley-Hamilton, the endomorphism $(\mathrm{Res}_{\theta_i}\tilde{\nabla} - n1_M)^{-1}$ of $M\otimes_{\mathcal{O}(S)}\kappa(S)$ is a polynomial in $\mathrm{Res}_{\theta_i}\tilde{\nabla} - n1_M$ with coefficients in K, hence induces an endomorphism of M). This shows that

$$m(x-\theta_i)^{-n} \in \tilde{\nabla}\left(\prod_{k=1}^{r}(x-\theta_k)\frac{d}{dx}\right)(M(x-\theta_i)^{-n}) + (x-\theta_i)^{-n+1}M[x]\,.$$

By virtue of the Mittag-Leffler decomposition

$$\begin{aligned} M\left[x, \frac{1}{\prod_{i=1}^{r}(x-\theta_i)}\right] &= M\otimes\left(\mathcal{O}(S)\left[x, \frac{1}{\prod_{i=1}^{r}(x-\theta_i)}\right]\right) \\ &= M[x]\oplus\bigoplus_{i=1}^{r}\frac{1}{x-\theta_i}M\left[\frac{1}{x-\theta_i}\right], \end{aligned}$$

one concludes that

$$(6.4.4)\qquad \begin{aligned} &M\left[x, \frac{1}{\prod_{i=1}^{r}(x-\theta_i)}\right] \\ &\quad = \tilde{\nabla}\left(\prod_{i=1}^{r}(x-\theta_i)\frac{d}{dx}\right)M\left[x, \frac{1}{\prod_{i=1}^{r}(x-\theta_i)}\right] + M[x]. \end{aligned}$$

At this point the proof splits into two alternative arguments (6.5 and 6.6).

6.5 From

$$\tilde{\nabla}\left(\frac{1}{x}\prod_{i=2}^{r}\left(\frac{1}{x}-\frac{1}{\theta_i}\right)\frac{d}{d\frac{1}{x}}\right)M'\left[\frac{1}{x}\right]\subset M'\left[\frac{1}{x}\right],$$

we deduce

$$\tilde{\nabla}\left(x^{1-r}\prod_{i=1}^{r}(x-\theta_i)\frac{d}{dx}\right)M'\subset\bigoplus_{j=0}^{l}\frac{1}{x^j}M',$$

for sufficiently big $l \geq 1$. We consider the endomorphism $Res_\infty\tilde{\nabla}$ of M', which satisfies (6.4.1) above, and

$$\tilde{\nabla}\left(x\frac{d}{dx}\right)m+(Res_\infty\tilde{\nabla})m\in\frac{1}{x}M'\left[\left[\frac{1}{x}\right]\right], \tag{6.5.1}$$

or, equivalently

$$\tilde{\nabla}\left(x^{1-r}\prod_{i=1}^{r}(x-\theta_i)\frac{d}{dx}\right)m+(Res_\infty\tilde{\nabla})m\in\bigoplus_{j=1}^{l}\frac{1}{x^j}M', \tag{6.5.2}$$

for any $m \in M'$. So, for any $n \geq 0$ and $m \in M'$,

$$\begin{aligned}&\tilde{\nabla}\left(x^{1-r}\prod_{i=1}^{r}(x-\theta_i)\frac{d}{dx}\right)(x^{n+l}m)+x^{n+l}\\ &(Res_\infty\tilde{\nabla}-(n+l)1_{M'})m\in\bigoplus_{j=0}^{n-1}x^jM'\,.\end{aligned} \tag{6.5.3}$$

From the invertibility of $Res_\infty\tilde{\nabla}-(n+l)1_{M'}$, we deduce, that

$$M'[x]/\tilde{\nabla}\left(x^{1-r}\prod_{i=1}^{r}(x-\theta_i)\frac{d}{dx}\right)x^lM'[x]$$

is generated by $\bigoplus_{j=0}^{l-1}x^jM'$. So, for any $h \geq 0$,

$$x^{r-h-1}M'[x]/\tilde{\nabla}\left(\prod_{i=1}^{r}(x-\theta_i)\frac{d}{dx}\right)x^lM'[x]$$

is generated by $\bigoplus_{j=r-h-1}^{r+l-2}x^jM'$. We now pick l, h sufficiently big so that

$$x^lM'[x]\subset M[x]\subset x^{r-h-1}M'[x]\,.$$

This proves that $M[x]/\tilde{\nabla}(\prod_{i=1}^{r}(x-\theta_i)\frac{d}{dx})M[x]$ is finitely generated over $\mathcal{O}(S)$. A fortiori $\Gamma(X,\mathcal{E})/\nabla(\frac{d}{dx})\Gamma(X,\mathcal{E})$ is finitely generated over $\mathcal{O}(S)$.

6.6 Alternative conclusion of proof

We come back to (6.4.4), and derive the coherence of $\mathrm{Coker}_{\Gamma(X,\mathcal{E})}\nabla(\frac{d}{dx})$ using another argument which will be of importance in the next section.

Let us first remark that

(6.6.1)
$$M[x] \cap \tilde{\nabla}\left(\prod_{i=1}^{r}(x-\theta_i)\frac{d}{dx}\right) M\left[x, \frac{1}{\prod_{i=1}^{r}(x-\theta_i)}\right] = \tilde{\nabla}\left(\prod_{i=1}^{r}(x-\theta_i)\frac{d}{dx}\right) M[x].$$

Indeed, if an element of $M[x, \frac{1}{\prod_{i=1}^{r}(x-\theta_i)}]$ has a pole of order $n \geq 0$ at θ_i, with Mittag-Leffler component $m(x-\theta_i)^{-n}$, and is sent to $M[x]$ by $\tilde{\nabla}(\prod_{i=1}^{r}(x-\theta_i)\frac{d}{dx})$, then $m \in \mathrm{Ker}(\mathrm{Res}_{\theta_i}\tilde{\nabla} - n1_M)$, in view of (6.4.3); hence $n \in \mathrm{Im}\,\tau \cap \mathbf{Z} = \{0\}$. This argument also shows that

(6.6.2)
$$\mathrm{Ker}_{M[x]}\tilde{\nabla} = \mathrm{Ker}_{M\left[x, \frac{1}{\prod_{i=1}^{r}(x-\theta_i)}\right]}\nabla.$$

Putting (6.4.4) and (6.6.1) together, we see that multiplication by $\prod_{i=1}^{r}(x-\theta_i)^{-1}$ induces an isomorphism

(6.6.3)
$$M[x]/\tilde{\nabla}\left(\prod_{i=1}^{r}(x-\theta_i)\frac{d}{dx}\right) M[x] \cong M\left[x, \frac{1}{\prod_{i=1}^{r}(x-\theta_i)}\right] \Big/ \nabla\left(\frac{d}{dx}\right) M\left[x, \frac{1}{\prod_{i=1}^{r}(x-\theta_i)}\right].$$

By symmetry $x \longmapsto \frac{1}{x}$, we also have

(6.6.2′)
$$\mathrm{Ker}_{M'\left[\frac{1}{x}\right]}\tilde{\nabla} = \mathrm{Ker}_{M'\left[x, \frac{1}{\prod_{i=1}^{r}(x-\theta_i)}\right]}\nabla.$$

(6.6.3′)
$$M'\left[\frac{1}{x}\right] \Big/ \tilde{\nabla}\left(\frac{1}{x}\prod_{i=2}^{r}\left(\frac{1}{x}-\frac{1}{\theta_i}\right)\frac{d}{d\frac{1}{x}}\right) M'\left[\frac{1}{x}\right] \cong M\left[x, \frac{1}{\prod_{i=1}^{r}(x-\theta_i)}\right] \Big/ \nabla\left(\frac{d}{dx}\right) M\left[x, \frac{1}{\prod_{i=1}^{r}(x-\theta_i)}\right].$$

This can be summarized in the following diagram of isomorphisms.

$$\begin{array}{ccc}
(\mathrm{Co})\mathrm{Ker}_{M[x]}\tilde{\nabla}\left(\prod_{i=1}^{r}(x-\theta_i)\frac{d}{dx}\right) & & (\mathrm{Co})\mathrm{Ker}_{M'\left[\frac{1}{x}\right]}\tilde{\nabla}\left(\frac{1}{x}\prod_{i=2}^{r}\left(\frac{1}{x}-\frac{1}{\theta_i}\right)\frac{d}{d\frac{1}{x}}\right) \\
\searrow & & \swarrow \cdot(-1)^r x^{r-1}\prod_{i=2}^{r}\theta_i \\
& (\mathrm{Co})\mathrm{Ker}_{M\left[x,\frac{1}{x}\right]=M'\left[x,\frac{1}{x}\right]}\tilde{\nabla}\left(\prod_{i=1}^{r}(x-\theta_i)\frac{d}{dx}\right) & \\
& \downarrow \cdot \prod_{i=1}^{r}(x-\theta_i)^{-1} & \\
& (\mathrm{Co})\mathrm{Ker}_{M\left[x,\frac{1}{\prod_{i=1}^{r}(x-\theta_i)}\right]}\nabla\left(\frac{d}{dx}\right) &
\end{array}$$

Diagram 6.6.4

In terms of the open covering $\mathcal{U} = \{U_0, U_1\}$ of $\mathbf{P}^1_S$ with $U_0 := \mathbf{A}^1_S$, $U_1 := \mathbf{P}^1_S \setminus (\{0\} \times S)$, $(X \subset U_{01} := \mathbf{A}^1_S \setminus (\{0\} \times S))$, the previous diagram may be rewritten as follows.

$$\begin{array}{ccc}
\underline{h}^i\overline{f}_{|U_0*}(\Omega^{\cdot}_{U_0/S}(\log(U_0 \setminus X)) \otimes \tilde{\mathcal{E}}_{|U_0}) & & \underline{h}^i\overline{f}_{|U_1*}(\Omega^{\cdot}_{U_1/S}(\log(U_1 \setminus X)) \otimes \tilde{\mathcal{E}}_{|U_1}) \\
\searrow & & \swarrow \\
& \underline{h}^i\overline{f}_{|U_{01}*}(\Omega^{\cdot}_{U_{01}/S}(\log(U_{01} \setminus X)) \otimes \tilde{\mathcal{E}}_{|U_{01}}) & \\
& \downarrow & \\
& \underline{h}^i f_*(\Omega^{\cdot}_{X/S} \otimes \mathcal{E}) & \\
& \downarrow\cong & \\
& R^i_{DR} f_*(\mathcal{E}, \nabla) &
\end{array}$$

Diagram 6.6.5

For $i = 0, 1$, these maps are isomorphisms of $\mathcal{O}_S$-modules (in fact, isomorphisms in $\mathbf{MIC}(S)$). Let us now compute $\mathbf{R}^i\overline{f}_*(\Omega^{\cdot}_{\mathbf{P}^1_S/S}(\log Z) \otimes \tilde{\mathcal{E}})$, using the (2×2) Čech bicomplex $C^{\cdot}(\mathcal{U}, \Omega^{\cdot}_{\mathbf{P}^1_S/S}(\log Z) \otimes \tilde{\mathcal{E}})$ of $\mathcal{O}_S$-modules. The differentials $d_2^{0,0}$ and $d_2^{1,0}$ are given by $\tilde{\nabla}$, while $d_1^{0,0}$ (resp. $d_1^{0,1}$) is induced by $(j_0^*, -j_1^*)$ (resp. $(-j_0^*, j_1^*)$), where j_k stands for the inclusion $U_{01} \subset U_k$. For $k = 0, 1$, (6.6.4) identifies $\mathrm{Ker} d_2^{k,0}$ (resp. $\mathrm{Coker} d_2^{k,0}$) with $2 - k$ copies of $R^0_{DR} f_*(\mathcal{E}, \nabla)$ (resp. $R^1_{DR} f_*(\mathcal{E}, \nabla)$), and for the first spectral sequence of the bicomplex, we find $\mathbf{I}_2^{0,0} \cong R^0_{DR} f_*(\mathcal{E}, \nabla)$, $\mathbf{I}_2^{0,1} \cong R^1_{DR} f_*(\mathcal{E}, \nabla)$, $\mathbf{I}_2^{1,0} = \mathbf{I}_2^{1,1} = 0$. Applying [CE, Chapter XV, Section 6, case 4], we conclude:

Proposition 6.6.6 *The morphism of $\mathcal{O}_S$-modules*

$$\mathbf{R}^i \overline{f}_*(\Omega^{\cdot}_{\mathbf{P}^1_S/S}(\log Z) \otimes \tilde{\mathcal{E}}) \longrightarrow \mathbf{R}^i \overline{f}_*(j_*(\Omega^{\cdot}_{X/S} \otimes \mathcal{E})) \cong R^i_{DR} f_*(\mathcal{E}, \nabla)\ ,$$

induced by the open S-embedding $j : X \hookrightarrow \mathbf{P}^1_S$, is an isomorphism.

Since $\mathbf{R}^i \overline{f}_*(\Omega^{\cdot}_{\mathbf{P}^1_S/S}(\log Z) \otimes \tilde{\mathcal{E}})$ is the abutment of a spectral sequence with $E_1^{pq} = R^q \overline{f}_*(\Omega^p_{\mathbf{P}^1_S/S}(\log Z) \otimes \tilde{\mathcal{E}})$, the coherence of $R^i_{DR} f_*(\mathcal{E}, \nabla)$ now follows from the stability of coherence under direct image by the projective morphism $\overline{f}$.

6.7 Base change On combining 6.1 and 5.5, we obtain that *for any fibered square of smooth K-varieties*

$$\begin{array}{ccc} X^\sharp & \xrightarrow{u^\sharp} & X \\ f^\sharp \downarrow & & \downarrow f \\ S^\sharp & \xrightarrow{u} & S \end{array}$$

with f a rational elementary fibration, and for any regular object $(\mathcal{E}, \nabla)$ of $\mathbf{MIC}(X)$, *the base change morphism*

$$u^* R^i_{DR} f_*(\mathcal{E}, \nabla) \longrightarrow R^i_{DR} f^\sharp_* u^{\sharp *}(\mathcal{E}, \nabla)$$

is an isomorphism for $i = 0, 1$.

In fact it is immediate to generalize 6.1 (coherence) and this base change property to the case where f is a coordinatized elementary fibration.

§7 Regularity and exponents of the cokernel of a regular connection

7.1 In this section, we establish that *for any rational elementary fibration*

(7.1.1)
$$\begin{array}{ccccc} X & & \overset{j}{\hookrightarrow} & & \overline{X} \\ & {\scriptstyle f}\searrow & & \swarrow{\scriptstyle \overline{f}} & \\ & & S & & \end{array}$$

and for any locally free $\mathcal{O}_X$-module of finite rank with regular integrable connection $(\mathcal{E}, \nabla)$, the Gauss-Manin connection on $R^i_{DR} f_*(\mathcal{E}, \nabla)$ is regular for $i = 0, 1$; moreover, if the exponents of $(\mathcal{E}, \nabla)$ belong to a certain $\mathbf{Q}$-space Δ ($\mathbf{Q} \subset \Delta \subset K$), so do the exponents of $R^i_{DR} f_*(\mathcal{E}, \nabla)$.

We notice that the case $i = 0$ is an easy consequence of (4.1.2).

7.2 By Theorem (I.5.7) and by base change (6.7), we may assume that S is the complement of a finite set of points Σ in a smooth projective *curve* $\overline{S}$. By the classical theory of embedded resolution of curves (Zariski), there is a smooth projective surface $\overline{\overline{X}}$ containing $\overline{X}$, such that $\overline{Z} := \overline{\overline{X}} \setminus X$ is a union $\bigcup_{i=1}^{s} \overline{Z}_j$ of smooth connected curves with normal crossings, and a projective morphism $\overline{\overline{f}} : \overline{\overline{X}} \longrightarrow \overline{S}$ extending $\overline{f}$. By rearranging the $\overline{Z}_j$, we may assume that $\overline{Z}_j \cap \overline{X} \cong Z_j$, for $i \leq r$, while $\overline{\overline{f}}^{-1}(\Sigma) = \sum_{j=r+1}^{s} e_j \overline{Z}_j$, for some positive integers e_j. Following Katz [Ka2], we shall construct a locally free extension of $R^i_{DR} f_*(\mathcal{E}, \nabla)$ on $\overline{S}$ and an extension of the Gauss-Manin connection with logarithmic poles along Σ.

7.3 Let us set

$$\Omega^p_{\overline{\overline{X}}/\overline{S}}(\log \overline{Z}/\Sigma) = \wedge^p_{\mathcal{O}_{\overline{\overline{X}}}}(\Omega^1_{\overline{\overline{X}}}(\log \overline{Z}))/\overline{\overline{f}}^* \Omega^1_{\overline{S}}(\log \Sigma)).$$

Let $(\overset{\approx}{\mathcal{E}}, \overset{\approx}{\nabla})$ be a locally free extension of $(\mathcal{E}, \nabla)$ on $\overline{\overline{X}}$ with logarithmic poles along $\overline{Z}$ (Section 8). Let us filter the logarithmic De Rham complex $\Omega^{\cdot}_{\overline{\overline{X}}}(\log \overline{Z}) \otimes_{\mathcal{O}_{\overline{\overline{X}}}} \overset{\approx}{\mathcal{E}}$ by

$$F^p = \text{image of } \overline{\overline{f}}^* \Omega^p_{\overline{S}}(\log \Sigma) \otimes \Omega^{\cdot - p}_{\overline{\overline{X}}}(\log \overline{Z}) \otimes_{\mathcal{O}_{\overline{\overline{X}}}} \overset{\approx}{\mathcal{E}},$$

so that

$$gr^p = F^p/F^{p+1} = \overline{\overline{f}}^* \Omega^p_{\overline{S}}(\log \Sigma) \otimes_{\mathcal{O}_{\overline{\overline{X}}}} \Omega^{\cdot - p}_{\overline{\overline{X}}/\overline{S}}(\log \overline{Z}/\Sigma) \otimes_{\mathcal{O}_{\overline{\overline{X}}}} \overset{\approx}{\mathcal{E}}.$$

We define

$$R^i_{DR} \overline{\overline{f}}_*(\overset{\approx}{\mathcal{E}}, \overset{\approx}{\nabla}) = \mathbf{R}^i \overline{\overline{f}}_*(\Omega^{\cdot}_{\overline{\overline{X}}/\overline{S}}(\log \overline{Z}/\Sigma) \otimes_{\mathcal{O}_{\overline{\overline{X}}}} \overset{\approx}{\mathcal{E}})/\{\mathcal{O}_{\overline{S}} - \text{torsion}\},$$

which is a locally free $\mathcal{O}_{\overline{S}}$-module of finite rank because $\overline{\overline{f}}$ is projective and $\overline{S}$ is a curve. Moreover,

$$\mathbf{R}^i \overline{\overline{f}}_*(\Omega^{\cdot}_{\overline{\overline{X}}/\overline{S}}(\log \overline{Z}/\Sigma) \otimes_{\mathcal{O}_{\overline{\overline{X}}}} \overset{\approx}{\mathcal{E}})_{|S} \cong \mathbf{R}^i \overline{f}_*(\Omega^{\cdot}_{\overline{X}/\overline{S}}(\log Z)$$

$$\otimes_{\mathcal{O}_{\overline{X}}} \overset{\approx}{\mathcal{E}}_{|\overline{X}}) \cong R^i_{DR} f_*(\mathcal{E}, \nabla),$$

according to 6.6.6, so that $R^i_{DR}\overline{\overline{f}}_*(\overset{\approx}{\mathcal{E}}, \overset{\approx}{\nabla})$ is a locally free extension of $R^i_{DR} f_*(\mathcal{E}, \nabla)$. On the other hand, an extension of the Gauss-Manin connection

$$\mathbf{R}^i\overline{\overline{f}}_*(\Omega^{\cdot}_{\overline{\overline{X}}/\overline{S}}(\log \overline{Z}/\Sigma) \otimes_{\mathcal{O}_{\overline{\overline{X}}}} \overset{\approx}{\mathcal{E}}) \longrightarrow \Omega^1_{\overline{S}}(\log \Sigma)$$

$$\otimes_{\mathcal{O}_{\overline{S}}} \mathbf{R}^i\overline{\overline{f}}_*(\Omega^{\cdot}_{\overline{\overline{X}}/\overline{S}}(\log \overline{Z}/\Sigma) \otimes_{\mathcal{O}_{\overline{\overline{X}}}} \overset{\approx}{\mathcal{E}})$$

is provided by the coboundary map of the long exact sequence

$$0 \longrightarrow gr^1 \longrightarrow F^0/F^2 \longrightarrow gr^0 \longrightarrow 0$$

(since $\Omega^1_{\overline{S}}(\log \Sigma) \otimes_{\mathcal{O}_{\overline{S}}} \mathbf{R}^i\overline{\overline{f}}_*(\Omega^{\cdot}_{\overline{\overline{X}}/\overline{S}}(\log \overline{Z}/\Sigma) \otimes_{\mathcal{O}_{\overline{\overline{X}}}} \overset{\approx}{\mathcal{E}})$ may be identified with $\mathbf{R}^{i+1}\overline{\overline{f}}_*(gr^1)$). That map factors through torsion, and induces an extension of the Gauss-Manin connection on $R^i_{DR}\overline{\overline{f}}_*(\overset{\approx}{\mathcal{E}}, \overset{\approx}{\nabla})$, with logarithmic poles along Σ.

7.4 Katz's argument [Ka2, 7] applies without change to our situation and shows that the indicial polynomial of $R^i_{DR}\overline{\overline{f}}_*(\overset{\approx}{\mathcal{E}}, \overset{\approx}{\nabla})$ at $s \in \Sigma$ divides

$$\prod_{j\in J} \prod_{f_j=0}^{e_j-1} P_j(e_j X - f_j), \tag{7.4.1}$$

where P_j is the indicial polynomial of $(\overset{\approx}{\mathcal{E}}, \overset{\approx}{\nabla})$ at $\overline{Z}_j$, $\overline{\overline{f}}^{-1}(s) = \sum_{j\in J} e_j \overline{Z}_j$. This completes the proof of the statement in 7.1.

Remarks 7.5 1) A closer examination of Katz's argument allows us to replace (7.4.1) by the least common multiple of the $P_j(e_j X - f_j)$, for $j \in J$ and $f_j \in \{0, 1, \ldots, e_j - 1\}$.

2) If $K \subset \mathbf{C}$ and τ is the canonical section $K/\mathbf{Z} \longrightarrow K$, defined by $Re(\tau(z)) \in [0, 1[$, and if $(\overset{\approx}{\mathcal{E}}, \overset{\approx}{\nabla})$ is the τ-extension of $(\mathcal{E}, \nabla)$ on $\overline{\overline{X}}$, then 7.4 shows that $R^i_{DR}\overline{\overline{f}}_*(\overset{\approx}{\mathcal{E}}, \overset{\approx}{\nabla})$ is the τ-extension of $R^i_{DR} f_*(\mathcal{E}, \nabla)$ on $\overline{S}$; indeed $Re(z) \in [0, 1[\Rightarrow Re(\frac{z+f_j}{e_j}) \in [0, 1[$.

3) The arguments 7.2, 7.3, 7.4 work as well for any elementary fibration, not necessarily rational, over a curve.

§8 Proof of the main theorems: finiteness, regularity, monodromy, base change (in the regular case)

Let i be a non-negative integer. Let $f : X \longrightarrow S$ be a smooth morphism of smooth K-varieties, such that *the Artin subset* $A_i(f)$ *is* S *itself* (*cf.* (1.8)). Let $\mathcal{E}$ be a coherent $\mathcal{O}_X$-module endowed with an integrable *regular* connection ∇.

Theorem 8.1 (Finiteness theorem). $R^i_{DR} f_*(\mathcal{E}, \nabla)$ *is a coherent* $\mathcal{O}_S$*-module.*

Theorem 8.2 (Regularity theorem). *The Gauss-Manin connection* $\aleph$ *on* $R^i_{DR} f_*(\mathcal{E}, \nabla)$ *is regular.*

Theorem 8.3 (Monodromy theorem). *If the exponents of* ∇ *all belong to some* $\mathbf{Q}$*-subspace* Δ *of K* ($\Delta \supset \mathbf{Q}$), *so do the exponents of* $\aleph$.

Theorem 8.4 (Base change theorem). *For any smooth* K*-variety* $S^\sharp$ *and any morphism* $u : S^\sharp \longrightarrow S$, *the base change morphism*

$$u^* R^i_{DR} f_*(\mathcal{E}, \nabla) \longrightarrow R^i_{DR} f^\sharp_* u^{\sharp *}(\mathcal{E}, \nabla)$$

attached to the fibered square

$$\begin{array}{ccc} X^\sharp & \xrightarrow{u^\sharp} & X \\ f^\sharp \downarrow & & \downarrow f \\ S^\sharp & \xrightarrow{u} & S \end{array}$$

is an isomorphism.

Proofs: Let us say that $(\mathcal{E}, \nabla)$ satisfies property $\mathcal{P}_\Delta$ if and only if $\mathcal{E}$ is coherent, ∇ is regular, and all exponents of ∇ belong to Δ. Then $\mathcal{P}_\Delta$ is a strongly exact property (I.5.2,I.6.5.3), local for the étale topology (I.5.3,I.6.5.4). By Lemma 3.2 on dévissage, in order to prove Theorems 8.1, 8.2, 8.3, it suffices to consider the case when f is a rational elementary fibration and $i = 0$ or 1, which was settled in Sections 6 and 7.

As for the base change theorem, one reduces as in (5.8.2) to the case when f is a tower of coordinatized elementary fibrations. The associated Leray spectral sequences allow us to further reduce to the case when f is a single coordinatized elementary fibration, which was settled in 6.7. □

Appendix C: Berthelot's comparison theorem on $\mathcal{O}_X$- vs. $\mathcal{D}_X$-linear duals

We reproduce in this section the proof of a very powerful theorem of Berthelot, in its relative version[1]. Let $f : X \longrightarrow S$ be a smooth morphism of smooth K-varieties of relative dimension d, and let $\omega_{X/S} := \Omega^d_{X/S}$.

We recall a few basic definitions from the theory of $\mathcal{D}$-modules.

$\mathcal{D}_{X/S}$ (resp. $\mathcal{D}_{X/S,n}$) = the sheaf of differential operators (resp. of order $\leq n$) on X relative to S, with its two structures of $\mathcal{O}_X$-module, via left and right multiplication. For any $\mathcal{O}_X$-module $\mathcal{E}$, in the tensor product $\mathcal{D}_{X/S} \otimes_{\mathcal{O}_X} \mathcal{E}$ (resp. $\mathcal{E} \otimes_{\mathcal{O}_X} \mathcal{D}_{X/S}$) the right (resp. left) structure of $\mathcal{O}_X$-module of $\mathcal{D}_{X/S}$ is used.

For any sheaf of rings $\mathcal{A}$ on X, let $D^b(\mathcal{A})$ be the bounded derived category of left $\mathcal{A}$-modules.

$D^b_c(\mathcal{D}_{X/S})$ = the full triangulated subcategory of $D^b(\mathcal{D}_{X/S})$, consisting of objects with $\mathcal{D}_{X/S}$-coherent cohomology. It is well-known that in this framework the relative De Rham functor may be viewed as a derived functor

$$DR_{X/S} = R\underline{Hom}_{\mathcal{D}_{X/S}}(\mathcal{O}_X, -) : D^b(\mathcal{D}_{X/S}) \longrightarrow D^b(f^{-1}\mathcal{O}_S), \mathcal{E} \longmapsto \Omega^{\cdot}_{X/S} \otimes \mathcal{E}.$$

In the next lemma, for a left $\mathcal{D}_{X/S}$-module $\mathcal{E}$, we form the two tensor products $\mathcal{D}_{X/S} \otimes_{\mathcal{O}_X} \mathcal{E}$ and $\mathcal{E} \otimes_{\mathcal{O}_X} \mathcal{D}_{X/S}$. On the first, we have a left (resp. right) $\mathcal{D}_{X/S}$-module structure given by

$$Q(P \otimes e) = QP \otimes e,$$

(resp.

$$(P \otimes e)D = PD \otimes e - P \otimes De),$$

for sections P, Q of $\mathcal{D}_{X/S}$, e of $\mathcal{E}$ and D of the relative tangent space. On the second tensor product, we have instead a left (resp. right) $\mathcal{D}_{X/S}$-module structure given by

$$D(e \otimes P) = De \otimes P + e \otimes DP,$$

(resp.

$$(e \otimes P)Q = e \otimes PQ),$$

One checks that both $\mathcal{D}_{X/S} \otimes_{\mathcal{O}_X} \mathcal{E}$ and $\mathcal{E} \otimes_{\mathcal{O}_X} \mathcal{D}_{X/S}$ are $\mathcal{D}_{X/S}$-bi-modules. Berthelot [Be2, 1.3.1] noticed the existence of the following *transposition isomorphism*.

---

[1] We are indebted to M. Cailotto and L. Fiorot for providing us very good notes of two lectures held by Berthelot at Padova University in September 1998. P. Berthelot kindly gave us his permission to include them in this book.

Lemma C.1 *Let $\mathcal{E}$ be a left $\mathcal{D}_{X/S}$-module. There is a unique isomorphism of $\mathcal{D}_{X/S}$-bi-modules*

$$\mathcal{D}_{X/S} \otimes_{\mathcal{O}_X} \mathcal{E} \xrightarrow{\cong} \mathcal{E} \otimes_{\mathcal{O}_X} \mathcal{D}_{X/S}$$

sending $1 \otimes e$ to $e \otimes 1$, for any section e of $\mathcal{E}$.

Theorem C.2 *Let $\mathcal{F} \in D^b_c(\mathcal{D}_{X/S})$.*

(i) *There is a canonical morphism in $D^b(\mathcal{D}_{X/S})$*

$$\mathcal{F}^\vee := R\underline{Hom}_{\mathcal{O}_X}(\mathcal{F}, \mathcal{O}_X) \longrightarrow R\underline{Hom}_{\mathcal{D}_{X/S}}(\mathcal{F}, \mathcal{D}_{X/S} \otimes \omega^{-1}_{X/S}[d]) =: \mathcal{F}^*_{X/S}$$

(ii) *If $\mathcal{F}$ has $\mathcal{O}_X$-coherent cohomology, the previous morphism is an isomorphism.*

(iii) *In the general case, the morphism in (i) induces an isomorphism*

$$DR_{X/S}(\mathcal{F}^\vee) \xrightarrow{\cong} DR_{X/S}(\mathcal{F}^*_{X/S}).$$

Proof: We make use of the isomorphism, valid for $\mathcal{E}, \mathcal{F}, \mathcal{G}$ in $D^b(\mathcal{D}_{X/S})$,

$$R\underline{Hom}_{\mathcal{D}_{X/S}}(\mathcal{E} \otimes^{\mathbf{L}}_{\mathcal{O}_X} \mathcal{F}, \mathcal{G}) \cong R\underline{Hom}_{\mathcal{D}_{X/S}}(\mathcal{E}, R\underline{Hom}_{\mathcal{O}_X}(\mathcal{F}, \mathcal{G}))$$

in the special case of $\mathcal{E} = \mathcal{O}_X$ and $\mathcal{G} = \mathcal{D}_{X/S} \otimes \omega^{-1}_{X/S}[d]$ so as to obtain

$$\begin{aligned}\mathcal{F}^*_{X/S} &= R\underline{Hom}_{\mathcal{D}_{X/S}}(\mathcal{F}, \mathcal{D}_{X/S} \otimes \omega^{-1}_{X/S}[d]) \\ &\cong R\underline{Hom}_{\mathcal{D}_{X/S}}(\mathcal{O}_X, R\underline{Hom}_{\mathcal{O}_X}(\mathcal{F}, \mathcal{D}_{X/S} \otimes \omega^{-1}_{X/S}[d])).\end{aligned}$$

We have a canonical morphism

$$\text{(C.2.1)} \qquad \begin{array}{c} R\underline{Hom}_{\mathcal{O}_X}(\mathcal{F}, \mathcal{O}_X) \otimes_{\mathcal{O}_X} (\mathcal{D}_{X/S} \otimes \omega^{-1}_{X/S}[d]) \\ \downarrow \\ R\underline{Hom}_{\mathcal{O}_X}(\mathcal{F}, \mathcal{D}_{X/S} \otimes \omega^{-1}_{X/S}[d]). \end{array}$$

This in turn induces another canonical morphism

$$\text{(C.2.2)} \qquad \begin{array}{c} R\underline{Hom}_{\mathcal{D}_{X/S}}(\mathcal{O}_X, R\underline{Hom}_{\mathcal{O}_X}(\mathcal{F}, \mathcal{O}_X) \otimes_{\mathcal{O}_X} (\mathcal{D}_{X/S} \otimes \omega^{-1}_{X/S}[d])) \\ \downarrow \\ R\underline{Hom}_{\mathcal{D}_{X/S}}(\mathcal{O}_X, R\underline{Hom}_{\mathcal{O}_X}(\mathcal{F}, \mathcal{D}_{X/S} \otimes \omega^{-1}_{X/S}[d])) \cong \mathcal{F}^*_{X/S}. \end{array}$$

By Lemma C.1 applied to the left $\mathcal{D}_{X/S}$-module $\mathcal{E} = R\underline{Hom}_{\mathcal{O}_X}(\mathcal{F}, \mathcal{O}_X)$, the top term of (C.2.2) may be rewritten as

$$R\underline{Hom}_{\mathcal{D}_{X/S}}(\mathcal{O}_X, \mathcal{D}_{X/S} \otimes R\underline{Hom}_{\mathcal{O}_X}(\mathcal{F}, \mathcal{O}_X) \otimes_{\mathcal{O}_X} \omega^{-1}_{X/S}[d])$$

but, $\mathcal{O}_X$ being $\mathcal{D}_{X/S}$-coherent, this equals

$$\begin{aligned} &R\underline{Hom}_{\mathcal{D}_{X/S}}\left(\mathcal{O}_X, \mathcal{D}_{X/S}\right) \otimes R\underline{Hom}_{\mathcal{O}_X}(\mathcal{F}, \mathcal{O}_X) \otimes_{\mathcal{O}_X} \omega_{X/S}^{-1}[d] \\ &\quad \cong DR_{X/S}(\mathcal{D}_{X/S}) \otimes \mathcal{F}^\vee \otimes \omega_{X/S}^{-1}[d] \cong \mathcal{F}^\vee \end{aligned}$$

since $DR_{X/S}(\mathcal{D}_{X/S}) \cong \omega_{X/S}[-d]$. All in all, we obtain the canonical morphism (i), which is an isomorphism in the case of coherent $\mathcal{O}_X$-modules (because then (C.2.1) is an isomorphism).

From (i), we get the induced map of relative De Rham complexes appearing in (iii). To prove that this map is an isomorphism, we may work locally. We may then assume that $\mathcal{F}$ is free of finite rank over $\mathcal{D}_{X/S}$, and we are reduced to the case of $\mathcal{F} = \mathcal{D}_{X/S}$.

We then have

$$\begin{aligned} \mathcal{D}_{X/S}^\vee &= R\underline{Hom}_{\mathcal{O}_X}(\mathcal{D}_{X/S}, \mathcal{O}_X) \\ &= R\varprojlim R\underline{Hom}_{\mathcal{O}_X}(\mathcal{D}_{X/S,n}, \mathcal{O}_X) \\ &= \varprojlim \mathcal{P}_{X/S}^n =: \mathcal{P}_{X/S}^\infty \end{aligned}$$

where the next to last equality holds true because the projective system $\mathcal{P}_{X/S}^n = R\underline{Hom}_{\mathcal{O}_X}\left(\mathcal{D}_{X/S,n}, \mathcal{O}_X\right)$ has surjective transition morphisms. On the other side of (i) we get

$$\begin{aligned} R\underline{Hom}_{\mathcal{O}_X}\left(\mathcal{D}_{X/S}, \mathcal{D}_{X/S} \otimes \omega_{X/S}^{-1}[d]\right) &\cong R\varprojlim R\underline{Hom}_{\mathcal{O}_X}\left(\mathcal{D}_{X/S,n}, \mathcal{D}_{X/S} \otimes \omega_{X/S}^{-1}[d]\right) \\ &\cong R\varprojlim \Big(R\underline{Hom}_{\mathcal{O}_X}\left(\mathcal{D}_{X/S,n}, \mathcal{O}_X\right) \otimes_{\mathcal{O}_X} \\ &\qquad (\mathcal{D}_{X/S} \otimes \omega_{X/S}^{-1}[d])\Big) \\ &\cong \varprojlim \left(\mathcal{P}_{X/S}^n \otimes_{\mathcal{O}_X} \left(\mathcal{D}_{X/S} \otimes \omega_{X/S}^{-1}[d]\right)\right). \end{aligned}$$

The map (C.2.1) may then be spelled out as the natural morphism

$$\mathcal{P}_{X/S}^\infty \otimes_{\mathcal{O}_X} \mathcal{D}_{X/S} \otimes \omega_{X/S}^{-1}[d] \longrightarrow \varprojlim p(\mathcal{P}_{X/S}^n \otimes_{\mathcal{O}_X} (\mathcal{D}_{X/S} \otimes \omega_{X/S}^{-1}[d])).$$

On applying the functor $R\underline{Hom}_{\mathcal{D}_{X/S}}(\mathcal{O}_X, -) = DR_{X/S}(-) = \Omega^{\cdot}_{X/S} \otimes -$ we get the morphism (i) of the two duals

$$\begin{array}{ccc} \mathcal{D}_{X/S}^\vee & & (\mathcal{D}_{X/S})_{X/S}^* \\ \| & & \| \\ \Omega^{\cdot}_{X/S} \otimes \mathcal{P}_{X/S}^\infty \otimes_{\mathcal{O}_X} \mathcal{D}_{X/S} \otimes \omega_{X/S}^{-1}[d] & \longrightarrow & \Omega^{\cdot}_{X/S} \otimes \varprojlim \left(\mathcal{P}_X^n \otimes_{\mathcal{O}_X} \left(\mathcal{D}_{X/S} \otimes \omega_{X/S}^{-1}[d]\right)\right) \end{array}$$

(in particular, those duals are interpreted as relative De Rham complexes).

Until now, only the left structure of $\mathcal{D}_{X/S} \otimes \omega^{-1}_{X/S}$ as a $\mathcal{D}_{X/S}$-module has been used. We now use the right structure for calculating the relative De Rham functor

$$\begin{array}{ccc} DR_{X/S}(\mathcal{D}^{\vee}_{X/S}) & & DR_{X/S}((\mathcal{D}_{X/S})^{*}_{X/S}) \\ \| & & \| \\ \Omega^{\cdot}_{X/S} \otimes \mathcal{P}^{\infty}_{X/S} \otimes_{\mathcal{O}_X} \mathcal{D}_{X/S} \otimes \omega^{-1}_{X/S} \otimes \Omega^{\cdot}_{X/S} & \longrightarrow & \Omega^{\cdot}_{X/S} \otimes \varprojlim p(\mathcal{P}^{n}_{X/S} \otimes_{\mathcal{O}_X} (\mathcal{D}_{X/S} \otimes \omega^{-1}_{X/S})) \otimes \Omega^{\cdot}_{X/S} \end{array}$$

The above is a map $L^{\cdot\cdot} \longrightarrow K^{\cdot\cdot}$ of bicomplexes whose cohomology may be calculated from either one of the two spectral sequences. The second spectral sequence immediately gives the identification

$$_{II}E_2^{p,d}(L^{\cdot\cdot}) = \Omega^{p}_{X/S} \otimes \mathcal{P}^{\infty}_{X/S}$$

and $_{II}E_2^{p,q}(L^{\cdot\cdot}) = 0$ for $q \neq d$, since $\mathcal{D}_{X/S} \otimes \omega^{-1}_{X/S} \otimes \Omega^{\cdot}_{X/S} \cong \mathcal{O}_X[-d]$. The same fact, after we express $\varprojlim(\mathcal{P}^{n}_{X/S} \otimes_{\mathcal{O}_X} (\mathcal{D}_{X/S} \otimes \omega^{-1}_{X/S}))$ as $\underline{Hom}_{\mathcal{O}_X}(\mathcal{D}_{X/S}, \mathcal{D}_{X/S} \otimes \omega^{-1}_{X/S})$ to write $\lim p(\mathcal{P}^{n}_{X/S} \otimes_{\mathcal{O}_X} (\mathcal{D}_{X/S} \otimes \omega^{-1}_{X/S})) \otimes \Omega^{\cdot}_{X/S} = \underline{Hom}_{\mathcal{O}_X}(\mathcal{D}_{X/S}, \mathcal{D}_{X/S} \otimes \omega^{-1}_{X/S} \otimes \Omega^{\cdot}_{X/S})$, shows that

$$_{II}E_2^{p,d}(K^{\cdot\cdot}) = \Omega^{p}_{X/S} \otimes \underline{Hom}_{\mathcal{O}_X}(\mathcal{D}_{X/S}, \mathcal{O}_X)$$

while $_{II}E_2^{p,q}(K^{\cdot\cdot}) = 0$ for $q \neq d$, and our conclusion follows. □

Appendix D: Introduction to Dwork's algebraic dual theory

D.1 Serre duality for an elementary fibration We start with an elementary fibration as in (1.1.1). We denote by $\mathcal{J} = \mathcal{J}_Z$ the invertible sheaf of $\mathcal{O}_{\overline{X}}$-ideals of Z in $\overline{X}$. The dualizing sheaf for the morphism $\overline{f}$ is the $\mathcal{O}_{\overline{X}}$-module $\omega = \Omega^1_{\overline{X}/S}$. For any scheme Y, let $\mathcal{A}b(Y)$ (resp. $Mod(Y)$, resp. $Coh(Y)$) be the category of abelian sheaves on Y (resp. of $\mathcal{O}_Y$-modules, resp. of coherent $\mathcal{O}_Y$-modules). We also set $D^b_c(Y) := D^b_{Coh(Y)}(Mod(Y))$, $D^b(Y) := D^b(Mod(Y))$.

We consider the exact functor

$$\text{(D.1.1)} \qquad \begin{aligned} Mod(\overline{X}) &\longrightarrow Mod(\overline{X}), \\ \mathcal{E} &\longmapsto \mathcal{E}(*Z) = \varinjlim_{s} \mathcal{J}_Z^{-s} \otimes_{\mathcal{O}_{\overline{X}}} \mathcal{E}\ . \end{aligned}$$

For $\mathcal{E}$ coherent $\mathcal{E}(*Z) = j_*(j^{-1}(\mathcal{E}))$. For any $\mathcal{E}$ in $Coh(\overline{X})$,

$$\hat{\mathcal{E}} = \varprojlim_{N} \mathcal{E}/\mathcal{J}_Z^N \mathcal{E}$$

is a coherent $\hat{\mathcal{O}}_{\overline{X}|Z}$-module, which we rather regard as an object of $Mod(\overline{X})$, in general not even quasi-coherent. So again we have an exact functor

$$\text{(D.1.2)} \qquad \begin{aligned} \hat{:} Coh(\overline{X}) &\longrightarrow Mod(\overline{X})\ , \\ \mathcal{E} &\longmapsto \hat{\mathcal{E}} = \varprojlim_{N} \mathcal{E}/\mathcal{J}_Z^N \mathcal{E}\ . \end{aligned}$$

We assume from now on in this section that $\mathcal{E}$ is a coherent and locally free $\mathcal{O}_{\overline{X}}$-module. Grothendieck-Serre duality [H1, VII, Corollary 4.3] gives, for any $\mathcal{E}$ as before, an isomorphism

$$R\overline{f}_*(\omega \otimes \mathcal{E}^\vee)[1] \xrightarrow{\cong} R\underline{Hom}_S(R\overline{f}_*\mathcal{E}, \mathcal{O}_S).$$

We apply [CE, XV, Theorem 5.11] with $r = 2$, $q = -1$, $q' = 0$ to the spectral sequence

$$E_2^{u,v} := \underline{Ext}^u_S(R^{-v}\overline{f}_*\mathcal{E}, \mathcal{O}_S) \Rightarrow R^{u+v+1}\overline{f}_*(\omega \otimes \mathcal{E}^\vee)$$

and deduce exact sequences

$$\text{(D.1.3)} \qquad E_2^{-2,0} = 0 \longrightarrow E_2^{0,-1} \xrightarrow{\sim} H^{-1} \longrightarrow E_2^{-1,0} = 0,$$

$$\text{(D.1.4)} \qquad E_2^{-1,0} = 0 \longrightarrow E_2^{1,-1} \longrightarrow H^0 \longrightarrow E_2^{0,0} \longrightarrow E_2^{2,-1} \longrightarrow H^1 = 0.$$

So we get exact sequences:

(D.1.5) $$0 \longrightarrow \underline{Hom}_S(R^1\overline{f}_*\mathcal{E}, \mathcal{O}_S) \xrightarrow{\sim} R^0\overline{f}_*(\omega \otimes \mathcal{E}^\vee) \longrightarrow 0,$$

and

(D.1.6) $$\begin{aligned} 0 &\longrightarrow \underline{Ext}^1_S(R^1\overline{f}_*\mathcal{E}, \mathcal{O}_S) \longrightarrow R^1\overline{f}_*(\omega \otimes \mathcal{E}^\vee) \\ &\longrightarrow \underline{Hom}_S(R^0\overline{f}_*\mathcal{E}, \mathcal{O}_S) \longrightarrow \underline{Ext}^2_S(R^1\overline{f}_*\mathcal{E}, \mathcal{O}_S) \longrightarrow 0. \end{aligned}$$

On interchanging $\mathcal{E}$ and $\omega \otimes \mathcal{E}^\vee$, we obtain the exact sequences

(D.1.7) $$0 \longrightarrow \underline{Hom}_S(R^1\overline{f}_*(\omega \otimes \mathcal{E}^\vee), \mathcal{O}_S) \xrightarrow{\sim} R^0\overline{f}_*\mathcal{E} \longrightarrow 0,$$

and

(D.1.8) $$\begin{aligned} 0 &\longrightarrow \underline{Ext}^1_S(R^1\overline{f}_*(\omega \otimes \mathcal{E}^\vee), \mathcal{O}_S) \longrightarrow R^1\overline{f}_*\mathcal{E} \\ &\longrightarrow \underline{Hom}_S(R^0\overline{f}_*(\omega \otimes \mathcal{E}^\vee), \mathcal{O}_S) \\ &\longrightarrow \underline{Ext}^2_S(R^1\overline{f}_*(\omega \otimes \mathcal{E}^\vee), \mathcal{O}_S) \longrightarrow 0. \end{aligned}$$

Let us explain in concrete terms the pairing

(D.1.9) $$R^0\overline{f}_*(\omega \otimes \mathcal{E}^\vee) \times R^1\overline{f}_*(\mathcal{E}) \longrightarrow \mathcal{O}_S.$$

This pairing factors through cup-product and an isomorphism $R^1\overline{f}_*(\omega) \xrightarrow{\mathrm{Tr}} \mathcal{O}_S$ that may be described as follows. After finite étale base-change $S' \longrightarrow S$, we may assume that Z is a disjoint union of components Z_i, for $i = 1, \ldots, r, \infty$, isomorphic to S via g. In terms of the relatively affine covering $\{U_j = \overline{X} \setminus \bigcup_{i \neq j} Z_i\}_j$ of $\overline{X}$, we may represent a global section η of the $\mathcal{O}_S$-module $R^1\overline{f}_*(\omega)$ as the class of an (alternating) cocycle $\eta_{j,k} \in \Gamma(U_{j,k}, \omega)$. We may write $\eta_{j,k} = \eta_k - \eta_j$, for $\eta_j \in \Gamma(U_j, j_*(\omega_{|X}))$. Then

$$\mathrm{Tr}(\eta) = \sum_i \mathrm{Res}_{Z_i} \eta_i.$$

We now replace, in the previous pairing, $\mathcal{E}$ by $\mathcal{E} \otimes \mathcal{J}_Z^N$ for sufficiently big N, so that $R^0\overline{f}_*(\mathcal{E} \otimes \mathcal{J}_Z^N) = 0$ and $R^1\overline{f}_*(\omega \otimes \mathcal{E}^\vee \otimes \mathcal{J}_Z^{-N}) = 0$. The sequence D.1.8 then gives an isomorphism

(D.1.10) $$0 \longrightarrow R^1\overline{f}_*(\mathcal{E} \otimes \mathcal{J}_Z^N) \xrightarrow{\sim} \underline{Hom}_S(R^0\overline{f}_*(\omega \otimes \mathcal{E}^\vee \otimes \mathcal{J}_Z^{-N}), \mathcal{O}_S) \longrightarrow 0.$$

Taking into account (D.1.5) that now gives

(D.1.11) $$0 \longrightarrow \underline{Hom}_S(R^1\overline{f}_*(\mathcal{E} \otimes \mathcal{J}_Z^N), \mathcal{O}_S) \xrightarrow{\sim} R^0\overline{f}_*(\omega \otimes \mathcal{E}^\vee \otimes \mathcal{J}_Z^{-N}) \longrightarrow 0,$$

we obtain, for sufficiently big N, perfect pairings

$$\langle\, ,\, \rangle_N : R^0\overline{f}_*(\omega \otimes \mathcal{E}^\vee \otimes \mathcal{J}_Z^{-N}) \times R^1\overline{f}_*(\mathcal{E} \otimes \mathcal{J}_Z^N) \longrightarrow \mathcal{O}_S\,. \tag{D.1.12}$$

In order to take limits for $N \to \infty$ in (D.1.12), we consider the inverse system of exact 6-terms rows $(\mathcal{S}_N, \varphi_{M,N})_{M\geq N}$

$$\begin{aligned} \mathcal{S}_N = (0 &\longrightarrow R^0\overline{f}_*\mathcal{E} \longrightarrow R^0\overline{f}_*(\mathcal{E}/\mathcal{J}_Z^N\mathcal{E}) \\ &\longrightarrow R^1\overline{f}_*(\mathcal{J}_Z^N\mathcal{E}) \longrightarrow R^1\overline{f}_*\mathcal{E} \longrightarrow 0). \end{aligned} \tag{D.1.13}$$

Since when $S = \operatorname{Spec} A$ is affine the projective system of A-modules $(\Gamma(S, R^0\overline{f}_*(\mathcal{E}/\mathcal{J}_Z^N\mathcal{E})))_N$ satisfies the Mittag-Leffler condition, on taking the projective limit of $(\mathcal{S}_N, \varphi_{M,N})_{M\geq N}$, we obtain an exact rows

$$\begin{aligned} 0 &\longrightarrow R^0\overline{f}_*\mathcal{E} \longrightarrow R^0\overline{f}_*(\hat{\mathcal{E}}) \longrightarrow \\ &\varprojlim_N R^1\overline{f}_*(\mathcal{J}_Z^N\mathcal{E}) \longrightarrow R^1\overline{f}_*\mathcal{E} \longrightarrow 0\,. \end{aligned} \tag{D.1.14}$$

Similarly, we obtain a commutative diagram with exact rows

$$\begin{array}{ccccccccc} 0 \to & R^0\overline{f}_*\mathcal{E} & \to R^0\overline{f}_*\hat{\mathcal{E}} & \to \varprojlim_N R^1\overline{f}_*(\mathcal{J}_Z^N\mathcal{E}) & \to R^1\overline{f}_*\mathcal{E} \to & 0 \\ & \downarrow & \downarrow & \downarrow & \downarrow & \\ 0 \to & R^0\overline{f}_*(\mathcal{E}(*Z)) & \to R^0\overline{f}_*(\hat{\mathcal{E}}(*Z)) & \to \varprojlim_N R^1\overline{f}_*(\mathcal{J}_Z^N\mathcal{E}) & \to \quad 0 \to & 0. \end{array} \tag{D.1.15}$$

Therefore,

$$\begin{aligned} R^0\overline{f}_*(\hat{\mathcal{E}}(*Z))/R^0\overline{f}_*(\mathcal{E}(*Z)) &\xrightarrow{\sim} \varprojlim_N R^1\overline{f}_*(\mathcal{J}_Z^N\mathcal{E}) \\ &\xrightarrow{\sim} \underline{Hom}_S(R^0\overline{f}_*(\omega \otimes \mathcal{E}^\vee(*Z)), \mathcal{O}_S). \end{aligned} \tag{D.1.16}$$

If $R^1\overline{f}_*\mathcal{E} = 0$, we also get

$$R^0\overline{f}_*\hat{\mathcal{E}}/R^0\overline{f}_*\mathcal{E} \xrightarrow{\sim} R^0\overline{f}_*(\hat{\mathcal{E}}(*Z))/R^0\overline{f}_*(\mathcal{E}(*Z))\,. \tag{D.1.17}$$

We say that an $\mathcal{O}_S$-linear pairing $\mathcal{E} \times \mathcal{F} \longrightarrow \mathcal{O}_S$ of $\mathcal{O}_S$-modules is *left-perfect* if the corresponding $\mathcal{O}_S$-linear map $\mathcal{F} \longrightarrow \underline{Hom}_S(\mathcal{E}, \mathcal{O}_S)$ is an isomorphism. We again prefer to regard the composite isomorphism in (D.1.16) as a left-perfect pairing

$$\langle\, ,\, \rangle : R^0\overline{f}_*(\omega \otimes \mathcal{E}^\vee(*Z)) \times R^0\overline{f}_*\hat{\mathcal{E}}(*Z)/R^0\overline{f}_*\mathcal{E}(*Z) \longrightarrow \mathcal{O}_S\,. \tag{D.1.18}$$

We point out that in general the sheaf of $\mathcal{O}_S$-modules $R^0\overline{f}_*\hat{\mathcal{E}}$ is *not* quasi-coherent. We now assume that $\mathcal{E}$ is endowed with a relative connection

$$\nabla : \mathcal{D}_{\overline{X}/S} \longrightarrow \underline{End}_{\overline{f}^{-1}(\mathcal{O}_S)}(\mathcal{E}(*Z)) \tag{D.1.19}$$

with poles along Z, which makes $\mathcal{E}(*Z)$ into a left $\mathcal{D}_{\overline{X}/S}$-module. From this we deduce a similar structure on

$$\mathrm{Coker}(\mathcal{E}(*Z) \to \hat{\mathcal{E}}(*Z)) \cong \mathrm{Coker}(\mathcal{E} \to \hat{\mathcal{E}}) .$$

From the dual connection $(\mathcal{E}^\vee = \underline{Hom}_{\mathcal{O}_{\overline{X}}}(\mathcal{E}, \mathcal{O}_{\overline{X}}), \nabla^\vee)$ (with poles along Z), we get a structure of left $\mathcal{D}_{\overline{X}/S}$-module on $\mathcal{E}^\vee(*Z)$, and from the natural right structure of ω, we deduce a structure of right $\mathcal{D}_{\overline{X}/S}$-module on $\omega \otimes \mathcal{E}^\vee(*Z)$. For the previous structures and for sections L of $R^0\overline{f}_*\mathcal{D}_{\overline{X}/S}$, η of $R^0\overline{f}_*(\omega \otimes \mathcal{E}^\vee(*Z))$, and h of $R^0\overline{f}_*\hat{\mathcal{E}}(*Z)$, we have

$$\langle \eta L, h\rangle = \langle \eta, Lh\rangle \ , \tag{D.1.20}$$

as can be immediately checked via the previous residue description of Serre duality.

Proposition D.1.21 *(Dwork duality principle for an elementary fibration.) Let $\mathcal{E}$ be a coherent and locally free $\mathcal{O}_{\overline{X}}$-module and assume $\mathcal{E}(*Z)$ carries a structure of a left $\mathcal{D}_{\overline{X}/S}$-module. Let*

$$\begin{array}{c} DR(\overline{X}/S, \mathcal{E}(*Z)) := [\mathcal{E}(*Z) \longrightarrow \omega \otimes \mathcal{E}(*Z)] \\ \qquad\qquad\qquad\qquad\qquad\quad 0 \qquad\qquad\quad 1 \end{array}$$

be the De Rham complex of the $\mathcal{D}_{\overline{X}/S}$-module $\mathcal{E}(*Z)$ and let

$$\begin{array}{l} DW(\overline{X}/S, \mathcal{E}^\vee) := \\ \quad [R^0\overline{f}_*(\hat{\mathcal{E}}^\vee(*Z))/R^0\overline{f}_*(\mathcal{E}^\vee(*Z)) \to R^0\overline{f}_*(\omega \otimes \hat{\mathcal{E}}^\vee(*Z))/R^0\overline{f}_*(\omega \otimes \mathcal{E}^\vee(*Z))] \\ \qquad\qquad 0 \qquad\qquad\qquad\qquad\qquad\qquad\qquad\qquad 1 \end{array}$$

be the complex of Dwork's algebraic dual theory *(for $\mathcal{E}$), where $\mathcal{E}^\vee$ denotes the dual $\mathcal{O}_{\overline{X}}$-module of $\mathcal{E}$ with the dual connection. There is a canonical isomorphism*

$$\begin{aligned} DW(\overline{X}/S, \mathcal{E}^\vee) &\xrightarrow{\sim} \underline{Hom}_S(\overline{f}_*DR(\overline{X}/S, \mathcal{E}(*Z))[1], \mathcal{O}_S) \\ &= R\underline{Hom}_{\mathcal{O}_S}(Rf_*DR(X/S, \mathcal{E}_{|X})[1], \mathcal{O}_S) . \end{aligned} \tag{D.1.21.1}$$

In particular, we have a canonical isomorphism

$$H^0(DW(\overline{X}/S, \mathcal{E}^\vee)) \xrightarrow{\sim} \underline{Hom}_S(R^1_{DR}f_*((\mathcal{E}, \nabla)_{|X}), \mathcal{O}_S) ,$$

and an exact sequence

$$0 \longrightarrow \underline{Ext}^1_S(R^1_{DR} f_*((\mathcal{E},\nabla)_{|X}), \mathcal{O}_S) \longrightarrow H^1(DW(\overline{X}/S,\mathcal{E}^\vee))$$
$$\longrightarrow \underline{Hom}_S(R^0_{DR} f_*((\mathcal{E},\nabla)_{|X}), \mathcal{O}_S)$$
$$\longrightarrow \underline{Ext}^2_S(R^1_{DR} f_*((\mathcal{E},\nabla)_{|X}), \mathcal{O}_S) \longrightarrow 0.$$

If, in particular, the $\mathcal{O}_S$-module $R^1_{DR}(f_*(\mathcal{E},\nabla)_{|X})$ is locally free, we have a canonical isomorphism

$$H^1(DW(\overline{X}/S,\mathcal{E}^\vee)) \xrightarrow{\sim} \underline{Hom}_S(R^0_{DR} f_*((\mathcal{E},\nabla)_{|X}), \mathcal{O}_S)\,.$$

Proof: The main point that still requires a proof is the last equality in (D.1.21.1). Now $Rf_*DR(X/S,\mathcal{E}_{|X}) = f_*DR(X/S,\mathcal{E}_{|X})$ because f is affine, and $\underline{Hom}_{\mathcal{O}_S}(f_*DR(X/S,\mathcal{E}_{|X}),\mathcal{O}_S) = R\underline{Hom}_{\mathcal{O}_S}(Rf_*DR(X/S,\mathcal{E}_{|X}),\mathcal{O}_S)$ because the terms of $DR(X/S,\mathcal{E}_{|X})$ are $\mathcal{O}_S$-locally free. Granting that, the last two formulas follow by application of [CE, XV, Theorem 5.11] with $r=2$, $q=-1$, $q'=0$ to the spectral sequence

$$E_2^{u,v} := \underline{Ext}^u_S(R^{-v}_{DR} f_*(\mathcal{E},\nabla)_{|X}), \mathcal{O}_S) \Rightarrow H^{u+v+1}(DW(\overline{X}/S,\mathcal{E}^\vee))\,.$$

We deduce exact sequences (D.1.3)

$$E_2^{-2,0} = 0 \longrightarrow E_2^{0,-1} = \underline{Hom}_S(R^1_{DR} f_*((\mathcal{E},\nabla)_{|X}), \mathcal{O}_S)$$
$$\xrightarrow{\sim} H^0(DW(\overline{X}/S,\mathcal{E}^\vee)) \longrightarrow E_2^{-1,0} = 0\,,$$

and (D.1.4)

$$\begin{aligned} E_2^{-1,0} = 0 &\longrightarrow E_2^{1,-1} \\ &= \underline{Ext}^1_S(R^1_{DR} f_*((\mathcal{E},\nabla)_{|X}), \mathcal{O}_S) \longrightarrow H^0 \\ &= H^1(DW(\overline{X}/S,\mathcal{E}^\vee)) \longrightarrow E_2^{0,0} \\ &= \underline{Hom}_S(R^0_{DR} f_*((\mathcal{E},\nabla)_{|X}), \mathcal{O}_S) \longrightarrow E_2^{2,-1} \\ &= \underline{Ext}^2_S(R^1_{DR} f_*((\mathcal{E},\nabla)_{|X}), \mathcal{O}_S) \longrightarrow H^1 \\ &= H^2(DW(\overline{X}/S,\mathcal{E}^\vee)) = 0. \end{aligned}$$

The conclusion follows. □

D.2 Cohomology with compact supports. It is tempting to interpret the results of the previous section in terms of Deligne's theory of direct images with proper supports for coherent $\mathcal{O}_X$-modules [H1, Appendix]. In this section we relax the conditions on the diagram of K-varieties

(D.2.0)
$$\begin{array}{ccccc} X & \overset{j}{\hookrightarrow} & \overline{X} & \hookleftarrow & Z \\ & \overset{f}{\searrow} & \downarrow \overline{f} & \swarrow & \\ & & S & & \end{array}$$

by requiring only that f be smooth of relative dimension d, that $\overline{f}$ be proper, and that j be an open immersion of complement $Z = \overline{X} \setminus j(X)$, the closed subscheme defined by the ideal sheaf $\mathcal{J}_Z$.

For any category $\mathcal{C}$, let *Pro* $\mathcal{C}$ denote the category of (Artin-Rees) pro-objects of $\mathcal{C}$ [G2]. Objects of *Pro* $\mathcal{C}$ are projective systems indexed by $\mathbf{Z}$ of objects of $\mathcal{C}$, and morphisms are defined by

$$Hom_{Pro\,\mathcal{C}}((A_n)_{n\in\mathbf{Z}}, (B_n)_{n\in\mathbf{Z}}) = \varinjlim_{k\in\mathbf{Z}} Hom((A_n)_k, (B_n))$$

where $(A_n)_k$ denotes the pro-object obtained from (A_n) by shifting k places to the right (*i.e.* its n-th component is A_{n+k}), and the second Hom means morphisms of diagrams. We point out that, for any $\mathcal{C}$, the category *Pro Pro* $\mathcal{C}$ is equivalent to Pro $\mathcal{C}$.

Deligne [*loc. cit.*] defines an exact functor that he denotes $j_!$, and we denote here $j_!^{Del}$,

(D.2.1)
$$\begin{aligned} j_!^{Del} : Coh(X) &\longrightarrow ProCoh(\overline{X}), \\ \mathcal{E} &\longmapsto \text{“}\varprojlim_N\text{”}\, \mathcal{J}_Z^N \overline{\mathcal{E}} \,, \end{aligned}$$

independent of the coherent extension $\overline{\mathcal{E}}$ of $\mathcal{E}$ to $\overline{X}$. We prefer to define our functor "direct image with compact supports" (already in derived categories) as the composition of $R \varprojlim$ and Deligne's $j_!^{Del}$, that is we set

(D.2.2)
$$\begin{aligned} j_! : D^b_c(X) &\longrightarrow D^b(\overline{X}), \\ \mathcal{E} &\longmapsto R \varprojlim_N (\mathcal{J}_Z^N \overline{\mathcal{E}}) \,, \end{aligned}$$

where for a bounded complex $\cdots \longrightarrow \mathcal{E}_i \longrightarrow \mathcal{E}_{i+1} \longrightarrow \cdots$ of coherent $\mathcal{O}_X$-modules that represents $\mathcal{E}$, $\overline{\mathcal{E}}$ is represented an arbitrary extension to a bounded complex $\cdots \longrightarrow \overline{\mathcal{E}}_i \longrightarrow \overline{\mathcal{E}}_{i+1} \longrightarrow \cdots$ of coherent $\mathcal{O}_{\overline{X}}$-modules and $\mathcal{O}_{\overline{X}}$-linear maps. The exact functor "extension by zero" of [H1, II, Ex. 1.19] will be denoted by $j_!^{top} : \mathcal{A}b(X) \longrightarrow \mathcal{A}b(\overline{X})$.

When $\overline{X}$ is *smooth*, we may deduce from $j_!$, a triangulated functor

(D.2.3)
$$j_! : D^b_{Coh(X)}(\mathcal{D}_X) \longrightarrow D^b(\mathcal{D}_{\overline{X}}) \,,$$

where $D^b_{Coh(X)}(\mathcal{D}_X)$ denotes the bounded derived category of the category of $\mathcal{O}_X$-coherent $\mathcal{D}_X$-modules, and $D^b(\mathcal{D}_{\overline{X}})$ the bounded derived category of $\mathcal{D}_{\overline{X}}$-modules. If $\mathcal{E}$ is an object of $D^b_{Coh(X)}(\mathcal{D}_X)$ and $\overline{\mathcal{E}}$ denotes any extension of $\mathcal{E}$ to an object of

$D^b_{Coh(\overline{X})}(\mathcal{D}_{\overline{X}})$, the completion $\hat{\overline{\mathcal{E}}}$ of $\overline{\mathcal{E}}$ along Z is a well defined object of $D^b(\mathcal{D}_{\overline{X}})$. We may calculate ($\overline{X}$ smooth) $j_!\mathcal{E}$ from the exact triangle in $D^b(\mathcal{D}_{\overline{X}})$

(D.2.3.1)
$$\begin{array}{ccccc} & & \overline{\mathcal{E}} & & \\ & \nearrow & & \searrow & \\ j_!(\mathcal{E}) & & \overset{[+1]}{\longleftarrow} & & \hat{\overline{\mathcal{E}}}. \end{array}$$

In general, we set $Rf_! = R\overline{f}_* \circ j_! : D^b_c(X) \longrightarrow D^b(S)$. The interpretation of the functor $\varprojlim$ as a direct image of a morphism of topoi (*cf.* [BO, Proof of 7.20], *e.g.*) and [SGA 4, Example V, Proposition 5.4] show that

$$Rf_!(\mathcal{E}) = R\overline{f}_*(R\varprojlim(\mathcal{J}_Z^N\overline{\mathcal{E}})) = R\varprojlim(R\overline{f}_*(\mathcal{J}_Z^N\overline{\mathcal{E}})).$$

The definition of $\overline{f}^!\mathcal{F}$, for $\mathcal{F} \in D^b_c(S)$, is given in [H1, VII.3.4]. We are not too interested in what it is, except later when $\overline{f}$ will be assumed to be smooth and $\overline{f}^!\mathcal{F} = \overline{f}^{-1}\mathcal{F} \otimes \Omega^d_{\overline{X}/S}[d]$. In any case, $f^!\mathcal{F} = (\overline{f}^!\mathcal{F})_{|X} = \overline{f}^{-1}\mathcal{F} \otimes \Omega^d_{X/S}[d]$.

Lemma D.2.4 *$\forall \mathcal{E} \in D^b_c(X)$, $\mathcal{F} \in D^b_c(\overline{X})$, $\mathcal{G} \in D^b_c(S)$ we have in $D^b(\overline{X})$ the canonical isomorphism*

$$j_!R\underline{Hom}_X(\mathcal{E}, j^*\mathcal{F}) \xrightarrow{\sim} R\underline{Hom}_{\overline{X}}(Rj_*\mathcal{E}, \mathcal{F}),$$

and in $D^b(S)$ the canonical isomorphism

$$Rf_!R\underline{Hom}_X(\mathcal{E}, f^!\mathcal{G}) \xrightarrow{\sim} R\underline{Hom}_S(Rf_*\mathcal{E}, \mathcal{G}) .$$

Proof: The second assertion follows from the first, by taking $\mathcal{F} = \overline{f}^!\mathcal{G}$ and applying the functor $R\overline{f}_*$. In fact, by [H1, VII, Corollary 4.3]

$$R\overline{f}_*R\underline{Hom}_{\overline{X}}(Rj_*\mathcal{E}, \overline{f}^!\mathcal{G}) \xrightarrow{\sim} R\underline{Hom}_S(Rf_*\mathcal{E}, \mathcal{G}) .$$

So we only have to prove the first assertion. Now, let $\overline{\mathcal{E}}$ be a coherent extension of $\mathcal{E}$ and $\mathcal{E}^\vee = R\underline{Hom}(\mathcal{E}, \mathcal{O}_X)$, so that $\overline{\mathcal{E}}^\vee = R\underline{Hom}(\overline{\mathcal{E}}, \mathcal{O}_{\overline{X}})$ is a coherent extension of $\mathcal{E}^\vee$. From Deligne's adjunction formula [H1, Appendix, Proposition 4] we deduce

$$\underline{Hom}(j_!^{Del}\mathcal{E}^\vee, \mathcal{O}_{\overline{X}}) = j_*\underline{Hom}(\mathcal{E}^\vee, \mathcal{O}_X) ,$$

and, (since for any injective $\mathcal{O}_{\overline{X}}$-module $\mathcal{I}$, and any $\mathcal{O}_{\overline{X}}$-module $\mathcal{H}$, $\underline{Hom}_{\overline{X}}(\mathcal{H}, \mathcal{I})$ is flasque [SGA 4, Example V, Proposition 4.10], while $j_!^{Del}$ is exact)

$$R\underline{Hom}(j_!^{Del}\mathcal{E}^\vee, \mathcal{O}_{\overline{X}}) = Rj_*R\underline{Hom}(\mathcal{E}^\vee, \mathcal{O}_X) = Rj_*\mathcal{E} .$$

We now apply to the previous formula the functor $R\underline{Hom}(-, \mathcal{F})$ to obtain

$$R\underline{Hom}(Rj_*\mathcal{E}, \mathcal{F}) = R\lim_{\substack{\leftarrow \\ N}} R\underline{Hom}((\mathcal{J}_Z^N \overline{\mathcal{E}}^\vee)^\vee, \mathcal{F})$$

$$= R\lim_{\substack{\leftarrow \\ N}} \mathcal{J}_Z^N \overline{\mathcal{E}}^\vee \overset{\mathbf{L}}{\otimes} \mathcal{F} = j_! R\underline{Hom}_X(\mathcal{E}, j^*\mathcal{F}) \ .$$

□

Corollary D.2.5 $\forall \mathcal{E} \in D_c^b(X)$,

$$j_!\mathcal{E}^\vee = (Rj_*\mathcal{E})^\vee \ .$$

Let $\mathcal{E}$ be a coherent $\mathcal{O}_X$-module and $\overline{\mathcal{E}}$ be an arbitrary coherent extension of it to $\overline{X}$. The exact sequence of projective systems

$$0 \longrightarrow \{\mathcal{J}_Z^N \overline{\mathcal{E}}\}_N \longrightarrow \overline{\mathcal{E}} \longrightarrow \{\overline{\mathcal{E}}/\mathcal{J}_Z^N \overline{\mathcal{E}}\}_N \longrightarrow 0$$

shows that

$$H^0(j_!\mathcal{E}) = j_!^{top}(\mathcal{E}) H^1(j_!\mathcal{E}) = \mathrm{Coker}(\overline{\mathcal{E}} \longrightarrow \hat{\overline{\mathcal{E}}}). \tag{D.2.6}$$

We are now interested in introducing, for any base K-scheme S and any smooth S-scheme Y, a triangulated category $D^b(Coh(Y), \mathrm{diff.op.})$, intermediate between $D_c^b(Y)$ and $D^b(\mathcal{A}b(Y))$, where De Rham complexes relative to S of $\mathcal{D}_Y$-modules naturally live, and with the further property that $j_!$ makes sense as a triangulated functor from $D^b(Coh(X), \mathrm{diff.op.})$ to, say, $D^b(\mathcal{A}b(\overline{X}))$. Our construction is similar to the one proposed by Herrera and Lieberman [HL], but replaces their use of a Yoneda pairing of hyperext functors by Berthelot's recent duality result in the theory of $\mathcal{D}$-modules, *cf.* Appendix C[(1)].

Definition D.2.7 Let S be a K-scheme. For any smooth S-scheme Y, let $\mathcal{C}$ be any of the categories $Mod(Y)$, $Coh(Y)$, $Pro\ Coh(Y)$. Let $C^b(\mathcal{C}, \mathrm{diff.op.})$ be the (abelian) category of bounded complexes of objects of $\mathcal{C}$, with $\mathcal{O}_S$-linear differential operators of order ≤ 1 as differentials. Morphisms of two such complexes are supposed to be $\mathcal{O}_Y$-linear maps commuting with the differentials.

Definition D.2.8 Let $Y \longrightarrow S$ and $\mathcal{C}$ be as in (D.2.7).

i) Two morphisms $f, g : \mathcal{F}^\cdot \longrightarrow \mathcal{G}^\cdot$ in $C^b(\mathcal{C}, \mathrm{diff.op.})$ are homotopic if there exist $\mathcal{O}_Y$-linear maps $s^i : \mathcal{F}^i \longrightarrow \mathcal{G}^{i-1}$ such that $f^i - g^i = s^{i+1} \circ d^i_{\mathcal{F}} + d^{i-1}_{\mathcal{G}} \circ s^i$, $\forall i \in \mathbf{Z}$.
ii) We define an additive category $K^b(\mathcal{C}, \mathrm{diff.op.})$ whose objects are the objects of $C^b(\mathcal{C}, \mathrm{diff.op.})$, and whose morphisms are the homotopy classes of morphisms of $C^b(\mathcal{C}, \mathrm{diff.op.})$.

---

(1) See also [Sa] for a similar construction.

iii) $K^b(\mathcal{C}, \text{diff.op.})$ is a triangulated category, where a triangle is distinguished if it comes from the mapping cone of a morphism in $C^b(\mathcal{C}, \text{diff.op.})$.

We now recall Grothendieck's linearization of differential operators [G2]. Let $Y \longrightarrow S$ and $\mathcal{C}$ be as in (D.2.7). We denote by $P^{\nu}_{Y/S}(i)$ the i-th infinitesimal neighborhood of the diagonal $Y \longrightarrow Y \times_S \cdots \times_S Y$ ($\nu + 1$ factors) and by $\mathcal{P}^{\nu}_{Y/S}(i)$ the $\mathcal{O}_Y$-module $p_{1*}\mathcal{O}_{P^{\nu}_{Y/S}(i)}$, where $p_k : P^{\nu}_{Y/S}(i) \longrightarrow Y$ denotes the projection on the k-th factor. Besides its canonical $\mathcal{O}_Y$-module structure (the "extreme left structure"), $\mathcal{P}^{\nu}_{Y/S}(i)$ has ν additional such structures, associated to each of its other projections. We only need to use the canonical and the "extreme right structure", associated to $p_{\nu+1}$. For any object $\mathcal{E}$ of $\mathcal{C}$, we set $\mathcal{P}^{\nu}_{Y/S}(i)(\mathcal{E}) = p_{1*}p^*_{\nu+1}\mathcal{E} = \mathcal{P}^{\nu}_{Y/S}(i) \otimes_{\mathcal{O}_Y} \mathcal{E}$, using the extreme right structure to take the tensor product. A tensor product of the form $\mathcal{E} \otimes_{\mathcal{O}_Y} \mathcal{P}^{\nu}_{Y/S}(i)$, would instead mean that the extreme right structure on $\mathcal{P}^{\nu}_{Y/S}(i)$ is used.

Given an object $\mathcal{E}^{\cdot}$ of $C^b(\mathcal{C}, \text{diff.op.})$, Grothendieck constructs functorially a bicomplex

$$Q^*(\mathcal{E}^{\cdot}) = Q^*_{Y/S}(\mathcal{E}^{\cdot}) = (\mathcal{P}^{\nu}_{Y/S}(i)(\mathcal{E}^j))$$

indexed by ν and j in the category *Pro* ($\mathcal{C}$) (the index i refers to inverse systems indexed by $i \in \mathbf{N}$), such that the bicomplex $\varprojlim_i Q^*(\mathcal{E}^{\cdot})(i)$ is a resolution of $\mathcal{E}^{\cdot}$ in $\mathcal{A}b(Y)$ (resp. *Pro* $\mathcal{A}b(Y)$), if $\mathcal{C} = Mod(Y)$ or $Coh(Y)$ (resp. *Pro* $Coh(Y)$).

Remark D.2.9 If $f, g \in Hom_{C^b(\mathcal{C},\text{diff.op.})}(\mathcal{F}^{\cdot}, \mathcal{G}^{\cdot})$ are homotopic via $s^i : \mathcal{F}^i \longrightarrow \mathcal{G}^{i-1}$, the maps $Q^0(f)$ and $Q^0(g) \in Hom_{Pro\mathcal{C}}(Q^0(\mathcal{F}^{\cdot}), Q^0(\mathcal{G}^{\cdot}))$ are homotopic via $Q^0(s^i) : Q^0(\mathcal{F}^i) \longrightarrow Q^0(\mathcal{G}^{i-1})$. So, Q^0 induces a functor $K^b(\mathcal{C}, \text{diff.op.}) \longrightarrow K^b(Pro\mathcal{C})$.

Definition D.2.10 [(1)]. Let $Y \longrightarrow S$ and $\mathcal{C}$ be as in (D.2.7). Let $\mathcal{S}$ be the family of morphisms f of $K^b(\mathcal{C}, \text{diff.op.})$ such that $Q^0(f)$ (hence $Q^{\nu}(f)$, $\forall \nu$) is a quasi-isomorphism in $K^b(\text{Pro}\mathcal{C})$.

It is easy to show that $\mathcal{S}$ is a multiplicative system in $K^b(\mathcal{C}, \text{diff.op.})$.

Definition D.2.11 Let $Y \longrightarrow S$ and $\mathcal{C}$ be as in (D.2.7). Let $\mathcal{S}$ be as in (D.2.10). We define the triangulated category $D^b(\mathcal{C}, \text{diff.op.})$ as the localization of $K^b(\mathcal{C}, \text{diff.op.})$ with respect to the multiplicative system $\mathcal{S}$.

It is easily seen that $D^b(\mathcal{C})$ is a full triangulated subcategory of $D^b(\mathcal{C}, \text{diff.op.})$. Since Grothendieck's construction gives a resolution of objects of $C^b(\mathcal{C}, \text{diff.op.})$ as abelian sheaves, there is a canonical triangulated functor $D^b(\mathcal{C}, \text{diff.op.}) \longrightarrow D^b(\mathcal{A}b(Y))$ (resp. $D^b(\mathcal{C}, \text{diff.op.}) \longrightarrow D^b(Pro\ \mathcal{A}b(Y))$) if $\mathcal{C} = Mod(Y)$ or $Coh(Y)$ (resp. *Pro* $Coh(Y)$).

Lemma D.2.12 *Let $\mathcal{E}$, $\mathcal{F}$ be coherent $\mathcal{O}_X$-modules and $\mathcal{E} \xrightarrow{L} \mathcal{F}$ be an $\mathcal{O}_S$-linear differential operator. Let $\overline{\mathcal{F}}$ be a coherent extension of $\mathcal{F}$ to $\overline{X}$, contained in $j_*(\mathcal{F})$.*

---

(1) We are indebted to P. Berthelot for suggesting localization of $K^b(\mathcal{C}, \text{diff.op.})$ with respect to this class of morphisms.

Then there exists a coherent extension $\overline{\mathcal{E}}$ *of* $\mathcal{E}$ *to* $\overline{X}$, *contained in* $j_*(\mathcal{E})$ *and an* $\mathcal{O}_S$*-linear differential operator* $\overline{\mathcal{E}} \xrightarrow{\overline{L}} \overline{\mathcal{F}}$ *extending* L.

Proof: Certainly $j_*(L) : j_*(\mathcal{E}) \longrightarrow j_*(\mathcal{F})$ is an $\mathcal{O}_S$-linear differential operator. Let $\overline{\mathcal{E}}'$ be any coherent extension of $\mathcal{E}$ to $\overline{X}$, contained in $j_*(\mathcal{E})$. We may take $\overline{\mathcal{E}} = \mathcal{J}_Z^N \overline{\mathcal{E}}'$ for sufficiently big N, and $\overline{L} = j_*(L)_{|\overline{\mathcal{E}}}$. □

The previous lemma shows that, when $\overline{f} : \overline{X} \longrightarrow S$ is smooth, given a bounded complex $\mathcal{E}^{\cdot} = \cdots \longrightarrow \mathcal{E}^i \xrightarrow{L^i} \mathcal{E}^{i+1} \dots$ of coherent $\mathcal{O}_X$-modules and $\mathcal{O}_S$-linear differential operators, there exists a bounded complex $\overline{\mathcal{E}}^{\cdot} = \cdots \longrightarrow \overline{\mathcal{E}}^i \xrightarrow{\overline{L}^i} \overline{\mathcal{E}}^{i+1} \dots$ in $C^b(Coh(\overline{X}), \text{diff.op.})$ extending $\mathcal{E}^{\cdot}$. Given homotopic maps $\varphi, \psi : \mathcal{E}^{\cdot} \longrightarrow \mathcal{F}^{\cdot}$ in $C^b(Coh(X), \text{diff.op.})$, we may extend $\mathcal{E}^{\cdot}$ and $\mathcal{F}^{\cdot}$ to objects of $C^b(Coh(\overline{X}), \text{diff.op.})$ and φ, ψ to homotopic maps $\overline{\varphi}, \overline{\psi} : \overline{\mathcal{E}}^{\cdot} \longrightarrow \overline{\mathcal{F}}^{\cdot}$ in $C^b(Coh(\overline{X}), \text{diff.op.})$.

We note that a differential operator of order d between $\mathcal{O}_{\overline{X}}$-modules $\overline{\mathcal{E}} \xrightarrow{\overline{L}} \overline{\mathcal{F}}$, induces a differential operator of projective systems $\{\mathcal{J}_Z^N \overline{\mathcal{E}}\}_N \longrightarrow \{\mathcal{J}_Z^{N-d} \overline{\mathcal{F}}\}_N$. For $\mathcal{E}^{\cdot}$ in $C^b(Coh(X), \text{diff.op.})$ and any extension $\overline{\mathcal{E}}^{\cdot}$ of $\mathcal{E}^{\cdot}$ in $C^b(Coh(\overline{X}), \text{diff.op.})$, "$\varprojlim_N$"$\mathcal{J}_Z^N \overline{\mathcal{E}}^{\cdot}$ is an object of $C^b(ProCoh(\overline{X}), \text{diff.op.})$, independent of the chosen extension $\overline{\mathcal{E}}^{\cdot}$ of $\mathcal{E}^{\cdot}$. So, we get a functor

$$j_!^{Del} : C^b(Coh(X), \text{diff.op.}) \longrightarrow C^b(Pro\ Coh(\overline{X}), \text{diff.op.})\ .$$

It is easily seen that $j_!^{Del}\mathcal{E}^{\cdot}$ up to homotopy depends only upon the homotopy class of $\mathcal{E}^{\cdot}$, so that

$$j_!^{Del} : K^b(Coh(X), \text{diff.op.}) \longrightarrow K^b(Pro\ Coh(\overline{X}), \text{diff.op.})\ .$$

Lemma D.2.13 *Assume* $\overline{f} : \overline{X} \longrightarrow S$ *is smooth. Let* $\varphi : \mathcal{E}^{\cdot} \longrightarrow \mathcal{F}^{\cdot}$ *be a morphism in* $C^b(Coh(X), \text{diff.op.})$, *representing an element in the multiplicative system* $\mathcal{S}$ *of morphisms of* $K^b(Coh(X), \text{diff.op.})$. *Then*

$$j_!^{Del}(\varphi) : j_!^{Del}(\mathcal{E}^{\cdot}) \longrightarrow j_!^{Del}(\mathcal{F}^{\cdot})$$

is a quasi-isomorphism in $K^b(ProCoh(\overline{X}), \text{diff.op.})$. *We conclude that* $j_!^{Del}$ *defines a triangulated functor*

$$j_!^{Del} : D^b(Coh(X), \text{diff.op.}) \longrightarrow D^b(Pro\ Coh(\overline{X}), \text{diff.op.}).$$

Proof: Let $\overline{\varphi} : \overline{\mathcal{E}}^{\cdot} \longrightarrow \overline{\mathcal{F}}^{\cdot}$ be an extension of φ to a morphism of $C^b(Coh(\overline{X}), \text{diff.op.})$. By application of Grothendieck's linearization functor, we get a morphism of bicomplexes

$$Q^0_{\overline{X}/S}(\overline{\varphi}) : Q^0_{\overline{X}/S}(\overline{\mathcal{E}}^{\cdot}) \longrightarrow Q^0_{\overline{X}/S}(\overline{\mathcal{F}}^{\cdot}) \tag{D.2.13.1}$$

extending

$$\mathcal{Q}^0_{X/S}(\varphi) : \mathcal{Q}^0_{X/S}(\mathcal{E}^{\cdot}) \longrightarrow \mathcal{Q}^0_{X/S}(\mathcal{F}^{\cdot}).$$

The functor "$\varprojlim_N$" $\mathcal{J}_Z^N(-)$ also applies to the morphism (D.2.13.1): we obtain

$$j_!^{Del}(\mathcal{Q}^0_{X/S}(\varphi)) : j_!^{Del}(\mathcal{Q}^0_{X/S}(\mathcal{E}^{\cdot})) \longrightarrow j_!^{Del}(\mathcal{Q}^0_{X/S}(\mathcal{F}^{\cdot}))$$

where, for example, $j_!^{Del}(\mathcal{Q}^0_{X/S}(\mathcal{E}^{\cdot}))$ is $(\mathcal{J}_Z^N \mathcal{Q}^0_{\overline{X}/S}(\overline{\mathcal{E}}^{\cdot})) = (\mathcal{J}_Z^N \otimes \mathcal{P}^1_{\overline{X}/S}(i)(\overline{\mathcal{E}}^j))$, a complex of pro-objects (each indexed by $(N, i) \in \mathbf{N} \times \mathbf{N}$) in $Coh(\overline{X})$. The difference between the left and the right structure of $\mathcal{P}^1_{\overline{X}/S}(i)$ gives

$$\mathcal{J}_Z^N \otimes \mathcal{P}^1_{\overline{X}/S}(i)(\overline{\mathcal{E}}^j) = \mathcal{P}^1_{\overline{X}/S}(i)(\overline{\mathcal{E}}^j) \otimes \mathcal{J}_Z^{N+i} = \mathcal{P}^1_{\overline{X}/S}(i)(\mathcal{J}_Z^{N+i}\overline{\mathcal{E}}^j),$$

so that we may write $\mathcal{Q}^0_{\overline{X}/S}(j_!^{Del}\mathcal{E}^{\cdot}) = j_!^{Del}\mathcal{Q}^0_{X/S}(\mathcal{E}^{\cdot})$. More precisely,

$$j_!^{Del}(\mathcal{Q}^0_{X/S}(\varphi)) : j_!^{Del}(\mathcal{Q}^0_{X/S}(\mathcal{E}^{\cdot})) \longrightarrow j_!^{Del}(\mathcal{Q}^0_{X/S}(\mathcal{F}^{\cdot}))$$

identifies with

$$\mathcal{Q}^0_{\overline{X}/S}(j_!^{Del}(\varphi)) : \mathcal{Q}^0_{\overline{X}/S}(j_!^{Del}\mathcal{E}^{\cdot}) \longrightarrow \mathcal{Q}^0_{\overline{X}/S}(j_!^{Del}\mathcal{F}^{\cdot}).$$

We recall that Deligne's functor $j_!^{Del} : Coh(X) \longrightarrow Coh(\overline{X})$ (hence $j_!^{Del}$: $Pro\ Coh(X) \longrightarrow Pro\ Coh(\overline{X})$) is exact. So, if $\mathcal{Q}^0_{X/S}(\varphi)$ is a quasi-isomorphism in $K^b(Pro\ Coh(X))$, then $j_!^{Del}(\mathcal{Q}^0_{X/S}(\varphi)) = \mathcal{Q}^0_{\overline{X}/S}(j_!^{Del}(\varphi))$ is a quasi-isomorphism in $K^b(Pro\,Coh(\overline{X}))$, and the assertion is proven. □

Definition D.2.14 We define

$$Rj_! : D^b(Coh(X), \text{diff.op.}) \longrightarrow D^b(\mathcal{A}b(\overline{X}))$$

as the composition of

$$D^b(Coh(X), \text{diff.op.}) \xrightarrow{j_!^{Del}} D^b(Pro\ Coh(\overline{X}), \text{diff.op.})$$
$$\longrightarrow D^b(Pro\mathcal{A}b(\overline{X})) \xrightarrow{R\varprojlim} D^b(\mathcal{A}b(\overline{X})).$$

We also define

$$Rf_! = R\overline{f}_* \circ Rj_! : D^b(Coh(X), \text{diff.op.}) \longrightarrow D^b(\mathcal{A}b(S)).$$

Obviously, $Rj_!$ takes its values in the derived category of the category of sheaves of $\overline{f}^{-1}\mathcal{O}_S$-modules on $\overline{X}$, a triangulated subcategory of $D^b(\mathcal{A}b(\overline{X}))$. As a consequence, $Rf_!$ may be viewed as a triangulated functor

$$Rf_! = R\overline{f}_* \circ Rj_! : D^b(Coh(X), \text{diff.op.}) \longrightarrow D^b(S) .$$

Let $\overline{f}$ be smooth and $\mathcal{E}^\cdot$ be an object of $C^b(Coh(X), \text{diff.op.})$, and let $\overline{\mathcal{E}}^\cdot$ be an extension of $\mathcal{E}^\cdot$ to an object of $C^b(Coh(\overline{X}), \text{diff.op.})$. We may compute $Rj_!(\mathcal{E}^\cdot)$ from the exact triangle in $D^b(\mathcal{A}b(\overline{X}))$

(D.2.15)
$$\begin{array}{ccccc} & & \overline{\mathcal{E}}^\cdot & & \\ & \nearrow & & \searrow & \\ Rj_!(\mathcal{E}^\cdot) & & \overset{[+1]}{\longleftarrow} & & \hat{\overline{\mathcal{E}}}^\cdot . \end{array}$$

We note that if $\mathcal{E}$ is an object of $D^b_{Coh(X)}(\mathcal{D}_X)$ (D.2.3),

(D.2.15.1)
$$DR_{\overline{X}/S}(j_!\mathcal{E}) = Rj_! DR_{X/S}(\mathcal{E}) .$$

Definition D.2.16 Let S (and therefore X) be smooth. For $\mathcal{E}$ in $D^b_{Coh(X)}(\mathcal{D}_X)$ we define the direct image with compact supports by $f : X \longrightarrow S$ as $f_!\mathcal{E} := Rf_! DR_{X/S}(\mathcal{E})[d] \in D^b(\mathcal{D}_S)$. For a coherent object $(\mathcal{E}, \nabla)$ of $\mathbf{MIC}(X)$, we define the q-th De Rham cohomology sheaf with compact supports of $(\mathcal{E}, \nabla)$ relative to S, denoted $R^q_{DR} f_!(\mathcal{E}, \nabla)$ or $H^q_{DR,c}(X/S, (\mathcal{E}, \nabla))$, as the object $H^q(Rf_! DR(X/S, (\mathcal{E}, \nabla)))$ of $\mathbf{MIC}(S)$.

The following is our main duality statement. It is a corollary of Berthelot's duality theorem proven in Appendix C, combined with the previous definition. It allows us to regard Dwork's duality principle for elementary fibrations as a version in algebraic coherent cohomology of the familiar Poincaré duality.

We recall that, $f : X \longrightarrow S$ being smooth, the direct image by f of $\mathcal{D}_X$-modules may be expressed via the classical Gauss-Manin connection on relative De Rham cohomology:

$$\begin{aligned} f_+ : D^b(\mathcal{D}_X) &\longrightarrow D^b(\mathcal{D}_S) \\ \mathcal{E} &\longmapsto Rf_* DR_{X/S}(\mathcal{E})[d] \ . \end{aligned}$$

Theorem D.2.17 *(Main Duality Theorem) Let $f : X \longrightarrow S$ be a smooth map of smooth algebraic K-varieties. Assume that f can be embedded in a diagram (D.2.0), with $\overline{f}$ smooth and proper. Then for any object $\mathcal{E}$ of $D^b_{Coh(X)}(\mathcal{D}_X)$, we have a canonical isomorphism in $D^b(\mathcal{D}_S)$*

$$f_!\mathcal{E} \xrightarrow{\cong} (f_+\mathcal{E}^*_{X/S})^\vee .$$

Proof: By (D.2.5),

$$j_!(\mathcal{E}^\vee) = (Rj_*\mathcal{E})^\vee .$$

By Berthelot's theorem C.1, we have a natural morphism in $D^b(\mathcal{D}_{\overline{X}})$

$$(Rj_*\mathcal{E})^\vee \longrightarrow (Rj_*\mathcal{E})^*_{\overline{X}/S} ,$$

and therefore a morphism

$$j_!(\mathcal{E}^\vee) \longrightarrow (Rj_*\mathcal{E})^*_{\overline{X}/S} ,$$

inducing an isomorphism of De Rham complexes in $D^b(\mathcal{D}_S)$ (*cf.* (D.2.15.1))

$$Rj_! DR_{X/S}(\mathcal{E}^\vee) = DR_{\overline{X}/S}(j_!\mathcal{E}^\vee) \xrightarrow{\cong} DR_{\overline{X}/S}((Rj_*\mathcal{E})^*_{\overline{X}/S}).$$

We now apply to the previous isomorphism the functor $R\overline{f}_*$ and obtain to the r.h.s.

$$R\overline{f}_* DR_{\overline{X}/S}((Rj_*\mathcal{E})^*_{\overline{X}/S}) \cong \overline{f}_+((Rj_*\mathcal{E})^*_{\overline{X}/S}) \cong (\overline{f}_+ Rj_*\mathcal{E})^*_{S/S} \cong (f_+\mathcal{E})^\vee .$$

□

We point out that the formula obtained in our duality theorem with $\vee$ replaced by $*$ has been taken as the *definition* of the algebraic direct image with compact support of a $\mathcal{D}$-module by previous authors [Me 5, I.5.2.4], [Bo, VI.4.0, eq. 1]. Our construction of $f_!$ has the advantage of allowing explicit computations of De Rham relative cohomology groups, especially in the case of an elementary fibration, and should offer an alternative proof of the results in Section 8. We start by showing how Dwork's duality principle for elementary fibrations may be restated as a version in coherent cohomology of the familiar Poincaré duality. Some explicit computations due to Dwork, in the framework of his dual theory, will be given as examples in the following sections.

Proposition D.2.18 *(Dwork-Poincaré duality for an elementary fibration.) Let $f : X \longrightarrow S$ be an elementary fibration. Let $(\mathcal{E}, \nabla)$ be a coherent object of* $\mathbf{MIC}(X)$ *and $(\mathcal{E}^\vee, \nabla^\vee)$ be its dual. There is a canonical isomorphism*

$$Rf_! DR(X/S, (\mathcal{E}^\vee, \nabla^\vee))[2] \xrightarrow{\sim} R\underline{Hom}_S(Rf_* DR(X/S, (\mathcal{E}, \nabla)), \mathcal{O}_S).$$

In particular, we have a canonical isomorphism in $\mathbf{MIC}(S)$

$$H^1_{DR,c}(X/S, (\mathcal{E}^\vee, \nabla^\vee)) \xrightarrow{\sim} \underline{Hom}_{\mathcal{O}_S}(H^1_{DR}(X/S, (\mathcal{E}, \nabla)), \mathcal{O}_S),$$

and an exact sequence in $\mathbf{MIC}(S)$

$$0 \longrightarrow \underline{Ext}^1_S(H^1_{DR}(X/S, (\mathcal{E}, \nabla)), \mathcal{O}_S) \longrightarrow H^2_{DR,c}(X/S, (\mathcal{E}^\vee, \nabla^\vee)) \longrightarrow$$
$$\underline{Hom}_S(H^0_{DR}(X/S, (\mathcal{E}, \nabla)), \mathcal{O}_S) \longrightarrow \underline{Ext}^2_S(H^1_{DR}(X/S, (\mathcal{E}, \nabla)), \mathcal{O}_S) \longrightarrow 0.$$

Corollary D.2.19

(i) *If* $\underline{Ext}^1_S(H^1_{DR}(X/S, (\mathcal{E}, \nabla)), \mathcal{O}_S) = 0$, *then* $H^2_{DR,c}(X/S, (\mathcal{E}^\vee, \nabla^\vee))$ *is coherent (hence locally free).*

(ii) *If* $\underline{Ext}^1_S(H^1_{DR}(X/S, (\mathcal{E}, \nabla)), \mathcal{O}_S) = 0 = \underline{Ext}^2_S(H^1_{DR}(X/S, (\mathcal{E}, \nabla)), \mathcal{O}_S)$ *(e.g., if* $H^1_{DR}(X/S, (\mathcal{E}, \nabla))$ *is coherent) we also have a canonical isomorphism of coherent* (4.2.1) *objects of* **MIC**(S)

$$H^2_{DR,c}(X/S, (\mathcal{E}^\vee, \nabla^\vee)) \xrightarrow{\sim} \underline{Hom}_{\mathcal{O}_S}(H^0_{DR}(X/S, (\mathcal{E}, \nabla)), \mathcal{O}_S).$$

(iii) *If* $\underline{Ext}^1_S(H^1_{DR}(X/S, (\mathcal{E}, \nabla)), \mathcal{O}_S) = 0 = \underline{Ext}^2_S(H^1_{DR}(X/S, (\mathcal{E}, \nabla)), \mathcal{O}_S)$ *and* $H^1_{DR,c}(X/S, (\mathcal{E}^\vee, \nabla^\vee))$ *is coherent, then* $H^1_{DR}(X/S, (\mathcal{E}, \nabla))$ *is coherent, too.*

We now show that the discussion of this section in fact interprets and generalizes Dwork's duality principle (D.1.21).

Theorem D.2.20 *In the situation of* (D.2.18), *let* $\overline{\mathcal{E}}$ *be any coherent extension of* $\mathcal{E}$ *to* $\overline{X}$, *contained in* $j_*\mathcal{E}$, *and let* $\omega = \Omega^1_{\overline{X}/S}$. *Then* $Rf_!DR(X/S, (\mathcal{E}, \nabla))[1]$ *is quasi-isomorphic to*

$$\begin{aligned} DW(\overline{X}/S, \overline{\mathcal{E}}) = [R^0\overline{f}_*(\hat{\overline{\mathcal{E}}}(*Z))/R^0\overline{f}_*(\overline{\mathcal{E}}(*Z)) &\longrightarrow \\ 0 \quad & \\ R^0\overline{f}_*(\omega \otimes \hat{\overline{\mathcal{E}}}(*Z))/R^0\overline{f}_*(\omega \otimes \overline{\mathcal{E}}(*Z))]. & \\ 1 \quad & \end{aligned}$$

Proof: (L. Fiorot) We may extend $\mathcal{E}^{\cdot} := DR(X/S, (\mathcal{E}, \nabla))$ to $\overline{X}$ by

$$\overline{\mathcal{E}}^{\cdot} := [\underset{0}{\overline{\mathcal{E}}} \longrightarrow \underset{1}{\omega \otimes \overline{\mathcal{E}} \otimes \mathcal{J}_Z^{-m}}]$$

for a sufficiently high m. By (D.2.15) $Rj_!(\mathcal{E}^{\cdot})$ is the mapping fiber of the morphism $\overline{\mathcal{E}}^{\cdot} \longrightarrow \hat{\overline{\mathcal{E}}}^{\cdot}$, independent of the choice of $\overline{\mathcal{E}}$ and m. So,

$$j_!(\mathcal{E}^{\cdot})[1] = [\underset{-1}{\overline{\mathcal{E}}} \longrightarrow \underset{0}{\hat{\overline{\mathcal{E}}} \oplus (\omega \otimes \overline{\mathcal{E}} \otimes \mathcal{J}_Z^{-m})} \longrightarrow \underset{1}{\omega \otimes \hat{\overline{\mathcal{E}}} \otimes \mathcal{J}_Z^{-m}}]$$

with the obvious differential. Since $j_*(\mathcal{E}) = \overline{\mathcal{E}}(*Z) = \varinjlim \overline{\mathcal{E}}_i$ where the direct limit runs over the directed set of extensions $\overline{\mathcal{E}}_i$ of $\mathcal{E}$ to $\overline{X}$, that are coherent submodules of $j_*(\mathcal{E})$, considering extensions of $\mathcal{E}^{\cdot}$ as above with $\overline{\mathcal{E}}$ replaced by $\overline{\mathcal{E}}_i$ (and suitable $m = m_i$) and taking direct limits, we obtain

$$Rj_!(\mathcal{E}^{\cdot})[1] = [\underset{-1}{\overline{\mathcal{E}}(*Z)} \longrightarrow \underset{0}{\hat{\overline{\mathcal{E}}}(*Z) \oplus (\omega \otimes \overline{\mathcal{E}}(*Z))} \longrightarrow \underset{1.}{\omega \otimes \hat{\overline{\mathcal{E}}}(*Z)}]$$

So,

$$Rf_!(\mathcal{E}^\cdot)[1] = R\overline{f}_*(j_!(\mathcal{E}^\cdot))[1] = R\overline{f}_*[\overline{\mathcal{E}}(*Z) \longrightarrow \hat{\overline{\mathcal{E}}}(*Z) \oplus (\omega \otimes \overline{\mathcal{E}}(*Z)) \longrightarrow \omega \otimes \hat{\overline{\mathcal{E}}}(*Z)].$$

Since the objects of $[\overline{\mathcal{E}}(*Z) \longrightarrow \hat{\overline{\mathcal{E}}}(*Z) \oplus (\omega \otimes \overline{\mathcal{E}}(*Z)) \longrightarrow \omega \otimes \hat{\overline{\mathcal{E}}}(*Z)]$ are f_*-acyclic, we obtain

$$Rf_!(\mathcal{E}^\cdot)[1] = [\overline{f}_*\overline{\mathcal{E}}(*Z) \longrightarrow \overline{f}_*(\hat{\overline{\mathcal{E}}}(*Z)) \oplus \overline{f}_*(\omega \otimes \overline{\mathcal{E}}(*Z)) \longrightarrow \overline{f}_*(\omega \otimes \hat{\overline{\mathcal{E}}}(*Z))] \,.$$

It is immediate to check that the map of complexes

$$\begin{array}{ccccccc} \Big[\overline{f}_*\overline{\mathcal{E}}(*Z) & \longrightarrow & \overline{f}_*(\hat{\overline{\mathcal{E}}}(*Z)) & \oplus & \overline{f}_*(\omega \otimes \overline{\mathcal{E}}(*Z)) & \longrightarrow & \overline{f}_*(\omega \otimes \hat{\overline{\mathcal{E}}}(*Z))\Big] \\ \downarrow 0 & & can \searrow & & \swarrow 0 & & can \downarrow \\ \Big[\ 0 & \longrightarrow & & \dfrac{\overline{f}_*(\hat{\overline{\mathcal{E}}}(*Z))}{\overline{f}_*(\overline{\mathcal{E}}(*Z))} & & \longrightarrow & \dfrac{\overline{f}_*(\omega \otimes \hat{\overline{\mathcal{E}}}(*Z))}{\overline{f}_*(\omega \otimes \overline{\mathcal{E}}(*Z))}\Big] \end{array}$$

is a quasi-isomorphism. □

D.3 Duality for rational elementary fibrations We now come back to the situation of section D.1 and consider the case of a rational elementary fibration as in (1.2.1) and (4.5.1). So now $\overline{X} = \mathbf{P}^1_S$, $\overline{f} = \mathrm{pr}$, and we have a canonical relative affine coordinate x. We let $(\mathcal{O}_{\overline{X}}, d)$ play the role of the $(\mathcal{E}, \nabla)$ of (D.1). We set

(D.3.1) $$\mathcal{R}(f) := R^0\overline{f}_*(\hat{\mathcal{O}}_{\overline{X}})/R^0\overline{f}_*(\mathcal{O}_{\overline{X}}) = R^0\overline{f}_*(\hat{\mathcal{O}}_{\overline{X}})/\mathcal{O}_S \,,$$

(D.3.2) $$\begin{aligned} \mathcal{R}'(f) &:= \lim_{s\to\infty} R^0\overline{f}_*(\hat{\mathcal{J}}_Z^{-s}) = R^0\overline{f}_*(\hat{\mathcal{O}}_{\overline{X}} \otimes_{\mathcal{O}_{\overline{X}}} j_*\mathcal{O}_X), \\ \mathcal{L}(f) &= R^0 f_*(\mathcal{O}_X), \\ \omega(f) &:= R^0 f_*(\omega_{|X}) = \mathcal{L}(f)dx, \\ \omega'(f) &:= R^0\overline{f}_*(\hat{\mathcal{O}}_{\overline{X}} \otimes_{\mathcal{O}_{\overline{X}}} j_*(\omega_{|X})) = \mathcal{R}'(f)dx, \\ \mathcal{D}(f) &:= R^0 f_*(\mathcal{D}_{X/S}). \end{aligned}$$

According to Section D.1, there exists an exact sequence of left $\mathcal{D}(f)$-modules (which *defines* the structure of left $\mathcal{D}(f)$-module of $\mathcal{R}(f)$)

(D.3.3) $$0 \longrightarrow \mathcal{L}(f) \longrightarrow \mathcal{R}'(f) \xrightarrow{\gamma_-(f)} \mathcal{R}(f) \longrightarrow 0 \,.$$

Proposition D.3.4 *There is a unique $\mathcal{O}_S$-linear pairing*

$$\langle\, , \rangle : \omega(f) \times \mathcal{R}(f) \longrightarrow \mathcal{O}_S \tag{D.3.4.1}$$

compatible with the Serre pairings

$$\langle\, , \rangle_N : R^0\overline{f}_*(\omega \otimes \mathcal{J}_Z^{-N}) \times R^1\overline{f}_*(\mathcal{J}_Z^N) \longrightarrow \mathcal{O}_S$$

and the natural morphisms

$$\mathcal{R}^0\overline{f}_*\hat{\mathcal{O}}_{\overline{X}} \longrightarrow R^0\overline{f}_*(\mathcal{O}_{\overline{X}}/\mathcal{J}_Z^N) \longrightarrow R^1\overline{f}_*(\mathcal{J}_Z^N)$$

(where $R^0\overline{f}_*\mathcal{O}_{\overline{X}} \to 0$) and

$$R^0\overline{f}_*(\omega \otimes \mathcal{J}_Z^{-N}) \longrightarrow \omega(f)\ .$$

We have, for any point $s \in S$:

i) *let $h \in \mathcal{R}(f)_s$; if $\langle \eta, h\rangle = 0$ in $\mathcal{O}_{S,s}$ for any $\eta \in \omega(f)_s$ then $h = 0$;*
ii) *let $\eta \in \omega(f)_s$; if $\langle \eta, h\rangle = 0$ in $\mathcal{O}_{S,s}$ for any $h \in \mathcal{R}(f)_s$ then $\eta = 0$;*
iii) *$\langle\, , \rangle$ establishes an isomorphism*

$$\mathcal{R}(f) \xrightarrow{\sim} \omega(f)^\vee = \underline{Hom}_{\mathcal{O}_S}(\omega(f), \mathcal{O}_S);$$

iv) *for any sections η of $\omega(f)$, h of $\mathcal{R}(f)$, L of $\mathcal{D}(f)$, we have*

$$\langle \eta L, h\rangle = \langle \eta, Lh\rangle.$$

Proof: It is a restatement of Dwork's duality principle for elementary fibrations, in the special case of $\mathcal{E} = \mathcal{O}_{\mathbf{P}^1_S}$, when $R^1\overline{f}_*\mathcal{O}_{\mathbf{P}^1_S} = 0$. □

There is a natural embedding of $\mathcal{O}_S$-modules

$$R^0\overline{f}_*(\hat{\mathcal{O}}_{\overline{X}}) \hookrightarrow R^0\overline{f}_*(\hat{\mathcal{O}}_{\overline{X}}(*Z)) = \mathcal{R}'(f)\ , \tag{D.3.5}$$

that "almost splits" the exact sequence (D.3.3) in the sense that $R^0 f_*(\mathcal{O}_X) \cap R^0\overline{f}_*(\hat{\mathcal{O}}_{\overline{X}}) = R^0\overline{f}_*(\mathcal{O}_{\overline{X}})$.

We define an $\mathcal{O}_S$-linear pairing

$$\begin{aligned} \langle\, , \rangle' : \omega'(f) \times \mathcal{R}'(f) &\longrightarrow \mathcal{O}_S, \\ ((\eta_i)_i, (\xi_i)_i) &\longmapsto \sum_{i=1}^{\infty} \mathrm{Res}_{Z_i} \xi_i \eta_i\ . \end{aligned} \tag{D.3.6}$$

where, for some open subset $U \subset S$ and some $N = 0, 1, \dots,$ η_i (resp. ξ_i), for $i = 1, \dots, r, \infty$, denotes a section of $\hat{\mathcal{J}}^{-N}\omega$ (resp. $\hat{\mathcal{J}}^{-N}$) over $g^{-1}(U) \cap Z_i$. For any section L of $\mathcal{D}_{X/S}$ over $f^{-1}(U)$, we then have

$$\langle (\eta_i L)_i, (\xi_i)_i\rangle' = \langle (\eta_i)_i, (L\xi_i)_i\rangle'\ . \tag{D.3.6.1}$$

Under the pairing $\langle\, , \rangle'$ $\omega(f) \perp \mathcal{L}(f)$. Therefore

Proposition D.3.7 *The pairing* $\langle\, ,\, \rangle$ *of* (D.3.4) *is induced by the pairing* $\langle\, ,\, \rangle'$ *of* (D.3.6), *in the sense that for any section* η *(resp.* ξ*) of* $\omega(f)$ *(resp.* $\mathcal{R}'(f)$*)*

$$\langle \eta, \gamma_-(f)(\xi)\rangle = \langle \eta, \xi\rangle'.$$

In order to follow faithfully Dwork's calculations, we now assume that one of the Z_i, say Z_1, has equation $x = 0$. We will identify $\omega'(f)$ with $\mathcal{R}'(f)$ via

$$\omega'(f) = \mathcal{R}'(f)\frac{dx}{x}\ , \quad \omega(f) = \mathcal{L}(f)\frac{dx}{x}.$$

To be completely explicit we recall Dwork's definition of the (logarithmic) *adjoint* operator L^* of L, with respect to the relative coordinate x, namely:

(D.3.8) $$\left(\sum_i a_i \left(x\frac{d}{dx}\right)^i\right)^* = \sum_i \left(-x\frac{d}{dx}\right)^i a_i$$

characterized by the property[(1)] that for any section $h\frac{dx}{x} = \eta$ of $\omega_{\eta X}$,

(D.3.9) $$\left(f\frac{dx}{x}\right)L = (L^* f)\frac{dx}{x}\ .$$

D.4 Explicit calculations Let the assumptions be as in (D.3); we assume moreover that $S = \operatorname{Spec} A$ is affine, and compute global sections of the $\mathcal{O}_S$-modules introduced above. We have:

$$R^0 f_*(\mathcal{O}_X)(S) = \mathcal{L} = A\left[x, \frac{1}{\prod_i (x-\theta_i)}\right],$$
$$\mathcal{D}(f)(S) = \mathcal{D} = \mathcal{L}[\frac{d}{dx}],$$
$$\mathcal{R}'(f)(S) = \mathcal{R}' = \oplus \mathcal{R}'_i, \text{ with}$$

(D.4.1) $$\mathcal{R}'_i = \begin{cases} A((x-\theta_i)) & \text{if } i = 1, \ldots, r, (\theta_1 = 0), \\ A((\frac{1}{x})) & \text{if } i = \infty. \end{cases}$$

Note that $\mathcal{L}$ is embedded in $\mathcal{R}'$ by the "diagonal" embedding $h \longmapsto (h, h, \ldots, h)$.

---

(1) The classical definition differs from ours in that it uses the basis dx of $\omega_{\eta X}$, so that, classically,

$$\left(\sum_i a_i \left(\frac{d}{dx}\right)^i\right)^* = \sum_i \left(-\frac{d}{dx}\right)^i a_i .$$

$\omega'(f)(S) = \mathcal{R}'(f)(S)dx,$
$\omega(f)(S) = \mathcal{L}dx,$
$R^0\overline{f}_*(\hat{\mathcal{O}}_{\overline{X}})(S) = \oplus_i R^0\overline{f}_*(\hat{\mathcal{O}}_{\overline{X}|\sigma_i(S)})(S)$where

(D.4.2) $$R^0\overline{f}_*(\hat{\mathcal{O}}_{\overline{X}|\sigma_i(S)})(S) = \begin{cases} A[[x-\theta_i]] & \text{if } i = 1,\dots,r, \\ A[[\frac{1}{x}]] & \text{if } i = \infty. \end{cases}$$

To give a useful (non-canonical) description of $\mathcal{R}(f)(S) = \mathcal{R}$, we split (D.3.3) ($\gamma_-(f) =: \gamma_-$), by a map

(D.4.3) $$\mathcal{R} \hookrightarrow \mathcal{R}'$$

taking $\xi \in \mathcal{R}$ to the unique $(\xi')_i \in R^0 f_*(\hat{\mathcal{O}}_{\overline{X}}) \cap \gamma_-^{-1}(\xi)$ such that $\xi_\infty \in \frac{1}{x}A[[\frac{1}{x}]]$. We will then (non-canonically) identify $\mathcal{R}$ with the A-submodule of $\mathcal{R}'$ direct sum $\oplus\mathcal{R}_i$ where

(D.4.4) $$\mathcal{R}_i = \begin{cases} A[[x-\theta_i]] & \text{if } i = 1,\dots,r, \\ \frac{1}{x}A[[\frac{1}{x}]] & \text{if } i = \infty. \end{cases}$$

The map

(D.4.5) $$\gamma_+ : \mathcal{R}' \longrightarrow \mathcal{L}$$

associated to this splitting of (D.3.3) takes $(\xi_i)_i$ to $\sum_i P_i(\xi_i) \in \mathcal{L}$, where $P_i : A((x-\theta_i)) \longrightarrow A[[x-\theta_i]]$, is the usual "principal part" at θ_i, for $i = 1,\dots,r$, while $P_\infty : A((\frac{1}{x})) \longrightarrow \frac{1}{x}A[[\frac{1}{x}]]$ takes $\sum_i a_i x^i$ to $\sum_{i\geq 0} a_i x^i$. This is easily seen because $\gamma_+(\mathcal{R}) = 0$ while $\gamma_{+|\mathcal{L}} = \mathrm{id}_{\mathcal{L}}$. So, as endomorphisms of $\mathcal{R}'$,

(D.4.6) $$\gamma_+ + \gamma_- = \mathrm{id}_{\mathcal{R}'}.$$

So we now have a decomposition

(D.4.7) $$\mathcal{R}' = \mathcal{R} \oplus \mathcal{L}\ ,$$

where the first projection is the map γ_-, which is a morphism of left $\mathcal{D}$-modules, while the second projection is the A-linear map γ_+. On the other hand the (diagonal) embedding $\mathcal{L} \hookrightarrow \mathcal{R}'$ is an embedding of left $\mathcal{D}$-modules. The pairing $\langle\ ,\ \rangle' : (\oplus_i\mathcal{R}'_i)\times(\oplus_i\mathcal{R}'_i) \longrightarrow A$, $((\eta_i)_i, (\xi_i)_i) \longmapsto \sum_i \mathrm{Res}_{\theta_i}\eta_i\xi_i\frac{dx}{x}$ of (D.3.6) then satisfies

(D.4.8) $$\langle (L\eta_i)_i, (\xi_i)_i\rangle' = \langle (\eta_i)_i, (L^*\xi_i)_i\rangle'\ ,$$

for any $(\eta_i)_i, (\xi_i)_i$ in $\mathcal{R}'$.

The subspaces $\mathcal{L}$ and $\mathcal{R}$ of $\mathcal{R}'$ are isotropic for $\langle\ ,\ \rangle'$. Hence $\langle\ ,\ \rangle'$ induces a pairing (that of Proposition D.3.4)

$$\text{(D.4.9)} \qquad \langle\ ,\ \rangle : \mathcal{L} \times \mathcal{R} \longrightarrow A$$

which identifies $\mathcal{R}$ with $Hom_A(\mathcal{L}, A)$. More precisely, for any $(\xi_i)_i$ in $\mathcal{R}'$ and η in $\mathcal{L}$, we have

$$\langle \eta, (\xi_i)_i \rangle' = \langle \eta, \gamma_-((\xi_i)_i) \rangle \ ,$$

and for any $(\xi_i)_i$ in $\mathcal{R}'$ and $(\eta_i)_i$ in $\mathcal{R}$, we have

$$\langle (\xi_i)_i, (\eta_i)_i \rangle' = \langle \gamma_+((\xi_i)_i), (\eta_i)_i \rangle \ .$$

Lemma D.4.10 *Let T be an A-linear endomorphism of $\mathcal{R}'$, stable on $\mathcal{L}$. Let T^* be an A-linear endomorphism of $\mathcal{R}'$ such that*

$$\text{(D.4.10.1)} \qquad \langle T(\eta_i)_i, (\xi_i)_i \rangle' = \langle (\eta_i)_i, T^*(\xi_i)_i \rangle' \ ,$$

for any $(\eta_i)_i$, $(\xi_i)_i$ in $\mathcal{R}'$. Then the (unique) endomorphism of $\mathcal{R}$, adjoint of $T_{|\mathcal{L}}$ under the pairing $\langle\ ,\ \rangle$, is

$$\text{(D.4.10.2)} \qquad \gamma_- \circ T^* \ .$$

We can now make completely explicit the duality of Proposition D.2.8, in the special case of a rational elementary fibration. Let $\underline{\lambda} = (\lambda_1, \ldots, \lambda_d)$ be étale coordinates on S, so that $(x, \lambda_1, \ldots, \lambda_d)$ are étale coordinates on X, and assume we are given an integrable connection ∇ on $\mathcal{E} = \mathcal{O}_X^\mu$

$$\text{(D.4.11)} \qquad \nabla = d_{X/K} - G_{1,\underline{0}} d_{X/K} x - \sum_{i=1}^{d} G_{0,\underline{1}_i} d_{X/K} \lambda_i ,$$

with $G_{1,\underline{0}}, G_{0,\underline{1}_i} \in M_{\mu \times \mu}(\mathcal{L})$, for $i = 1, \ldots, d$. The De Rham complex relative to S is

$$\begin{aligned} & DR(X/S, (\mathcal{E}, \nabla)) \\ \text{(D.4.12)} \qquad & = [\mathcal{O}_X^\mu \xrightarrow{d_{X/S} - G_{1,\underline{0}} d_{X/S} x} (\Omega^1_{X/S})^\mu] \\ & \cong [\mathcal{O}_X^\mu \xrightarrow{\frac{\partial}{\partial x} - G_{1,\underline{0}}} \mathcal{O}_X^\mu] \ . \end{aligned}$$

To be completely clear we specify that the first-order differential operator $\frac{\partial}{\partial x} - G_{1,\underline{0}}$ acts on $\mathcal{O}_X^\mu$ as

$$\begin{pmatrix} y_1 \\ \cdot \\ \cdot \\ y_\mu \end{pmatrix} \longmapsto \begin{pmatrix} \frac{\partial y_1}{\partial x} \\ \cdot \\ \cdot \\ \frac{\partial y_\mu}{\partial x} \end{pmatrix} - G_{1,\underline{0}} \begin{pmatrix} y_1 \\ \cdot \\ \cdot \\ y_\mu \end{pmatrix}.$$

We are interested in the two dual (up to a suitable shift) complexes of A-modules

$$\begin{aligned} &(Rf_* DR(X/S, (\mathcal{E}, \nabla)))(S) = \Gamma(X, DR(X/S, (\mathcal{E}, \nabla))) \\ \text{(D.4.13)} \quad &= [\mathcal{L}^\mu \overset{d_{X/S} - G_{1,\underline{0}} d_{X/S} x}{\longrightarrow} \mathcal{L}^\mu dx] \\ &\cong [\mathcal{L}^\mu \overset{\frac{\partial}{\partial x} - G_{1,\underline{0}}}{\longrightarrow} \mathcal{L}^\mu] \ , \end{aligned}$$

and

$$\begin{aligned} &\Gamma(S, Rf_! DR(X/S, (\mathcal{E}^\vee, \nabla^\vee)))[1] \\ &= [(\overline{f}_*(\hat{\mathcal{O}}_{\overline{X}|Z}(*Z))/\overline{f}_*(\mathcal{O}_{\overline{X}}(*Z)))^\mu \overset{\nabla^\vee}{\longrightarrow} \\ \text{(D.4.14)} \quad &(\overline{f}_*(\hat{\Omega}^1_{\overline{X}/S}(*Z))/\overline{f}_*(\Omega^1_{\overline{X}/S}(*Z)))^\mu](S) \cong [\mathcal{R}^\mu \overset{\gamma_- \circ (\frac{\partial}{\partial x} + G^t_{1,\underline{0}})}{\longrightarrow} \mathcal{R}^\mu] \\ &\cong [(\mathcal{R}'/\mathcal{L})^\mu \overset{\frac{\partial}{\partial x} + G^t_{1,\underline{0}}}{\longrightarrow} (\mathcal{R}'/\mathcal{L})^\mu], \end{aligned}$$

where $G^t_{1,\underline{0}}$ is the transposed matrix of $G_{1,\underline{0}}$ and again $\frac{\partial}{\partial x} + G^t_{1,\underline{0}}$ acts on $\mathcal{R}'^\mu$ as

$$\begin{pmatrix} y_1 \\ \cdot \\ \cdot \\ y_\mu \end{pmatrix} \longmapsto \begin{pmatrix} \frac{\partial y_1}{\partial x} \\ \cdot \\ \cdot \\ \frac{\partial y_\mu}{\partial x} \end{pmatrix} + G^t_{1,,\underline{0}} \begin{pmatrix} y_1 \\ \cdot \\ \cdot \\ y_\mu \end{pmatrix},$$

while $\gamma_- : \mathcal{R}'^\mu \longrightarrow \mathcal{R}^\mu$ operates componentwise.

We deduce the following down-to-earth description of De Rham cohomology groups.

Proposition D.4.15 *Let* $(\mathcal{E}, \nabla)$ *be an object of* $\mathbf{MIC}(X)$, *with* $\mathcal{E} \cong \mathcal{O}_X^\mu$ *and* ∇ *as in* (D.4.11). *Then*

(D.4.15.1)
$$H^0_{DR}(X/S, (\mathcal{E}, \nabla))(S) \cong \mathrm{Ker}_{\mathcal{L}^\mu} \left(\frac{\partial}{\partial x} - G_{1,\underline{0}} \right),$$

(D.4.15.2)
$$H^1_{DR}(X/S, (\mathcal{E}, \nabla))(S) \cong \mathcal{L}^\mu / \left(\frac{\partial}{\partial x} - G_{1,\underline{0}} \right) \mathcal{L}^\mu,$$

(D.4.15.3) $$H^1_{DR,c}(X/S,(\mathcal{E}^\vee,\nabla^\vee))(S) \cong \mathrm{Ker}_{\mathcal{R}^\mu}\gamma_- \circ \left(\frac{\partial}{\partial x} + G^t_{1,\underline{0}}\right),$$

(D.4.15.4) $$\begin{aligned} H^2_{DR,c}(X/S,(\mathcal{E}^\vee,\nabla^\vee))(S) &\cong \mathcal{R}^\mu/\left(\gamma_- \circ \left(\frac{\partial}{\partial} + G^t_{1,\underline{0}}\right)\mathcal{R}^\mu\right) \\ &\cong \mathcal{R}'^\mu/(\mathcal{L}^\mu + \left(\frac{\partial}{\partial x} + G^t_{1,\underline{0}}\right)\mathcal{R}'^\mu). \end{aligned}$$

The A-bilinear pairings of A/K-differential modules

(D.4.15.5) $$\langle\, ,\, \rangle : H^1_{DR}(X/S,(\mathcal{E},\nabla))(S) \times H^1_{DR,c}(X/S,(\mathcal{E}^\vee,\nabla^\vee))(S) \longrightarrow A$$

and

(D.4.15.6) $$\langle\, ,\, \rangle : H^0_{DR}(X/S,(\mathcal{E},\nabla))(S) \times H^2_{DR,c}(X/S,(\mathcal{E}^\vee,\nabla^\vee))(S) \longrightarrow A$$

are induced by the pairing (D.4.9).

Corollary D.4.16 *If $H^1_{DR,c}(X/S,(\mathcal{E}^\vee,\nabla^\vee))(S)$ is of finite type and $H^1_{DR}(X/S,(\mathcal{E},\nabla))(S)$ is projective, then* (D.4.15.5) *and* (D.4.15.6) *are perfect pairings of projective A-modules of finite type.*

Proof: The $\mathcal{O}_S$-module $H^1_{DR,c}(X/S,(\mathcal{E}^\vee,\nabla^\vee))$ is not quasi-coherent in general. Still, if $H^1_{DR,c}(X/S,(\mathcal{E}^\vee,\nabla^\vee))(S)$ is an A-module of finite type, the same holds for $H^1_{DR}(X/S,(\mathcal{E},\nabla))(S)$. Since $H^1_{DR}(X/S,(\mathcal{E},\nabla))$ is quasi-coherent, this implies that it is in fact coherent, and therefore locally free. So, $H^1_{DR,c}(X/S,(\mathcal{E}^\vee,\nabla^\vee))$ is coherent and locally free, too, and (D.4.15.5) is perfect. On the other hand, the local freeness of $H^1_{DR}(X/S,(\mathcal{E},\nabla))(S)$ implies that (D.4.15.6) is left-perfect. But $H^0_{DR}(X/S,(\mathcal{E},\nabla))$ is always coherent by (4.2.1), hence (D.4.15.6) is perfect. □

The A-modules $H^j_{DR}(X/S,(\mathcal{E},\nabla))$ are endowed with the S/K-connection $\aleph^j$ induced by

(D.4.17) $$\aleph\left(\frac{\partial}{\partial\lambda_i}\right) = \frac{\partial}{\partial\lambda_i} - G_{0,\underline{1}_i}\ ,\ i = 1,\ldots,d\ ,$$

on the quotient space (this makes sense because of the integrability of ∇). Similarly, the A-modules $H^j_{DR,c}(X/S,(\mathcal{E}^\vee,\nabla^\vee))$ are endowed with an integrable S/K-connection induced by

(D.4.18) $$\aleph^\vee\left(\frac{\partial}{\partial\lambda_i}\right) = \gamma_- \circ \left(\frac{\partial}{\partial\lambda_i} + G^t_{0,\underline{1}_i}\right),\ i = 1,\ldots,d\ .$$

D.5 Example: Gauss hypergeometric functions. We are here in the situation of the previous section, but we also fix $a, b, c \in K$ and assume

$$A = K\left[\lambda, \frac{1}{\lambda(1-\lambda)}\right],$$

$$S = \operatorname{Spec} A,$$

$$\theta_1 = 0, \quad \theta_2 = 1, \quad \theta_3 = \frac{1}{\lambda}.$$

We prefer to reindex our θ_i's on setting $\theta_i = i$, for $i = 0, 1, \frac{1}{\lambda}, \infty$.

$$\mathcal{L} = A\left[x, \frac{1}{x(1-x)(1-x\lambda)}\right],$$

$$h_{a,b,c}(\lambda, x) = x^b(1-x)^{c-b}(1-x\lambda)^{-a},$$

(in the sense of differential algebra)

$$\mathcal{E} = \mathcal{O}_X, \nabla = \nabla_{a,b,c} = h_{a,b,c}^{-1} \circ d_{X/K} \circ h_{a,b,c}.$$

We also reindex our spaces as

$$\mathcal{R} = \mathcal{R}_0 \oplus \mathcal{R}_1 \oplus \mathcal{R}_{\frac{1}{\lambda}} \oplus \mathcal{R}_\infty,$$

where

$$\mathcal{R}_0 = A[[x]], \mathcal{R}_1 = A[[1-x]], \mathcal{R}_{\frac{1}{\lambda}} = A[[1-x\lambda]]\ , \ \mathcal{R}_\infty = \frac{1}{x}A\left[\left[\frac{1}{x}\right]\right],$$

and similarly for $\mathcal{R}'$. We also set

$$T_0 = x, T_1 = 1-x, T_{\frac{1}{\lambda}} = 1-\lambda x, T_\infty = \frac{1}{x},$$

and

$$t_0 = \lambda, t_1 = \frac{\lambda}{\lambda - 1}, t_{\frac{1}{\lambda}} = \frac{1}{1-\lambda}, t_\infty = \frac{1}{\lambda}.$$

Using the identification

$$\mathcal{L} \longrightarrow \omega(f)$$

$$\eta \longmapsto \eta \frac{dx}{x},$$

the De Rham complex of $(\mathcal{O}_X, \nabla_{a,b,c})$ is identified with the complex

$$[\mathcal{O}_X \overset{L_{a,b,c}}{\longrightarrow} \mathcal{O}_X]$$
$$0 \qquad\qquad 1$$

where

$$L_{a,b,c} = x\frac{\partial}{\partial x} + c - a + \frac{b-c}{1-x} + \frac{a}{1-\lambda x}.$$

Therefore the relative De Rham cohomology groups $H^i_{DR}(X/S, (\mathcal{O}_X, \nabla_{a,b,c}))$, are the associated sheaves to the A-modules obtained as cohomology groups of the complex of A-modules

(D.5.1)
$$\left[A\left[x, \frac{1}{x(1-x)(1-x\lambda)}\right] \overset{L_{a,b,c}}{\longrightarrow} A\left[x, \frac{1}{x(1-x)(1-x\lambda)}\right]\right]$$
$$0 \qquad\qquad\qquad\qquad 1\,.$$

One can show by direct computation [Dw, Chapter 1] that, if

(D.5.2)
$$c-a\,,\quad c-b\,,\quad b\,,\quad a \notin \mathbf{Z}\,,$$

(D.5.3)
$$H^i_{DR}(X/S, (\mathcal{O}_X, \nabla_{a,b,c})) = \begin{cases} 0 & \text{if} i \neq 1, \\ A\,cl(1) \oplus A\,cl(\frac{1}{1-x}) & \text{if} i = 1, \end{cases}$$

where $cl(\eta)$ denotes the class of $\eta \in \mathcal{L}$ modulo $L_{a,b,c}\mathcal{L}$.

The dual theory of $\nabla_{a,b,c}$ involves the calculation of the dual connection

$$\nabla^\vee_{a,b,c} = h_{a,b,c} \circ d_{X/K} \circ h^{-1}_{a,b,c} = \nabla_{-a,-b,-c}$$

on $\mathcal{E}^\vee = \mathcal{O}_X$ (so that $L^*_{a,b,c} = L_{-a,-b,-c}$) and of the corresponding De Rham complex with compact supports, dual of (D.5.1),

(D.5.4)
$$[\mathcal{R} \overset{\gamma_- \circ L_{-a,-b,-c}}{\longrightarrow} \mathcal{R}].$$
$$1 \qquad\qquad 2$$

So,

(D.5.5)
$$H^i_{DR,c}(X/S, (\mathcal{O}_X, \nabla_{-a,-b,-c})) = \begin{cases} 0 & \text{if} i \neq 1, \\ \mathrm{Ker}_{\mathcal{R}}\gamma_- \circ L_{-a,-b,-c} & \text{if} i = 1. \end{cases}$$

The dual A-module $H^1_{DR,c}(X/S, (\mathcal{O}_X, \nabla_{-a,-b,-c}))$ of $H^1_{DR}(X/S, (\mathcal{O}_X, \nabla_{a,b,c}))$, may be explicitly computed in the following way. We examine the equation

(D.5.6) $$\gamma_- \circ L_{-a,-b,-c}\xi = 0\,,$$

for $\xi = (\xi_0, \xi_1, \xi_{\frac{1}{\lambda}}, \xi_\infty) \in \mathcal{R}$. Equation D.5.6 is equivalent to

(D.5.7) $$L_{-a,-b,-c}\xi_j = \gamma_+((L_{-a,-b,-c}\xi_i)_i) = \sum_i P_i(L_{-a,-b,-c}\xi_i) \quad \forall j\,.$$

Since ξ is going to be an element of $\mathcal{R}$, so that each ξ_i is formally holomorphic at i (at ∞, it even has a zero) consideration of the poles of $L_{-a,-b,-c}$ shows that the r.h.s. of (D.5.7) is

(D.5.8) $$\begin{aligned}&(c-b)\frac{\xi_1(1)}{1-x} - a\frac{\xi_{\frac{1}{\lambda}}\left(\frac{1}{\lambda}\right)}{1-x\lambda}\\ &= (b-c)\langle\frac{1}{1-x}, \xi\rangle\frac{1}{1-x} + a\lambda\langle\frac{1}{1-x\lambda}, \xi\rangle\frac{1}{1-x\lambda}.\end{aligned}$$

Equation D.5.7 then permits explicit calculations [Dw, 2.3.5] and provides v-adic growth estimates for the power series ξ_i for any discrete valuation v of the quotient field $K(\lambda)$ of A. We easily obtain:

Proposition D.5.9 **([Dw, 2.3])** *Assume v is of one of the following two types:*

a) *v is a discrete valuation of $K(\lambda)$ trivial on K*
b) *v induces the p-adic valuation on $\mathbf{Q}$, K is algebraic over $\mathbf{Q}$, $v(a)$, $v(b)$, $v(c) \geq 0$, $v(\sum_j a_j\lambda^j) = \max v(a_j) + jv(\lambda)$ for any polynomial $\sum_j a_j\lambda^j \in K[\lambda]$.*

Then, let $\xi \in H^1_{DR,c}(X/S, (\mathcal{O}_X, \nabla_{-a,-b,-c}))$; for any $i = 0, 1, \frac{1}{\lambda}, \infty$, ξ_i is a power series in T_i with coefficients in $K(\lambda)$ converging for $v(T_i) > \max(0, -v(t_i))$.

From a result of this type, one could deduce comparison theorems of algebraic versus p-adic analytic De Rham cohomology.

4 Complex and p-adic comparison theorems

Introduction

In this chapter, we show how to adapt Artin's strategy of proof of his comparison theorem for étale cohomology [SGA IV] in the De Rham context. The result is a particularly simple proof of the Grothendieck-Deligne comparison theorem (algebraic versus complex-analytic De Rham cohomology with regular coefficients [G1], [De]). As a corollary, we obtain an elementary proof of Riemann's existence theorem for coverings, in higher dimensions.

Not only does this proof not rely on resolution of singularities, but it also does not make use of moderate growth conditions, nor even of the notion of monodromy. This enables us to translate almost literally our argument into the p-adic setting, which yields a simple proof of the Kiehl-Baldassarri theorem (algebraic versus rigid-analytic De Rham cohomology with regular coefficients [Ki2], [B3]). We then use our study (II) of irregularity in several variables to prove that, in the p-adic setting, the comparison theorem extends to *irregular* connections (with the usual proviso on the field of definition).

Let us now describe in more detail the content of each section.

In the first section, we review complex-analytic and rigid-analytic connections and De Rham cohomology.

Section 2 generalizes and formalizes, in the abstract setting of differential algebra (2.1,2.4), the basic structure of some "comparison theorems" for De Rham cohomologies in one variable which one may find scattered in the literature (especially the literature on p-adic analysis).

Section 3, which relies on the technique of dévissage of (III), contains our proof of the *Grothendieck-Deligne comparison theorem* between algebraic and complex-analytic De Rham cohomologies with regular coefficients (3.1). For a discussion of the original proofs and of Mebkhout's contribution [Me3], we refer to the general introduction of this book.

In all these approaches (as well as their p-adic avatars [Ki2], [B2], [B3]), the central point remains the comparison between two analytic De Rham complexes, one meromorphic and one with essential singularities. This point does not arise (at least explicitly) in our proof. Indeed, by dévissage, we are able to reduce the question to a global comparison problem in one variable(1), so simple that it may be settled by the formal criteria of Section 2, together with Frobenius' convergence theorem of formal solutions of fuchsian differential equations.

---

(1)Let us mention that Mebkhout's strategy also uses an argument of projection of relative dimension 1, but in a local, analytic setting.

This proof is therefore analytical in nature (free from monodromy considerations, *cf.* (3.5)), and may be transferred into the p-adic setting. It also avoids any result of "GAGA type".

In Section 4, we prove in this way the *rigid-analytic* analogue of the Grothendieck-Deligne comparison theorem, under the assumption that the group generated by the exponents contains no p-adic Liouville number (4.1).

In the subsequent sections, we consider the case of *irregular* connections (in the rigid-analytic setting). Our main theorem is the following statement (*cf.* (6.1)), which was conjectured by the second author [B2]:

Let X be a smooth $\overline{\mathbf{Q}}$-variety and let $(\mathcal{E}, \nabla)$ be a coherent $\mathcal{O}_X$-module with integrable connection. Then for any $i \geq 0$,

$$(H^i_{DR}(X, (\mathcal{E}, \nabla))) \otimes_{\overline{\mathbf{Q}}} \mathbf{C}_p \cong H^i_{DR}(X^{an}_{\mathbf{C}_p}, (\mathcal{E}^{an}_{\mathbf{C}_p}, \nabla^{an}_{\mathbf{C}_p})).$$

We first treat the relative-dimension-one case (5.2) (a generalization of [B2] and [C]); here again, the statement is reduced to Clark's non-archimedean convergence theorem of formal solutions of possibly irregular differential equations, but the method of reduction is different from those used in [B2] or [C]. The strategy of proof of the theorem in arbitrary dimension uses refinements of our dévissages and our previous study of irregularity in several variables (II).

As a corollary, we obtain that over a smooth algebraic variety defined over $\overline{\mathbf{Q}} \subset \mathbf{C}_p$, the rigid GAGA functor

$$(\text{algebraic connections}) \longrightarrow (\text{analytic connections})$$

is fully faithful.

We close the chapter with a p-adic version of the Turrittin theorem in the relative case.

§1 Review of analytic connections and De Rham cohomology

1.1 Complex-analytic connections

1.1.1 Let $\mathcal{X}$ be a smooth complex analytic variety. The notion of an integrable connection on a $\mathcal{O}_{\mathcal{X}}$-module $\mathcal{E}$ is defined as in the algebraic situation (III.2.1.1) (without finiteness condition). It is equivalent to the notion of $\mathcal{D}_{\mathcal{X}}$-module. We denote by $\mathbf{MIC}(\mathcal{X})$ the category of $\mathcal{O}_{\mathcal{X}}$-modules with integrable connection. If $\mathcal{E}$ is coherent, it is then automatically locally free.

1.1.2 Inverse images are constructed as in the algebraic case. For a smooth morphism of smooth complex analytic varieties $f : \mathcal{X} \longrightarrow \mathcal{S}$, one constructs the direct image and the higher direct images $R^q_{DR} f_*(\mathcal{E}, \nabla) \in Ob\, \mathbf{MIC}(\mathcal{S})$ of any $(\mathcal{E}, \nabla) \in Ob\, \mathbf{MIC}(\mathcal{X})$, as in the algebraic situation (III.2.4). The formation of $R^q_{DR} f_*$ is compatible with localization on $\mathcal{S}$.

1.1.3 One associates functorially to any (smooth) algebraic **C**-variety X a (smooth) complex analytic variety X^{an}. The GAGA functor

$$\mathbf{MIC}(X) \longrightarrow \mathbf{MIC}(X^{an}),$$
$$(\mathcal{E}, \nabla) \longmapsto (\mathcal{E}^{an}, \nabla^{an})$$

commutes with inverse images (essentially because $\mathcal{O}_{X^{an}}$ is flat over $\mathcal{O}_X$), but not with direct images in general. In the sequel, we limit ourselves to the case of analytic connections on algebraic complex analytic varieties (*i.e.* objects of $\mathbf{MIC}(X^{an})$ for smooth algebraic varieties X over **C**).

1.1.4 The discussion of Čech spectral sequences in (III.2.5) carries over for analytic connections, except that "quasi-coherent" should be replaced by "coherent" (and using the fact that affine varieties are Stein, and Cartan's theorem B).

1.1.5 Flat base change (III.2.6) is more delicate in the analytic context: the interpretation of $(u^{\sharp\, an})^*$ as $(f^{\sharp\, an})^{-1}\mathcal{O}_{S^{\sharp\, an}} \otimes_{(u^{\sharp\, an})^{-1}(f^{an})^{-1}\mathcal{O}_{S^{an}}} (u^{\sharp\, an})^{-1}$ can only hold for a finite base change morphism u. For finite flat u, the argument of (III.2.6) (flat base change for ${}_I E_1^{p,q}$, using Čech complexes for affine covers as usual) carries over, and shows in particular that the formation of $R^q_{DR} f^{an}_*$ is compatible with étale localization (by factorizing an étale map as a composition of open immersions and finite flat morphisms[(1)]).

1.1.6 The computation (III.2.7) $R^j_{DR} f^{an}_*(\mathcal{E}, \nabla) \cong H^j(f^{an}_* DR_{X^{an}/S^{an}}(\mathcal{E}, \nabla))$ for affine f (and vanishing for $j > d$), also holds. (However, for affine (or Stein) S, the computation of that cohomology sheaf in terms of the j-th cohomology module of the global relative De Rham complex seems to require the *a priori* knowledge of its coherence; using Cartan's theorem B, one then shows that the presheaves of cohomology of $f^{an}_* DR_{X^{an}/S^{an}}(\mathcal{E}, \nabla)$ are sheaves.)

1.1.7 The first lemma of dévissage (III.3.2) also holds for properties $\mathcal{P}$ of modules with integrable connection on algebraic complex-analytic (with the same proof), if one replaces "quasi-coherent" by "coherent". For instance, let $\mathcal{P}$ = (analytic) coherence, *i.e.* $(\mathcal{E}, \nabla)$ on X^{an} satisfies $\mathcal{P}$ if and only if $\mathcal{E}$ is $\mathcal{O}_{X^{an}}$-coherent. This property is clearly strongly exact and local for the étale topology. Moreover, since any rational elementary fibration $f' : X' \longrightarrow S'$ is (locally on S') topologically trivial for the classical topology (III.1.2), and since for any $\zeta \in S'^{an}$ and any coherent $(\mathcal{E}', \nabla') \in Ob\, \mathbf{MIC}(X'^{an})$, $H^j(X'_\zeta, \mathcal{E}'^{\nabla'}_{|X'_\zeta})$ is finite-dimensional ($j = 0, 1$), it follows that $R^j_{DR} f'^{an}_*(\mathcal{E}', \nabla') \simeq R^j f'^{an}_*(\mathcal{E}'^{\nabla'_{|Der_{cont} X^{an}/S^{an}}}) \otimes_{\mathbf{C}} \mathcal{O}_{S'^{an}}$ is $\mathcal{O}_{S'^{an}}$-coherent [De, I, 2.28].

(1) A singular variety may appear in such a factorization. This does not make trouble since the $\mathcal{O}_S$-module $\mathbf{R}^q f_* DR_{X/S}(\mathcal{E}, \nabla)$ is well defined even if S is singular, provided f is smooth.

One concludes that *for any smooth morphism $f : X \longrightarrow S$ of smooth complex algebraic varieties of pure relative dimension d, for any $j \geq 0$, and any coherent $(\mathcal{E}, \nabla) \in Ob\ \mathbf{MIC}(X^{an})$, $R^j_{DR} f^{an}_*(\mathcal{E}, \nabla)_{|U^{an}}$ is coherent, with $U = A_j(f)$ if $j \leq d + \dim X$, and $U = S$ otherwise.*

1.1.8 However, we do not know whether the second lemma of dévissage (III.3.3) holds in the analytic category. While trying to adapt the given algebraic proof (or its mentioned alternative), one is confronted with the delicate question of comparing a De Rham complex with essential singularities along a divisor, with a meromorphic one.

1.1.9 Just as for the finiteness question (*cf.* (1.1.7)), the analytic situation shows a simpler behaviour than the algebraic situation with respect to base change, namely:

Let $f : X \longrightarrow S$ be a smooth morphism of smooth complex algebraic varieties, and let $(\mathcal{E}, \nabla)$ be a coherent object of $\mathbf{MIC}(X^{an})$. Then, for any smooth complex variety $S^\sharp$, any morphism $u : S^\sharp \longrightarrow S$, and any $i \geq 0$, the restriction to $(u^{-1}A_i(f))^{an}$ of the base change morphism

$$u^{an*} R^i_{DR} f^{an}_*(\mathcal{E}, \nabla) \longrightarrow R^i_{DR} f^{\sharp an}_*(u^{\sharp an*}(\mathcal{E}, \nabla))$$

is an isomorphism in $\mathbf{MIC}((u^{-1}A_i(f))^{an})$.

Indeed, replacing S by affine étale neighborhoods of the connected components of $A_i(f)$, we may assume that X admits a finite open affine cover $\{U_\alpha\}$ with the properties listed in definition (III.1.8). There is a morphism of Čech spectral sequences in $\mathbf{MIC}(S^{\sharp an})$ analogous to the one considered in (III.5.3), which allows us to reduce ourselves to the case when f is a tower of (coordinatized) elementary fibrations. The result is then an easy consequence of the topological local triviality of the fibration f, which implies that

$$u^{an*} R^i f^{an}_* \mathcal{E}^{\nabla_{|\underline{Der}_{cont} X^{an}/S^{an}}} \xrightarrow{\sim} R^i f^{\sharp an}_* \mathcal{E}^{\nabla^\sharp_{|\underline{Der}_{cont} X^{an}/S^{an}}},$$

is an isomorphism.

1.2 Rigid analytic connections

1.2.1 Similar comments apply to the rigid-analytic situation, over an algebraically closed field K of characteristic 0, complete with respect to a non-archimedean absolute value $|\ \ |$ (*e.g.* $K = \mathbf{C}_p$) ([Be1], [BGR]). Notions and comments (1.1.1) to (1.1.4) may be repeated in this context. One uses the fact that affine covers of X give rise to admissible covers of X^{an}, that affine varieties are quasi-Stein, and Kiehl's analogue of theorem B [Ki1].

1.2.2 The discussion (III.2.5) also carries over if $\{U_\alpha\}$ is chosen to be an admissible affinoid cover of X^{an}. This remark allows us to prove finite flat base change (using the first spectral sequence of (III.2.5.1) for such a cover, and flat base change for $f_{\underline{\alpha}*}(\Omega^j_{U_{\underline{\alpha}}}/S)$ with affinoid $f_{\underline{\alpha}}$). Therefore, the formation of $R^i_{DR}f^{an}_*$ is again compatible with étale localization. The previous discussion of $R^i_{DR}f^{an}_*$ for affine f carries over in the rigid-analytic context (on replacing "Stein" by "quasi-Stein", and Stein covers by admissible affinoid covers), but the question of coherence of $R^i_{DR}f^{an}_*(\mathcal{E}, \nabla)$ is more delicate.

In fact, since Dwork's early studies, it is well-known that one cannot expect generic finiteness of $R^1_{DR}f^{an}_*(\mathcal{E}^{an}, \nabla^{an})$ in general, due to "Liouville phenomena". As we shall see later, in the case of an elementary fibration, the problem disappears when *the roots* α_{ij} *of the indicial polynomials* ϕ_i *are non-Liouville* (in the sense that the series $\sum_{\substack{n\geq 0\\ n\neq -\alpha_{ij}}} \frac{x^n}{\alpha_{ij}+n}$ converges).

§2 Abstract comparison criteria

Let (C, ∂) be a differential $\mathbf{Q}$-algebra, A, B be differential sub-algebras, such that $A^\partial = C^\partial$ (equality of rings of constants), A is faithfully flat over A^∂, and $\partial_{|A}$ is surjective onto A.

Proposition 2.1 Let E be an $(A \cap B)[\partial]$-module, projective of finite rank μ as a module over $A \cap B$. Assume that E is solvable in A, i.e.

$$(E \otimes A)^\partial \otimes_{A^\partial} A \xrightarrow{\sim} E \otimes A.$$

Then $\varphi_0 : \mathrm{Ker}_E\partial \longrightarrow \mathrm{Ker}_{E\otimes B}\partial$ is an isomorphism and $\varphi_1 : \mathrm{Coker}_E\partial \longrightarrow \mathrm{Coker}_{E\otimes B}\partial$ is injective.

Proof: (i) Injectivity of φ_0. This follows trivially from the fact that, since E is flat over $A \cap B$, E injects into $E \otimes B$. (Similarly, we have a diagram of inclusions

$$\begin{array}{ccccc} & & E\otimes A & & \\ & \nearrow & & \searrow & \\ E & & & & E\otimes C; \\ & \searrow & & \nearrow & \\ & & E\otimes B & & \end{array}$$

since E is locally free $(A \cap B)$-module, $(E \otimes A) \cap (E \otimes B) = E$).

(ii) Injectivity of φ_1. Since E is solvable in A, which is faithfully flat over A^∂, the A^∂-module $(E \otimes A)^\partial$ is projective (of rank μ), and, locally with respect to the Zariski topology on $\mathrm{Spec}A^\partial$, the $A[\partial]$-module $E \otimes A$ is a sum of copies of A. Since $A^\partial = C^\partial$, this implies in turn. □

Lemma 2.2 *Under the previous assumptions,* $(E \otimes A)^\partial = (E \otimes C)^\partial$.

In order to complete the proof of (ii), we need the following.

Lemma 2.3 *Under the previous assumptions,* $\partial : E \otimes A \longrightarrow E \otimes A$ *is surjective.*

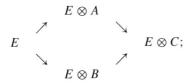

Proof: By assumption, $\partial : A \longrightarrow A$ is surjective. Since Zariski-locally on A^∂, $E \otimes A \cong A^\mu$ as $A[\partial]$-modules, it follows that $\partial : E \otimes A \longrightarrow E \otimes A$ is Zariski-locally (on A^∂) surjective, hence surjective. □

We now deduce the injectivity of φ_1 from Lemma 2.3. Let $e \in E \otimes B$ be such that $\partial e \in E$. By Lemma 2.3, there exists $e' \in E \otimes A$ satisfying $\partial e' = \partial e$. We have $e - e' \in (E \otimes C)^\partial = (E \otimes A)^\partial$, by Lemma 2.2. Hence $e \in (E \otimes A) \cap (E \otimes B) = E$.

(iii) Surjectivity of φ_0. It is a special case of the last argument, for $\partial e = 0$.

Proposition 2.4 *In addition to the previous assumptions, we assume that C (resp. A) is a product of differential* **Q***-algebras without zero divisors C_i (resp. $A_i \subset C_i$), $i = 0, 1, \ldots, r$ containing a differential subalgebra C'_i (resp. $A'_i := A_i \cap C'_i$). We also assume the existence for each i of a differential A'_i-subalgebra T_i of C_i subject to the following conditions:*

a) *$\partial : C_i \longrightarrow C_i$ is surjective and C_i is faithfully flat over C_i^∂;*
b) *the composed homomorphism of differential algebras $B \longrightarrow C \overset{\mathrm{pr}_i}{\longrightarrow} C_i$ factors through an injection $B \hookrightarrow C'_i$;*
c) *there is a decomposition $C_i = T_i \otimes_{A'_i} C_i^{log}$, a tensor product of differential algebras, where $C_i^{log} := \oplus_{j \geq 0} C'_i \partial_i^{-j} 1$, $\partial_i = u_i \partial$ for some unit u_i in A'_i, and $(1 = \partial_i^0 1, \partial_i^{-1} 1, \ldots)$ is a sequence of elements of A_i such that $\partial_i(\partial_i^{-j} 1) = \partial_i^{-j+1} 1$;*
d) *$C'_0 \subset A'_0 + \mathrm{pr}_0 B$, and, for $i > 0$, $C'_i \subset A'_i + \bigcap_{j<i} \mathrm{pr}_i(B \cap A_j)$ (we use b) to make sense of this intersection).*

Then $\varphi_1 : \mathrm{Coker}_E \partial \longrightarrow \mathrm{Coker}_{E \otimes B} \partial$ is surjective.

Proof: Since C_i is a domain, $A_i^{log} := A_i \cap C_i^{log} = \oplus_{j \geq 0} A'_i \partial_i^{-j} 1$; therefore $A_i = T_i \otimes_{A'_i} A_i^{log}$.

We remark that (on applying pr_i) E is solvable in A_i, which is faithfully flat over A_i^∂, and $A_i^\partial = C_i^\partial$. A fortiori, E is solvable in C_i. On the other hand, b) implies that $\mathrm{pr}_{i|B}$ induces an injection $A \cap B \hookrightarrow A'_i$. We have a diagram of inclusions

$$\begin{array}{ccccc} & & E \otimes A'_i & \longrightarrow & E \otimes A_i \\ & \nearrow & & & \\ E & & \downarrow & & \downarrow \\ & \searrow & & & \\ & & E \otimes C'_i & \longrightarrow & E \otimes C_i \end{array}$$

and $(E \otimes A_i) \cap (E \otimes C'_i) = E \otimes A'_i$ (see the argument in (2.1.i)). Moreover $B \cap A_i = B \cap A'_i$, which we shall view either as a subring of B or of C_i.

We also note that $E_{T_i} := E \otimes T_i$ is a $T_i[\partial]$-module projective of rank μ over T_i. The results (2.1), (2.2), (2.3) above apply on replacing E by E_{T_i} and the differential algebras C, A, B by C_i, A_i and T_i, respectively.

Let us fix an element $e = e_{-1}$ of $E \otimes B$. We look for an element $f \in E$ and an element $e' \in E \otimes B$ such that $e = f + \partial e'$. In view of this, we shall construct recursively, for $i = 0, \dots, r$, an element $e'_i \in E \otimes (\bigcap_{j<i} B \cap A_j)$, such that $e_i := e_{i-1} - \partial e'_i$ belongs to $E \otimes (\bigcap_{j\leq i} B \cap A_j)$. It will then suffice to take $e' := \sum_{i=0}^{r} e'_i$, and $f := e_r$ (which belongs to $E \otimes (\bigcap_{j=0}^{r} B \cap A_j) = E$).

Let us start with $e_{i-1} \in E \otimes (\bigcap_{j<i} B \cap A_j) \subset E \otimes C'_i$ (if $i = 0$ $e_{-1} \in E \otimes B \subset E \otimes C'_i$). We deduce the existence of an element $f_i \in E \otimes C_i = E_{T_i} \otimes_{A'_i} C_i^{log}$ such that $e_{i-1} = \partial f_i$. According to c), we may write $f_i = \sum_{j=0}^{N} f_{i,j} \partial_i^{-j} 1$, with $f_{i,j} \in E_{T_i} \otimes_{A'_i} C'_i$. Equation $e_{i-1} = \partial f_i$ decomposes into the following set of equations

$$\begin{aligned} \partial f_{i,N} &= 0, \\ \partial f_{i,N-1} &= -u_i^{-1} f_{i,N}, \\ &. \\ &. \\ \partial f_{i,1} &= -u_i^{-1} f_{i,2}, \\ \partial f_{i,0} &= -u_i^{-1} f_{i,1} + e_{i-1} \end{aligned}$$

from which we deduce, step by step, using Lemmata 2.2 and 2.3 for E_{T_i}, that $f_{i,N} \in (E_{T_i} \otimes_{A'_i} C'_i)^\partial = (E_{T_i})^\partial$, $f_{i,N-1}, \dots, f_{i,1} \in E_{T_i}$. According to condition d), $f_{i,0} = e'_i + f'_i$, with $e'_i \in E \otimes (\bigcap_{j<i} B \cap A_j)$, $f'_i \in E \otimes A'_i$. One then observes that the element $e_i := e_{i-1} - \partial e'_i$ of $E \otimes (\bigcap_{j<i} B \cap A_j)$ may also be written $u_i^{-1} f_{i,1} + \partial f'_i$, an element of E_{T_i}. Therefore $e_i \in E \otimes (\bigcap_{j\leq i} B \cap A_j)$. □

§3 Comparison theorem for algebraic vs. complex-analytic cohomology

Theorem 3.1 **([De, II, 6.13]).** *Let i be a non-negative integer, and let $f : X \longrightarrow S$ be a smooth morphism of smooth complex algebraic varieties, with $A_i(f) = S$ (cf. Definition* III.1.8). *Let $(\mathcal{E}, \nabla)$ be a coherent $\mathcal{O}_X$-module endowed with an integrable* regular *connection. Then the canonical morphism in* $\mathbf{MIC}(S^{an})$:

$$\varphi_i : (R^i_{DR} f_*(\mathcal{E}, \nabla))^{an} \longrightarrow (R^i_{DR} f^{an}_*(\mathcal{E}^{an}, \nabla^{an}))$$

is an isomorphism.

In the special case $S = \mathrm{Spec}\,\mathbf{C}$ and $(\mathcal{E}, \nabla) = (\mathcal{O}_X, d_{X/S})$, this gives (taking into account the analytic Poincaré lemma):

Corollary 3.2 **([G1]).** *For any smooth complex algebraic variety X, and any $i \geq 0$, $H^i_{DR}(X) \cong H^i_{DR}(X^{an}) \cong H^i(X(\mathbf{C}), \mathbf{C})$.*

3.3 Reduction to the case of a rational elementary fibration

3.3.1 If f is of relative dimension 0, the statement amounts to the fact that for étale finite f, the canonical morphism $(f_*\mathcal{E})^{an} \longrightarrow f_*^{an}\mathcal{E}^{an}$ is an isomorphism, which is an elementary result of GAGA type (one may reduce, as in (III.3.1.1), to the case when f is a Galois covering, and then remark that $(f^*f_*\mathcal{E})^{an} \xrightarrow{\sim} f^{an*}f_*^{an}\mathcal{E}^{an}$).

3.3.2 As in (III.3.2.1), one then reduces to the case when f has pure relative dimension $d \geq 1$, $S = A_i(f)$ is affine connected, and $0 \leq i \leq d + \dim X$. Moreover, since the statement is local on S^{an}, and since the formation of algebraic or analytic De Rham cohomology commutes with étale localization on S ((III.2.3), (1.1.5)), we may replace S, as in (III.3.2.2), by an affine connected étale neighborhood S' such that $X_{S'}$ admits a finite open cover $\{U_\alpha\}$ with the properties listed in definition (III.1.8).

There is a natural morphism of Čech spectral sequences in $\mathbf{MIC}(S^{an})$,

$$\begin{array}{ccc} \bigoplus_{\alpha_0<\cdots<\alpha_p}(R^{i-p}_{DR}f_{\underline{\alpha}*}(\mathcal{E},\nabla)_{|U_{\underline{\alpha}}})^{an} & \Rightarrow & (R^i_{DR}f_*(\mathcal{E},\nabla))^{an} \\ \downarrow \varphi_{i-p} & & \downarrow \varphi_i \\ \bigoplus_{\alpha_0<\cdots<\alpha_p}R^{i-p}_{DR}f^{an}_{\underline{\alpha}*}(\mathcal{E}^{an},\nabla^{an})_{|U^{an}_{\underline{\alpha}}} & \Rightarrow & R^i_{DR}f^{an}_*(\mathcal{E}^{an},\nabla^{an}). \end{array}$$

We argue by induction on i: since $A_{i-|\underline{\alpha}|+1}(f_{\underline{\alpha}}) = A_i(f) = S$, φ_{i-p} is an isomorphism for $p > 0$; hence we may assume that f itself is a tower of coordinatized elementary fibrations.

3.3.3 For f a tower of coordinatized elementary fibrations, we have a natural morphism of Leray spectral sequences, by which we are reduced to proving the comparison theorem for each level of the tower. Therefore we may assume that f itself is a coordinatized elementary fibration, taking into account the fact that regularity is preserved under direct images by coordinatized elementary fibrations (III.7). In fact, since we are free to replace S by an étale covering, we may even assume that $f = f' \circ \pi$, where f' is a rational elementary fibration and π is an étale covering. We then have $R^i_{DR}f_*(\mathcal{E},\nabla) = R^i_{DR}f'_* \circ \pi_*(\mathcal{E},\nabla)$ (resp. $R^i_{DR}f^{an}_*(\mathcal{E}^{an},\nabla^{an}) = R^i_{DR}f'^{an}_* \circ \pi^{an}_*(\mathcal{E}^{an},\nabla^{an})$), and we are reduced to proving the comparison theorem for a rational elementary fibration, and $i = 0, 1$.

At this point the proof splits into two alternative arguments.

3.4 First way: reduction to an ordinary linear differential system. We use the fact that both the source and the target of φ_i are locally free $\mathcal{O}_{S^{an}}$-modules of finite rank ((III.6) and (1.1.7), resp.). Therefore it suffices to check that the fibre $(\varphi_i)_s$ is an isomorphism for any $s \in S(\mathbf{C})$. By base change ((III.5.5), (1.1.9)), we are finally reduced to the case of an ordinary linear (regular) differential system, *i.e.* $S = \mathrm{Spec}\mathbf{C}$, $X = \mathbf{A}^1 \setminus \{\theta_1, \ldots, \theta_r\}$, and it is enough to deal with global sections. Hence, we have reduced Theorem 3.1 to the following special case.

Proposition 3.4.1 *Let $X = \mathrm{Spec}\mathbf{C}[x, \frac{1}{\prod_{i=1}^{r}(x-\theta_i)}]$, and let (E, ∇) be a projective $\mathcal{O}(X)$-module of finite rank μ, with a regular connection. Then*

$$\varphi_0 : \mathrm{Ker}_E \nabla \left(\frac{d}{dx}\right) \longrightarrow \mathrm{Ker}_{E \otimes \mathcal{O}(X^{an})} \nabla \left(\frac{d}{dx}\right)$$

and

$$\varphi_1 : \mathrm{Coker}_E \nabla \left(\frac{d}{dx}\right) \longrightarrow \mathrm{Coker}_{E \otimes \mathcal{O}(X^{an})} \nabla \left(\frac{d}{dx}\right)$$

are isomorphisms.

Notice that, since $\mathbf{C}[x, \frac{1}{\prod_{i=1}^{r}(x-\theta_i)}]$ is a principal Dedekind ring, the projective module E is necessarily free. A proof of the proposition can then be found in [Ad, Theorem 2 and Section 5, Remark 2]. We prefer to deduce Proposition 3.4.1 from the abstract comparison criteria of Section 2.

3.4.2 We set

$X = \mathrm{Spec}\mathbf{C}[x, \frac{1}{\prod_{i=1}^{r}(x-\theta_i)}]$,
$B := \mathcal{O}(X^{an})$,

and for $i = 1, \ldots, r$,

$C_i' := \bigcup_{\epsilon>0} \mathcal{O}(D(\theta_i, \epsilon) \setminus \{\theta_i\})$,
$\mathcal{M}_{\theta_i}$:= the field of germs of meromorphic functions at θ_i,
$A_i := \mathcal{M}_{\theta_i}[\log(x - \theta_i), (x - \theta_i)^{\alpha}]_{\alpha \in \mathbf{C}}$ (with the usual identification

$$(x - \theta_i)^{\alpha}(x - \theta_i)^{\beta} = (x - \theta_i)^{\alpha+\beta}),$$

$C_i = A_i C_i'$.

We adopt similar definitions for $i = 0$, upon replacing θ_i by ∞, and set $C = \prod_{i=0}^{r} C_i$, $A = \prod_{i=0}^{r} A_i$, $\partial = \frac{d}{dx}$. We have $A \cap B = (\prod_{i=0}^{r} \mathcal{M}_{\theta_i}) \cap B = \mathcal{O}(X) = \mathbf{C}[x, \frac{1}{\prod_{i=1}^{r}(x-\theta_i)}]$ (diagonally embedded in A), $A_i^{\partial} = C_i^{\partial} = \mathbf{C}$, $A_i' := A_i \cap C_i' = \mathcal{M}_{\theta_i}$.

3.4.3 We now exploit the *regularity* of the $(A \cap B)[\partial]$-module E in the guise that E is *solvable in* A, or, which amounts to the same, in each A_i (Fuchs-Frobenius, the point being that formal solution series of a differential operator at a regular singular point, are convergent). It remains to check that all assumptions in Section 2 are met. Both A and C are clearly faithfully flat over $A^{\partial} = C^{\partial} \cong \mathbf{C}^{r+1}$. Using the formula

$$x^{\alpha-1}\frac{\log^k x}{k!} = \partial\left(\frac{x^{\alpha}}{\alpha} \sum_{i+j=k} \left(-\frac{1}{\alpha}\right)^i \frac{\log^j x}{j!}\right),$$

for $\alpha \neq 0$, and

$$x^{-1}\frac{\log^k x}{k!} = \partial \frac{\log^{k+1} x}{(k+1)!} ,$$

one sees that ∂ acts surjectively both on A and on C. This proves the requirements for (2.1), and condition a) of (2.4) in addition. Condition b) is clear (by analytic continuation). Condition c) holds for $i \geq 1$ if we set $u_i = x - \theta_i$, $\partial_i^{-j} 1 = \frac{\log^j (x-\theta_i)}{j!}$, $T_i = \mathcal{M}_{\theta_i}[(x-\theta_i)^\alpha]_{\alpha \in \mathbf{C}}$. For $i = 0$, we adopt similar definitions, replacing θ_i by ∞.

Finally, d) holds in a stronger form. For any $i \in \{0, \dots, r\}$, $C_i' = A_i' + \bigcap_{j \neq i}(B \cap A_j)$, which reflects the decomposition of the Laurent expansion of an element of C_i', into its positive and negative parts, respectively. This achieves the proof of (3.4.1) and of the comparison theorem.

3.5 Second way: dealing with the relative situation. The proofs (1.1.7), (1.1.9) of coherence and base change for $R^1_{DR} f_*^{an}(\mathcal{E}^{an}, \nabla^{an})$, which are used in (3.4), rely on topological arguments (monodromy). We can avoid such arguments by handling directly the case of a rational elementary fibration in the following way.

3.5.1 For $\mathcal{S}$ compact in $S(\mathbf{C})$ and $\epsilon > 0$, we introduce the compact tube in $S^{an} \times (\mathbf{A}^1)^{an}$

$$T_{\mathcal{S},i,\epsilon} = \{(s,x) \ , \ s \in \mathcal{S} \ , \ |x - \theta_i(s)| \leq \epsilon \ \} ,$$

and

$$T^*_{\mathcal{S},i,\epsilon} = \{(s,x) \ , \ s \in \mathcal{S} \ , \ 0 < |x - \theta_i(s)| \leq \epsilon \ \} .$$

Let Δ be a $\mathbf{Q}$-subspace of $\mathbf{C}$ containing $\mathbf{Q}$ (in the sequel, Δ will be the $\mathbf{Q}$-space generated by 1 and the exponents of the canonical extension $\tilde{\mathcal{E}}$ of $\mathcal{E}$, taking into account theorem (III.8.3) in the course of dévissage).

We define

$C'_{\mathcal{S},i} = \bigcup_{\epsilon>0} \mathcal{O}(T^*_{\mathcal{S},i,\epsilon})$,
$A_{\mathcal{S},i} = \bigcup_{\epsilon>0} \mathcal{O}(T_{\mathcal{S},i,\epsilon})[(x-\theta_i)^\alpha, \log(x-\theta_i)]_{\alpha \in \Delta}$ (with the usual identification

$$(x-\theta_i)^\alpha (x-\theta_i)^\beta = (x-\theta_i)^{\alpha+\beta} \ ),$$

$C_{\mathcal{S},i} = C'_{\mathcal{S},i} A_{\mathcal{S},i}$,
$A'_{\mathcal{S},i} := C'_{\mathcal{S},i} \cap A_{\mathcal{S},i} = \bigcup_{\epsilon>0} \mathcal{O}(T_{\mathcal{S},i,\epsilon})[\frac{1}{x-\theta_i}]$.

We adopt similar definitions for $i = 0$, upon replacing θ_i by ∞, and set

$$C_{\mathcal{S}} = \prod_{i=0}^r C_{\mathcal{S},i} \ , \ A_{\mathcal{S}} = \prod_{i=0}^r A_{\mathcal{S},i} \ , \ \partial = \frac{d}{dx} .$$

Finally, we introduce the inverse image $\mathcal{X}$ of $\mathcal{S}$ in X^{an}, and set $B_{\mathcal{S}} = \mathcal{O}(\mathcal{X})$, diagonally embedded into $C_{\mathcal{S}}$.

One then checks as in (3.4.3) that $A_{\mathcal{S}}$ is faithfully flat over $A^{\partial}_{\mathcal{S}} = C^{\partial}_{\mathcal{S}} = \mathcal{O}(\mathcal{S})^{r+1}$, that $\partial_{|A_{\mathcal{S}}}$ is surjective onto $A_{\mathcal{S}}$, and that conditions a), ... , d) of (2.4) are fulfilled (with

$$u_i = x - \theta_i \ , \ \partial_i^{-j} 1 = \frac{\log^j (x - \theta_i)}{j!} \ ,$$

$$T_i = \oplus_{\alpha \in \Delta \cap (\mathrm{Im}\tau \setminus \{0\})} C'_{\mathcal{S},i} (x - \theta_i)^{\alpha} \ ) \ .$$

Note that $A_{\mathcal{S}} \cap B_{\mathcal{S}} = \mathcal{O}(\mathcal{S})[x, \frac{1}{\prod (x-\theta_i)}]$.

3.5.2 Because $\tilde{\mathcal{E}}$ is locally free, there exists a Zariski covering $\{U\}$ of S such that $\tilde{\mathcal{E}}_{|Z_i \cap f^{-1}(U)}$ is free over $\mathcal{O}_{Z_i \cap f^{-1}(U)}$ for every i, where $Z_i = div(x - \theta_i) \subset \mathbf{P}^1_S$ (Z_0 is the divisor at infinity). Then for every compact $\mathcal{S} \subset U(\mathbf{C})$, there exists $\epsilon_{\mathcal{S}} > 0$ such that $\tilde{\mathcal{E}} \otimes \mathcal{O}(T_{\mathcal{S},i,\epsilon})$ is free over $\mathcal{O}(T_{\mathcal{S},i,\epsilon})$ if $\epsilon \leq \epsilon_{\mathcal{S}}$. We denote by $E_{\mathcal{S}}$ the $\mathcal{O}(\mathcal{S})[x, \frac{1}{\prod (x-\theta_i)}, \frac{d}{dx}]$-module $\Gamma(\mathcal{E}) \otimes_{\mathcal{O}(S)} \mathcal{O}(\mathcal{S})$.

Proposition 3.5.3 *$E_{\mathcal{S}}$ is solvable in $A_{\mathcal{S}}$.*

Proof: We have to show that $E_{\mathcal{S}}$ is solvable in $A_{\mathcal{S},i}$, for every $i = 0, \dots, r$. Let $\underline{e}$ be a basis of $\tilde{\mathcal{E}} \otimes \mathcal{O}(T_{\mathcal{S},i,\epsilon})$ in which the connection may be written $\nabla((x - \theta_i)\frac{d}{dx})\underline{e} = \underline{e}G$, where G has entries analytic in $T_{\mathcal{S},i,\epsilon}$, and $G_{|Z_i}$ is a constant matrix with eigenvalues in Δ: in fact because $T_{\mathcal{S},i,\epsilon}$ is a compact tube, there exists a compact neighborhood $\mathcal{S}'$ of $\mathcal{S}$ such that $G \in M_{\mu}(\mathcal{O}(T_{\mathcal{S}',i,\epsilon}))$.

It suffices to show that the differential system $((x - \theta_i)\frac{d}{dx} + G)Y = 0$ has a solution of the form $Y = W \cdot (x - \theta_i)^{G_{|Z_i}}$, where $W \in GL_{\mu}(\mathcal{O}(T_{\mathcal{S},i,\epsilon'}))$, for some $0 < \epsilon' < \epsilon$. We follow the classical method used in the case $\mathcal{S}$ = one point, *cf. e.g.* [DGS, III, 8.5 and Appendix II]: a formal computation expresses the coefficients of the Taylor expansion of W at $x = \theta_i$ by a recursive formula, under the extra condition: $W_{|Z_i} = I$; one then estimates the growth of these coefficients (with respect to the sup-norm on $\mathcal{S}'$) as in *loc. cit.*, Appendix II, and one concludes that $W \in M_{\mu}(\mathcal{O}(T_{\mathcal{S},i,\epsilon''}))$, $0 < \epsilon'' < \epsilon$. Since $W_{|Z_i} = I$, it is clear that there exists ϵ', $0 < \epsilon' \leq \epsilon''$, such that $W \in GL_{\mu}(\mathcal{O}(T_{\mathcal{S},i,\epsilon'}))$, as wanted. This concludes the proof of (3.5.3). □

3.5.4 We can apply the criteria (2.1) and (2.4), and conclude that

$$(\mathrm{Co})\mathrm{Ker}_{E_{\mathcal{S}}} \nabla \left(\frac{d}{dx} \right) \cong (\mathrm{Co})\mathrm{Ker}_{\Gamma(\mathcal{X}, \mathcal{E}^{an})} \nabla \left(\frac{d}{dx} \right) .$$

Since $\mathcal{O}(\mathcal{S})$ is flat over $\mathcal{O}(S)$, this may be rewritten

$$(\mathrm{Co})\mathrm{Ker}_{\Gamma(X,\mathcal{E})} \nabla \left(\frac{d}{dx} \right) \otimes_{\mathcal{O}(S)} \mathcal{O}(\mathcal{S}) \cong (\mathrm{Co})\mathrm{Ker}_{\Gamma(\mathcal{X}, \mathcal{E}^{an})} \nabla \left(\frac{d}{dx} \right) ,$$

which holds for any small enough compact neighborhood $\mathcal{S}$ of any point of $S(\mathbf{C})$. Therefore the natural morphism

$$(R^i_{DR} f_*(\mathcal{E}, \nabla))^{an} \longrightarrow R^i_{DR} f^{an}_*(\mathcal{E}^{an}, \nabla^{an})$$

is an isomorphism for $i = 0, 1$, and this achieves the alternative proof of 3.1.

Corollary 3.6 *Let X be a smooth complex algebraic variety. The functor*

$$\{\text{regular connections on } X\} \longrightarrow \{\text{local systems on } X(\mathbf{C})\},$$
$$(\mathcal{E}, \nabla) \longmapsto (\mathcal{E}^{an})^{\nabla^{an}}$$

is fully faithful.

Proof: Let $(\mathcal{E}, \nabla)$, $(\mathcal{E}', \nabla')$ be two regular objects in $\mathbf{MIC}(X)$. Then (the internal Hom) $\underline{Hom}((\mathcal{E}, \nabla), (\mathcal{E}', \nabla'))$ is a regular object in $\mathbf{MIC}(X)$ (7.5). By 3.1, with $i = 0$, we have

$$\begin{aligned} Hom((\mathcal{E}, \nabla), (\mathcal{E}', \nabla')) &= H^0_{DR}(X, \underline{Hom}((\mathcal{E}, \nabla), (\mathcal{E}', \nabla'))) \\ &\cong H^0_{DR}(X^{an}, \underline{Hom}((\mathcal{E}^{an}, \nabla^{an}), (\mathcal{E}'^{an}, \nabla'^{an}))) \\ &\cong H^0(X(\mathbf{C}), \underline{Hom}_{loc.\ syst.}((\mathcal{E}^{an})^{\nabla^{an}}, (\mathcal{E}'^{an})^{\nabla'^{an}})) \\ &= Hom_{loc.\ syst.}((\mathcal{E}^{an})^{\nabla^{an}}, (\mathcal{E}'^{an})^{\nabla'^{an}})\,. \end{aligned}$$

□

Corollary 3.7 *Let X and $(\mathcal{E}, \nabla)$, $(\mathcal{E}', \nabla')$ be as before. Then the natural map*

$$Ext^1_{MIC}((\mathcal{E}, \nabla), (\mathcal{E}', \nabla')) \longrightarrow Ext^1_{loc.\ syst.}((\mathcal{E}^{an})^{\nabla^{an}}, (\mathcal{E}'^{an})^{\nabla'^{an}})$$

is a bijection.

Proof: Same argument as for (3.6), but with $i = 1$. □

§4 Comparison theorems for algebraic vs. rigid-analytic cohomology (regular coefficients)

In this section, K is an algebraically closed field of characteristic 0, complete with respect to a non-archimedean absolute value.

Theorem 4.1 *Let i be a non-negative integer, and let $f : X \longrightarrow S$ be a smooth morphism of smooth K-varieties, with $A_i(f) = S$. Let $(\mathcal{E}, \nabla)$ be a coherent $\mathcal{O}_X$-module endowed with an integrable regular connection, such that the additive subgroup of*

K generated by 1 *and the exponents of* ∇ *contains no Liouville number. Then the canonical morphism*

$$\varphi_i : (R^i_{DR} f_*(\mathcal{E}, \nabla))^{an} \longrightarrow R^i_{DR} f^{an}_*(\mathcal{E}^{an}, \nabla^{an})$$

is an isomorphism.

In the special case $S = \operatorname{Spec} K$ and $(\mathcal{E}, \nabla) = (\mathcal{O}_X, d_{X/S})$, this gives:

Corollary 4.2 **([Ki2]).** *For any smooth algebraic K-variety X, and any* $i \geq 0$, $H^i_{DR}(X) \cong H^i_{DR}(X^{an})$.

In the case $S = \operatorname{Spec} K$, Theorem 4.1 was already proven in [B3] using resolution of singularities.

4.3 Our proof of (4.1) is very close to the proof in the complex case (in its second form, *cf.* (3.5)). We reduce to the case of a rational elementary fibration, $i = 0, 1$, and apply the abstract criteria of Section 2. We introduce $\mathcal{S}, \mathcal{X}, T_{\mathcal{S},i,\epsilon}, A_{\mathcal{S},i}, C'_{\mathcal{S},i}, A_{\mathcal{S}}, C_{\mathcal{S}}, E_{\mathcal{S}}$, as in (3.5) (on replacing the word "compact" by "affinoid"). The only difference is that for the solvability of $E_{\mathcal{S}}$ in $A_{\mathcal{S}}$ - more specifically, in the non-archimedean estimates replacing [DGS, Appendix II] -, as well as for the stability of $A_{\mathcal{S},i}$ and $C'_{\mathcal{S},i}$ by integration, we have to *use the fact that* Δ (the **Q**-vector space generated by 1 and the exponents) *does not contain Liouville numbers*, which follows from the assumption in (4.1).

We conclude again that the natural morphism

$$(\mathrm{Co})\mathrm{Ker}_{\Gamma(X,\mathcal{E})} \nabla\left(\frac{d}{dx}\right) \otimes_{\mathcal{O}(S)} \mathcal{O}(\mathcal{S}) \longrightarrow (\mathrm{Co})\mathrm{Ker}_{\Gamma(\mathcal{X},\mathcal{E}^{an})} \nabla\left(\frac{d}{dx}\right),$$

is an isomorphism for any affinoid subspace $\mathcal{S}$ of U^{an}, where $\{U\}$ is a finite Zariski open covering of S as in (3.5.2).

4.4 Since $(\mathrm{Co})\mathrm{Ker}_{\Gamma(X,\mathcal{E})} \nabla(\frac{d}{dx})$ is finitely generated over $\mathcal{O}(S)$ (III.6), we see that $\mathcal{S} \longmapsto \mathrm{Ker}_{\Gamma(\mathcal{X},\mathcal{E}^{an})} \nabla(\frac{d}{dx})$ (resp. $\mathcal{S} \longmapsto \mathrm{Coker}_{\Gamma(\mathcal{X},\mathcal{E}^{an})} \nabla(\frac{d}{dx})$) is a coherent sheaf on U^{an}, which coincides with

$$R^0_{DR} f^{an}_{|f^{an-1}U^{an}*}(\mathcal{E}^{an}, \nabla^{an})_{|f^{an-1}U^{an}} \cong (R^0_{DR} f^{an}_*(\mathcal{E}^{an}, \nabla^{an}))_{|U^{an}}$$

(resp. $(R^1_{DR} f^{an}_*(\mathcal{E}^{an}, \nabla^{an}))_{|U^{an}}$). By pasting, we conclude that φ_i is an isomorphism for $i = 0, 1$ in our special situation. This achieves the proof of (4.1).

§5 Rigid-analytic comparison theorem in relative dimension one

5.1 On the coherence of the cokernel of a connection in the analytic situation.

Let us consider a rational elementary fibration

$$(5.1.0)\qquad \begin{array}{ccccc} X & \hookrightarrow & \mathbf{P}^1_S & \hookleftarrow & Z = (\coprod_{i=1}^r \sigma_i(S)) \coprod \sigma_\infty(S) \\ & \underset{f}{\searrow} & \downarrow^{\mathrm{pr}} & \swarrow & \\ & & S & & \end{array}$$

where $\sigma_\infty(S) = \infty \times S$, $\sigma_i : \zeta \longmapsto (\theta_i(\zeta), \zeta)$, $\theta_i \in \mathcal{O}(S)$, $\theta_i - \theta_j \in \mathcal{O}(S)^\times$. We assume that the base S is affine and smooth and consider a *cyclic* $(\mathcal{E}, \nabla)$, with associated differential polynomial $\Lambda \in \mathcal{O}(X)[\frac{d}{dx}]$, as in (III.4.4); in particular, we assume that S is localized in such a way that the leading coefficients of the indicial polynomials ϕ_i at $x = \theta_i$ are units in $\mathcal{O}(X)$. We denote by $f^{an} : X^{an} \longrightarrow S^{an}$ and $(\mathcal{E}^{an}, \nabla^{an}) \in Ob\mathbf{MIC}(X^{an})$ the corresponding analytic objects. Let also $\mathcal{S} \subset S^{an}$ be an affinoid subspace, and let $\mathcal{X} \longrightarrow \mathcal{S}$ denote the (quasi-Stein) morphism obtained from f^{an} by base change.

Proposition 5.1.1 *If the roots α_{ij} of the indicial polynomials ϕ_i are non-Liouville (i.e. if the power series $\sum_{\substack{n\geq 0\\ n\neq\alpha_{ij}}} \frac{x^n}{\alpha_{ij}+n}$ has a non-zero radius of convergence), then the natural map*

$$\mathrm{Coker}(\Gamma(X, \mathcal{E}) \xrightarrow{\nabla} \Gamma(X, \mathcal{E})dx) \otimes \mathcal{O}(\mathcal{S}) \longrightarrow \mathrm{Coker}(\Gamma(\mathcal{X}, \mathcal{E}^{an}) \xrightarrow{\nabla^{an}} \Gamma(\mathcal{X}, \mathcal{E}^{an})dx)$$

is surjective.

The rest of this section is devoted to the proof of this assertion.

5.1.2 There is a commutative diagram (induced by (III.4.4.1))

$$(5.1.2.1)\qquad \begin{array}{ccc} \mathcal{O}(\mathcal{X}) & \xrightarrow{\Lambda} & \mathcal{O}(\mathcal{X}) \\ \downarrow & & \downarrow \\ \Gamma(\mathcal{X}, \mathcal{E}^{an}) = \bigoplus_{k=0}^{\mu-1} \mathcal{O}(\mathcal{X})v_k & \xrightarrow{\nabla^{an}(\frac{d}{dx})} & \Gamma(\mathcal{X}, \mathcal{E}^{an}) = \bigoplus_{k=0}^{\mu-1} \mathcal{O}(\mathcal{X})v_k \end{array}$$

which gives rise to isomorphisms:

$$(\mathrm{Co})\mathrm{Ker}(\Gamma(\mathcal{X}, \mathcal{E}^{an}) \xrightarrow{\nabla} \Gamma(\mathcal{X}, \mathcal{E}^{an})dx) \cong (\mathrm{Co})\mathrm{Ker}_{\mathcal{O}(\mathcal{X})}\Lambda \cong (\mathrm{Co})\mathrm{Ker}_{\mathcal{O}(\mathcal{X})}\Lambda'$$

where $\Lambda' = P(x)\Lambda \in \mathcal{O}(\mathcal{S})[x, \frac{d}{dx}]$ as in (III.4.4).

5.1.3 We have a Mittag-Leffler decomposition (induced by (III.4.5.2))

(5.1.3.1) $$\mathcal{O}(\mathcal{X}) = \mathcal{O}(\mathbf{A}^{1\,an}_{\mathcal{S}}) \oplus \bigoplus_{i=1}^{r} \frac{1}{x-\theta_i}\mathcal{O}(\mathbf{P}^{1\,an}_{\mathcal{S}} \setminus \sigma_i(\mathcal{S})),$$

where

$$\mathcal{O}(\mathbf{A}^{1\,an}_{\mathcal{S}}) = \left\{ \sum_{n\geq 0} a_{n,\infty}x^n \in \mathcal{O}(\mathcal{S})[[x]]\,,\ \lim_{n\to\infty} ||a_{n,\infty}||^{1/n} = 0 \right\}$$

$$\mathcal{O}(\mathbf{P}^{1\,an}_{\mathcal{S}} \setminus \sigma_i(\mathcal{S})) = \left\{ \sum_{n\geq 0} \frac{a_{n,i}}{(x-\theta_i)^n} \in \mathcal{O}(\mathcal{S})\left[\left[\frac{1}{x-\theta_i}\right]\right],\ \lim_{n\to\infty} ||a_{n,i}||^{1/n} = 0 \right\}$$

for some (any) Banach norm $||\ \ ||$ on the Tate K-algebra $\mathcal{O}(\mathcal{S})$.

5.1.4 Let us restate more precisely the induction process in (III.4.6), assuming that the leading coefficients τ_∞, τ_i of the indicial polynomials ϕ_∞, ϕ_{θ_i} $(i = 1, \ldots, r)$ of Λ' are invertible in $\mathcal{O}(S)$. Let us write

$$P(x)\gamma_k = \sum_l b_{k,l,i}(x-\theta_i)^l = \sum_l b_{k,l}x^l,$$

with $b_{k,l,i}, b_{k,l} \in \mathcal{O}(S)$, and set

$$\kappa_i = \max\left(1, \max_{k,l} ||k!\tau_i^{-1}b_{k,l,i}||\right)\ ,\ \kappa_\infty = \max\left(1, \max_{k,l} ||k!\tau_\infty^{-1}b_{k,l}||\right).$$

We contend that for $n > 0$ (resp. $n \geq 0$)

$(*)_{n,i}$ $$\frac{1}{(x-\theta_i)^n} = \Lambda'\left(\sum_{l=1}^{n-r_i} \gamma_{n,l,i}\frac{1}{(x-\theta_i)^l}\right) + \sum_{k=1}^{M} \delta_{n,k,i}\frac{1}{(x-\theta_i)^k},$$

with

$$||\gamma_{n,l,i}||\ ,\ ||\delta_{n,l,i}|| \leq \kappa_i^{\max(0,n-M)} \prod_{m=M+1}^{n} \max\left(1, \left|\frac{\tau_i}{\phi_{\theta_i}(-m+r_i)}\right|\right)$$

(resp.

$(*)_{n,\infty}$ $$x^n = \Lambda'\left(\sum_{l=0}^{n-r_\infty} \gamma_{n,l,\infty}x^l\right) + \sum_{k=0}^{M} \delta_{n,k,\infty}x^k,$$

with

$$||\gamma_{n,l,\infty}|| \,,\ ||\delta_{n,l,\infty}|| \le \kappa_\infty^{\max(0,n-M)} \prod_{m=M+1}^{n} \max\left(1, \left|\frac{\tau_\infty}{\phi_\infty(-m+r_\infty)}\right|\right).$$

Indeed this is trivial for $n \le M$. For $n > M$, we write

$$\begin{aligned}\frac{1}{(x-\theta_i)^n} &= \Lambda'\left(\frac{1}{\tau_i}\cdot\frac{\tau_i}{\phi_{\theta_i}(-n+r_i)}\cdot\frac{1}{(x-\theta_i)^{n-r_i}}\right)\\ &+ \sum_{k=n-r_i-r_\infty}^{\max(M,n-r_i-r_\infty)} \beta_{n,k,i}\frac{1}{(x-\theta_i)^k}\\ &+ \sum_{k=\max(M,n-r_i-r_\infty)+1}^{n-1} \beta_{n,k,i}\frac{1}{(x-\theta_i)^k},\end{aligned}$$

with $|\beta_{n,k,i}| \le \kappa_i \max(1, |\frac{\tau_i}{\phi_{\theta_i}(-n+r_i)}|)$, and $(*)_{n,i}$ follows by induction, on applying $(*)_{<n,i}$ to the terms of the last sum (resp.

$$\begin{aligned}x^n = \Lambda'\left(\tau_\infty{}^{-1}\cdot\frac{\tau_\infty}{\phi_\infty(-n+r_\infty)}\cdot x^{n-r_\infty}\right) &+ \sum_{k=\max(0,n-r_\infty-\mu)}^{\max(M,n-r_\infty-\mu)} \beta_{n,k,\infty}x^k\\ + \sum_{k=\max(M,n-r_\infty-\mu)+1}^{n-1} \beta_{n,k,\infty}x^k,&\end{aligned}$$

with $|\beta_{n,k,\infty}| \le \kappa_\infty \max(1, |\frac{\tau_\infty}{\phi_\infty(-n+r_\infty)}|), \ldots$). On the other hand,

$$\prod_{m=M+1}^{n} \max\left(1, \left|\frac{\tau_i}{\phi_{\theta_i}(-m+r_i)}\right|\right)$$

grows at most exponentially with n, because the roots of the ϕ_{θ_i} are assumed to be non-Liouville (*cf.* [Ad, Section 2]). Hence $||\gamma_{n,l,i}||, ||\delta_{n,l,i}|| \le \kappa'^n$, for some $\kappa' \ge 1$.

Therefore, any $\sum_{n\ge 0} a_{n,\infty}x^n + \sum_i \sum_{n\ge 1} \frac{a_{n,i}}{(x-\theta_i)^n} \in \mathcal{O}(\mathcal{X})$ may be written formally $\Lambda' u + v$, where

$$u = \sum_{l\ge 0}\left(\sum_{n\ge l+r_\infty} a_{n,\infty}\gamma_{n,l,\infty}\right)x^l + \sum_i\sum_{l\ge 1}\left(\sum_{n\ge l+r_i} a_{n,i}\gamma_{n,l,i}\right)\frac{1}{(x-\theta_i)^l},$$

$$v = \sum_{k=0}^{M}\left(\sum_{n\ge 0} a_{n,\infty}\delta_{n,k,\infty}\right)x^k + \sum_i\sum_{k=1}^{M}\left(\sum_{n\ge 1} a_{n,i}\delta_{n,k,i}\right)\frac{1}{(x-\theta_i)^k},$$

and the previous estimates show that $u \in \mathcal{O}(\mathcal{X})$, $v \in \mathcal{O}(\mathcal{S})[x, \frac{1}{\prod_i (x-\theta_i)}]$.

This shows that the natural map

$$(\mathrm{Coker}_{\mathcal{O}(X)}\Lambda') \otimes_{\mathcal{O}(S)} \mathcal{O}(\mathcal{S}) \longrightarrow \mathrm{Coker}_{\mathcal{O}(\mathcal{X})}\Lambda'$$

is surjective, hence so is

$$(\mathrm{Coker}(\Gamma(X,\mathcal{E}) \xrightarrow{\nabla} \Gamma(X,\mathcal{E})dx) \otimes \mathcal{O}(\mathcal{S}))$$
$$\longrightarrow (\mathrm{Coker}(\Gamma(\mathcal{X},\mathcal{E}^{an}) \xrightarrow{\nabla^{an}} \Gamma(\mathcal{X},\mathcal{E}^{an})dx)).$$

Remark 5.1.5 As was already pointed out in the algebraic situation (III.4.7.iii), the proof does not use the full hypothesis that v is a cyclic vector on the whole of X (*i.e.* that one may take $P(x) = \prod_{i=1}^{r}(x-\theta_i)^{s_i}$). It would suffice to assume that v is a cyclic vector outside some divisor D whose Zariski closure in $\mathbf{P}^1_S$ is contained in X.

5.2 Comparison theorem for morphisms of relative dimension one

We now give a relative version of the main theorem of [B2]. This result, bound to rigid-analytic comparison in relative dimension 1, will be generalized to arbitrary dimensions in the next section.

Proposition 5.2.1 *Let $f : X \longrightarrow S$ be a smooth morphism of smooth algebraic K-varieties of relative dimension 1, and let $(\mathcal{E},\nabla)$ be a coherent $\mathcal{O}_X$-module endowed with an integrable connection. We assume that f and $(\mathcal{E},\nabla)$ may be defined over some algebraically closed subfield $K_0 \subset K$ which contains no Liouville number (e.g. $K_0 = \overline{\mathbf{Q}}$). Then there is a dense open subset $U \subset S$ such that for any $i \geq 0$, the restriction to U^{an} of the canonical morphism*

$$\varphi_i : (R^i_{DR} f_*(\mathcal{E},\nabla))^{an} \longrightarrow R^i_{DR} f^{an}_*(\mathcal{E}^{an},\nabla^{an})$$

is an isomorphism in $\mathbf{MIC}(U^{an})$.

Proof:

5.2.2 We may replace S by any dense open subset, and then by any finite flat S-scheme S' (indeed, the formation of algebraic and analytic De Rham cohomologies commutes with finite flat base change (III.2.6), (1.2.2), $\mathcal{O}_{S'^{an}}$ is faithfully flat over $\mathcal{O}_{S^{an}}$, and the functor "inverse image of $\mathcal{O}_{S^{an}}$-modules" on S^{an} is nothing but $-\otimes_{\mathcal{O}_{S^{an}}} \mathcal{O}_{S'^{an}}$). As we saw in (III.3.3.1), this allows us to assume that X admits a finite open cover $\{U_\alpha\}$ such that all $f_{\underline{\alpha}} := f_{|U_{\underline{\alpha}}=U_{\alpha_0}\cap\cdots\cap U_{\alpha_p}}$ are coordinate elementary fibrations for $p \leq 1+\dim X$.

Thanks to the morphism of Čech spectral sequences in $\mathbf{MIC}(S^{an})$,

$$\begin{array}{ccc} \oplus_{\alpha_0<\cdots<\alpha_p}(R^{i-p}_{DR} f_{\underline{\alpha}*}(\mathcal{E},\nabla)_{|U_{\underline{\alpha}}})^{an} & \Rightarrow & (R^i_{DR} f_*(\mathcal{E},\nabla))^{an} \\ \downarrow \varphi_{i-p} & & \downarrow \varphi_i \\ \oplus_{\alpha_0<\cdots<\alpha_p} R^{i-p}_{DR} f^{an}_{\underline{\alpha}*}(\mathcal{E}^{an},\nabla^{an})_{|U^{an}_{\underline{\alpha}}} & \Rightarrow & R^i_{DR} f^{an}_*(\mathcal{E}^{an},\nabla^{an}), \end{array}$$

we may assume that f itself is a coordinatized elementary fibration.

On replacing S by an étale covering and arguing as in (3.3.3), we reduce at last to the case when f is a rational elementary fibration, and $i = 0, 1$.

5.2.3 If $S = \operatorname{Spec} K$ (as in the situation of [B2]), we are left with an ordinary differential system in one variable. Hence the comparison theorem

$$H^i_{DR}(X, (\mathcal{E}, \nabla)) \cong H^i_{DR}(X^{an}, (\mathcal{E}^{an}, \nabla^{an}))$$

for a curve is equivalent to the following statement (whose proof on relies only the non-archimedean Turrittin theorem of [B1]). □

Proposition 5.2.4 *Let* $X = \operatorname{Spec} K[x, \frac{1}{\prod_{i=1}^r (x-\theta_i)}]$, *and let* (E, ∇) *be a projective* $\mathcal{O}(X)$*-module of finite rank* μ, *with a connection (not necessarily regular), such that the* $\mathbf{Q}$*-vector subspace* $\Delta \subset K$ *generated by all Turrittin exponents (at the* θ_i *and at* ∞*) contains no Liouville numbers. Let* X^{an} *denote the rigid analytic space associated to the* K*-scheme* X*. Then*

$$\varphi_0 : \operatorname{Ker}_E \nabla\left(\frac{d}{dx}\right) \longrightarrow \operatorname{Ker}_{E\otimes\mathcal{O}(X^{an})} \nabla\left(\frac{d}{dx}\right)$$

and

$$\varphi_1 : \operatorname{Coker}_E \nabla\left(\frac{d}{dx}\right) \longrightarrow \operatorname{Coker}_{E\otimes\mathcal{O}(X^{an})} \nabla\left(\frac{d}{dx}\right)$$

are isomorphisms.

Proof: This is obtained by choosing B, C_i' as in (3.4.2),

$$T_i = \mathcal{M}_{\theta_i}[(x-\theta_i)^\alpha, \exp(a(x-\theta_i)^{-\frac{k}{\mu!}})]_{\alpha\in\Delta, k\in\mathbf{Z}_{>0}, a\in K},$$

with the usual identification

$$\begin{aligned} &(x-\theta_i)^\alpha \exp(a(x-\theta_i)^{-\frac{k}{\mu!}}) \cdot (x-\theta_i)^\beta \exp(b(x-\theta_i)^{-\frac{k}{\mu!}}) \\ &= (x-\theta_i)^{\alpha+\beta} \exp((a+b)(x-\theta_i)^{-\frac{k}{\mu!}}), \\ A_i \quad &= T_i[\log(x-\theta_i)], \end{aligned}$$

and $C_i = A_i C_i'$ (for $i = 0$ we are tacitly replacing θ_i by ∞, as in (3.4.2)). We also set $C = \prod_{i=0}^r C_i$, $A = \prod_{i=0}^r A_i$, $\partial = \frac{d}{dx}$. □

Lemma 5.2.4.1 $\partial : A \longrightarrow A$ *is surjective.*

Proof: We tacitly replace A by A_i, and assume $\theta_i = 0$. We also set $z = x^{-\frac{1}{\mu^2!}}$, an element of A. We may work with d/dz instead of ∂. Any element of A is a finite sum of terms of the form $u\ z^\alpha\ \frac{\log^m z}{m!}\ e^{a\ z^{-k}}$, where $u \in \mathcal{M}_0[z]$ is a p-adic meromorphic function of z. We have to show that $\int u\ z^{\alpha}\ \frac{\log^m z}{m!}\ e^{a\ z^{-k}} dz$ is of the same form. We first note that if $k = 0$ or $a = 0$, the result is an easy consequence of the following formulas

$$z^{\alpha-1}\frac{\log^m z}{m!} = \partial\left(\frac{z^{\alpha}}{\alpha}\sum_{i+j=m}\left(-\frac{1}{\alpha}\right)^i \frac{\log^j z}{j!}\right),$$

for $\alpha \neq 0$, and

$$z^{-1}\frac{\log^m z}{m!} = \partial\frac{\log^{m+1} z}{(m+1)!}.$$

Let us then assume that $k > 0,\ a \neq 0$. By integration by parts

$$\begin{aligned}&\int u\ z^{\alpha}\ \frac{\log^m z}{m!}\ e^{a\ z^{-k}} dz \\ &= \frac{\log^m z}{m!}\int u\ z^{\alpha}\ e^{a\ z^{-k}} dz - \int z^{-1}\ \frac{\log^{m-1} z}{(m-1)!} \\ &\left(\int u\ z^{\alpha}\ e^{a\ z^{-k}} dz\right)\ dz\end{aligned}$$

and descending induction on m, we reduce to the case where $m = 0$. We then have to show that, for $\alpha \in \Delta$ and $a \in K$, there is a formal integral $y = u^{-1}z^{-\alpha}\ e^{-az^{-k}} \int u\ z^{\alpha}\ \ e^{a\ z^{-k}} dz$ which is a p-adic meromorphic function of z. We notice that y satisfies the inhomogenous differential equation

$$dy/dz = (-u^{-1}du/dz - \alpha/z\ +\ kaz^{-k-1})\ y\ +\ 1$$

hence the homogenous differential equation of second order

$$\begin{aligned}&d^2y/dz^2 + (u^{-1}du/dz + \alpha/z - kaz^{-k-1})dy/dz \\ &\quad + (d/dz(u^{-1}du/dz) - \alpha/z^2 + k(k+1)az^{-k-2})y = 0.\end{aligned}$$

Since u is meromorphic, $d/dz(u^{-1}du/dz)$ has at worst a double pole at 0, and since $k > 0$, we see that the indicial polynomial of this differential equation at 0 is

$$\phi_0(T) = ak(T - k - 1).$$

As a simple instance of Clark's theorem [Cl], we conclude that the differential equation possesses a meromorphic solution y. This achieves the proof of (5.2.4.1). □

To apply Proposition 2.4, we need to show that

$$A'_i := A_i \cap C'_i = \mathcal{M}_{\theta_i}. \tag{5.2.4.2}$$

We first show that

$$A_i \cap C'_i \subset \mathcal{M}_{\theta_i}[(x-\theta_i)^\alpha, \log(x-\theta_i)]_{\alpha\in\Delta}. \tag{5.2.4.3}$$

In fact, any element $y \in A_i$ may be expanded in convergent power series in $x - \zeta_i$, for $\zeta_i \neq \theta_i$ sufficiently close to θ_i:

$$y = y_0(x-\zeta_i) + \sum_{k=1}^{r} y_k(x-\zeta_i)\exp(a((x-\zeta_i)^{-\frac{k}{\mu!}} - (\zeta_i-\theta_i)^{-\frac{k}{\mu!}}),$$

where $y_k(x-\zeta_i)$ is the expansion at ζ_i of an element y_k of $\mathcal{M}_{\theta_i}[(x-\theta_i)^\alpha, \log(x-\theta_i)]_{\alpha\in\Delta}$. So, the radius of convergence of $y_k(x-\zeta_i)$ is of the form $C|\zeta_i-\theta_i|$, for a positive constant C depending on y but not on $|\zeta_i-\theta_i|$ [B1, Lemma 6]. On the other hand, the radius of convergence of $\exp(a((x-\zeta_i)^{-\frac{k}{\mu!}} - (\zeta_i-\theta_i)^{-\frac{k}{\mu!}})$ is of the form $C|\zeta_i-\theta_i|^{1+k}$. So, for sufficiently small value of $|\zeta_i-\theta_i|$, these radii are all distinct, and the radius of convergence of y is the infimum of them. Formula (5.2.4.3) follows. We now prove that

$$\mathcal{M}_{\theta_i}[(x-\theta_i)^\alpha, \log(x-\theta_i)]_{\alpha\in\Delta} \cap C'_i \subset \mathcal{M}_{\theta_i}. \tag{5.2.4.4}$$

This follows directly from Dwork's lemma [Ad, Lemma 1].

From (5.2.4.2) we deduce $A \cap B = (\prod_{i=0}^{r} \mathcal{M}_{\theta_i}) \cap B = \mathcal{O}(X)$ and $A_i^\partial = C_i^\partial = \mathbf{C}$. As for the solvability of (E, ∇) in A, it follows from the non-archimedean Turrittin theorem of [B1].

5.2.5 However we cannot reduce the case of a rational elementary fibration to the case of "$\mathbf{P}^1$ minus a few points" as in (3.4), for lack of having yet proven coherence and base change for $R^i_{DR} f^{an}_* (\mathcal{E}^{an}, \nabla^{an})$. Nevertheless, since we are free to replace S by a dense affine open subset, the argument of (III.3.3.3) together with the use of Čech spectral sequences as above allows us to assume that $(\mathcal{E}, \nabla)$ is cyclic of rank μ, with associated differential operator $\Lambda \in \mathcal{O}(X)[\frac{d}{dx}]$, and that the leading coefficients of the indicial polynomials of Λ are invertible in S. In particular, $\mathrm{Ker}_{\Gamma(X,\mathcal{E})}\nabla(\frac{d}{dx})$ and $\mathrm{Coker}_{\Gamma(X,\mathcal{E})}\nabla(\frac{d}{dx})$ are projective $\mathcal{O}(S)$-modules of finite rank. We notice that all previous reductions could be carried over K_0, and in particular the roots of the indicial polynomials belong to K_0. More generally, by Robba's result reproduced in (I, Appendix A), all Turrittin exponents (at θ_i and at ∞) belong to K_0; we denote by Δ the $\mathbf{Q}$-subspace of K_0 which they generate (together with 1).

5.2.6 Let $\mathcal{S} \subset S^{an}$ be an affinoid subspace, and let $\mathcal{X} \longrightarrow \mathcal{S}$ be the morphism obtained from f^{an} by base change. Since Δ does not contain any Liouville number, we are in the situation of (5.1.1). On the other hand, by the non-archimedean Turrittin theorem [B1], there is a Banach field K' which is a finite extension of the completion of the fraction field of $\mathcal{O}(\mathcal{S})$ with respect to the spectral norm, and which enjoys the following property. If we set $E = \Gamma(X, \mathcal{E}) \otimes_{\mathcal{O}(S)} \mathcal{O}(\mathcal{S})$, $B = \mathcal{O}(\mathcal{X})$, and choose $A \subset C$ as in (5.2.4), but with K replaced by K' (so that $A \cap B \cong \mathcal{O}(X) \otimes_{\mathcal{O}(S)} \mathcal{O}(\mathcal{S}) \subset \mathcal{O}(\mathcal{X})$, as follows combining the Mittag-Leffler decomposition (5.1.3.1) and $A \cap \mathcal{O}((\mathrm{Spec} K'[x, \frac{1}{\prod(x-\theta_i)}])^{an}) = K'[x, \frac{1}{\prod(x-\theta_i)}])$, then E is solvable in A.

Proposition 2.1 (and flatness of $\mathcal{O}(\mathcal{S})$ over $\mathcal{O}(S)$) shows that the natural morphism

$$\mathrm{Ker}_{\Gamma(X,\mathcal{E})} \nabla\left(\frac{d}{dx}\right) \otimes_{\mathcal{O}(S)} \mathcal{O}(\mathcal{S}) \longrightarrow \mathrm{Ker}_{\Gamma(\mathcal{X},\mathcal{E}^{an})} \nabla\left(\frac{d}{dx}\right),$$

(resp.

$$\mathrm{Coker}_{\Gamma(X,\mathcal{E})} \nabla\left(\frac{d}{dx}\right) \otimes_{\mathcal{O}(S)} \mathcal{O}(\mathcal{S}) \longrightarrow \mathrm{Coker}_{\Gamma(\mathcal{X},\mathcal{E}^{an})} \nabla\left(\frac{d}{dx}\right),$$

is an isomorphism (resp. is injective). On the other hand, we know by (5.1.1) that the latter morphism is surjective. As in (3.5.4), we conclude that φ_i is an isomorphism for $i = 0, 1$ in this situation, and this achieves the proof of (5.2.1).

Remark 5.2.7 i) In proving the surjectivity of φ_1, we cannot apply (2.4), because the last condition d) of (2.4) is not fulfilled.

ii) In the situation of (5.2.1), assume moreover that S is affine, f is a rational elementary fibration, $(\mathcal{E}, \nabla)$ is cyclic outside some divisor which does not meet Z, and that the leading coefficients of the indicial polynomials at the branches of Z are units in $\mathcal{O}(S)$. Then our argument shows that the conclusion of (5.2.1) holds with $U = S$ (*cf.* (5.1.5)).

5.3 Purity. We complement the previous analysis by the following purity statement.

Lemma 5.3.1 *Let $f : X \longrightarrow S$ be a rational elementary fibration, and let $(\mathcal{E}, \nabla)$ be a coherent $\mathcal{O}_X$-module endowed with an integrable connection. We assume that there is an open subset $U \subset S$ such that $S\backslash U$ is of codimension ≥ 2 in S and for any $i \geq 0$, the restriction to U^{an} of the canonical morphism*

$$\varphi_i : (R^i_{DR} f_*(\mathcal{E}, \nabla))^{an} \longrightarrow R^i_{DR} f^{an}_*(\mathcal{E}^{an}, \nabla^{an})$$

is an isomorphism of coherent objects in $\mathbf{MIC}(U^{an})$.
Then φ_i is in fact an isomorphism of coherent objects in $\mathbf{MIC}(S^{an})$.

Proof: We construct the natural commutative square

$$\begin{array}{ccc} X' = X_U & \overset{j'}{\hookrightarrow} & X \\ f' \downarrow & & \downarrow f \\ U & \overset{j}{\hookrightarrow} & S. \end{array}$$

Because $S\backslash U$ is of codimension ≥ 2 and $\mathcal{E}$ is locally free of finite rank, Lütkebohmert's theorem [Lü1] applies and we have

$$(\mathcal{E}^{an}, \nabla^{an}) \cong j_*^{\prime an} j^{\prime an*}(\mathcal{E}^{an}, \nabla^{an}) .$$

(By Zariski localization, this amounts to the fact that $j_*^{\prime an}\mathcal{O}_{X^{\prime an}} \cong \mathcal{O}_{X^{an}}$). For $i \neq 0, 1$, the cohomology modules vanish. The case $i = 0$ is easy:

$$R^0_{DR} f_*^{\prime an}(j^{\prime an*}(\mathcal{E}^{an}, \nabla^{an})) \cong (R^0_{DR} f_*^{\prime}(j^{\prime *}(\mathcal{E}, \nabla)))^{an}$$

and compatibility of the formation of De Rham cohomology with localization (III.2.6), (1.2.2), imply the isomorphism of coherent objects in $\mathbf{MIC}(S^{an})$

$$\begin{aligned} R^0_{DR} f_*^{an}(\mathcal{E}^{an}, \nabla^{an}) &\cong R^0_{DR} f_*^{an}(j_*^{\prime an} j^{\prime an*}(\mathcal{E}^{an}, \nabla^{an})) \\ &\cong j_*^{an} R^0_{DR} f_*^{\prime an}(j^{\prime an*}(\mathcal{E}^{an}, \nabla^{an})) \\ &\cong (j_* R^0_{DR} f_*^{\prime}(j^{\prime *}(\mathcal{E}, \nabla)))^{an} \cong (R^0_{DR} f_*(\mathcal{E}, \nabla))^{an}. \end{aligned}$$

Let us now consider the case $i = 1$. We first prove the coherence of $R^1_{DR} f_*^{an}(\mathcal{E}^{an}, \nabla^{an})$. For it, the same argument as the one used in the algebraic context (III.4.9) works: we may reduce to the case where $(\mathcal{E}, \nabla)$ is *simple*, and we have the alternative that

$$f^{an*} j_*^{an} R^0_{DR} f_*^{\prime an}(j^{\prime an*}(\mathcal{E}^{an}, \nabla^{an})) \cong (f^* j_* R^0_{DR} f_*^{\prime}(j^{\prime *}(\mathcal{E}, \nabla)))^{an}$$

is either $(\mathcal{E}^{an}, \nabla^{an})$ or 0. In both cases, the analytic version of the argument of *loc. cit.* applies, and shows in particular that

$$R^1_{DR} f_*^{an}(\mathcal{E}^{an}, \nabla^{an}) \cong j_*^{an} j^{an*} R^1_{DR} f_*^{an}(\mathcal{E}^{an}, \nabla^{an}).$$

The composition of this isomorphism with φ_1 can also be decomposed as

$$\begin{aligned} &(R^1_{DR} f_*(\mathcal{E}, \nabla))^{an} \to (j_* j^* R^1_{DR} f_*(\mathcal{E}, \nabla))^{an} \to j_*^{an}(j^* R^1_{DR} f_*(\mathcal{E}, \nabla))^{an} \\ &\to j_*^{an}(R^1_{DR} f_*^{\prime} j^{\prime *}((\mathcal{E}, \nabla)))^{an} \to j_*^{an} R^1_{DR} f_*^{\prime an} j^{\prime an*}(\mathcal{E}^{an}, \nabla^{an}) \\ &\to j_*^{an} j^{an*} R^1_{DR} f_*^{an}(\mathcal{E}^{an}, \nabla^{an}). \end{aligned}$$

Using our assumption, we see that all maps of this sequence are isomorphisms. Hence φ_1 is an isomorphism. □

§6 Comparison theorem for algebraic vs. rigid-analytic cohomology (irregular coefficients)

In this section, we tackle the problem of comparing algebraic and rigid-analytic De Rham cohomologies with possibly irregular coefficients, *i.e.* of extending (5.2.1) to the case of higher dimensional morphisms. We prove the following statement, which was conjectured in [B2]:

Theorem 6.1 *Let X be a smooth K_0-variety and let $(\mathcal{E}, \nabla)$ be a coherent $\mathcal{O}_X$-module with integrable connection (as before, K_0 denotes an algebraically closed subfield of the non-archimedean complete algebraically closed field K such that K_0 does not contain Liouville numbers, e.g. $K_0 = \overline{\mathbf{Q}}$). Then for any $i \geq 0$,*

$$H^i_{DR}(X, (\mathcal{E}, \nabla)) \otimes_{K_0} K \cong H^i_{DR}(X^{an}_K, (\mathcal{E}^{an}, \nabla^{an})) .$$

Our proof extends the strategy which we have used before in the case of regular coefficients: combining (5.2.1) and dévissage (III.3).

6.2 Let $f : X \longrightarrow S$ be a coordinatized elementary fibration, and let $(\mathcal{E}, \nabla)$ be a coherent object of $\mathbf{MIC}(X)$. We saw in (III.4) that $R^0_{DR} f_*(\mathcal{E}, \nabla)$ is locally free of finite rank, and that there exists a dense open subset $U \subset S$, depending on $(\mathcal{E}, \nabla)$, such that $R^1_{DR} f_*(\mathcal{E}, \nabla)_{|U}$ is locally free of finite rank (the other cohomology modules $R^i_{DR} f_*(\mathcal{E}, \nabla)_{|U}$ vanish). Since $U \neq S$ in general for irregular connections, this prevents us from using the first form of dévissage (III.3.2) in the study of $R^i_{DR} f_*(\mathcal{E}, \nabla)_{|U}$ for higher dimensional f. On the other hand, as was mentioned in (1.1.8), the analytic version of the second form of dévissage (III.3.3) is problematic. In order to tackle the problem of comparing algebraic and rigid-analytic De Rham cohomologies with irregular coefficients by an extension of our strategy, we thus need suitable refinements of the coherence results of (III.4).

More precisely, the proof of (6.1) relies upon the following two *key propositions*. The first one is purely algebraic, and makes essential use of the results of (II) about irregularity in several variables:

Proposition 6.3 *Let X be a smooth K_0-variety and let $(\mathcal{E}, \nabla)$ be a coherent $\mathcal{O}_X$-module with integrable connection. There exists a finite open affine cover (U_α) of X and, for each α, a coordinatized elementary fibration $f_\alpha : U_\alpha \longrightarrow S_\alpha$ such that $R^0_{DR} f_{\alpha *}((\mathcal{E}, \nabla)_{|U_\alpha})$ and $R^1_{DR} f_{\alpha *}((\mathcal{E}, \nabla)_{|U_\alpha})$ are coherent $\mathcal{O}_{S_\alpha}$-modules.*

The coherence of $R^0_{DR} f_{\alpha *}((\mathcal{E}, \nabla)_{|U_\alpha})$ is in fact automatic (see (III.4.2.1)), and is put here for ease of reference. In the irregular case, however, example (III.4.8) shows that the coherence of $R^1_{DR} f_{\alpha *}((\mathcal{E}, \nabla)_{|U_\alpha})$ is not automatic.

The second proposition is a variant of (5.2.1), (5.2.7.ii):

Proposition 6.4 *In the situation of* (6.3), *one can moreover choose f_α in such a way that for $j = 0, 1$, the canonical morphism*

$$R^j_{DR} f_{\alpha *}((\mathcal{E}, \nabla)_{|U_\alpha})^{an} \longrightarrow R^j_{DR} f^{an}_{\alpha *}((\mathcal{E}^{an}, \nabla^{an})_{|U^{an}_\alpha})$$

is an isomorphism in $\mathbf{MIC}(S^{an}_\alpha)$.

6.5 Let us deduce (6.1) from (6.3) and (6.4). We consider a cover (U_α) as in (6.3) and (6.4). For $\underline{\alpha} = (\alpha_0, \ldots, \alpha_p)$, $\alpha_0 < \cdots < \alpha_p$, we set $U_{\underline{\alpha}} = U_{\alpha_0} \cap \cdots \cap U_{\alpha_p}$.

There is a natural morphism of Čech spectral sequences,

$$\begin{array}{ccc} \oplus_{\alpha_0<\cdots<\alpha_p} H^{i-p}_{DR}(U_{\underline{\alpha}}, (\mathcal{E}, \nabla)) \otimes K & \Rightarrow & H^i_{DR}(X, (\mathcal{E}, \nabla)) \otimes K \\ \downarrow \varphi_{i-p} & & \downarrow \varphi_i \\ \oplus_{\alpha_0<\cdots<\alpha_p} H^{i-p}_{DR}(U^{an}_{\underline{\alpha}K}, (\mathcal{E}^{an}, \nabla^{an})) & \Rightarrow & H^i_{DR}(X^{an}_K, (\mathcal{E}^{an}, \nabla^{an})). \end{array}$$

By induction on i, we are left with proving the theorem for $X = U_\alpha$. On the other hand, there is a natural morphism of Leray spectral sequences

$$\begin{array}{ccc} \oplus_{0\le j\le i} H^{i-j}_{DR}(S_\alpha, R^j_{DR} f_{\alpha *}((\mathcal{E}, \nabla)_{|U_\alpha})) \otimes K & \Rightarrow & H^i_{DR}(X, (\mathcal{E}, \nabla)) \otimes K \\ \downarrow \varphi'_{i-j} & & \\ \oplus_{0\le j\le i} H^{i-j}_{DR}(S^{an}_{\alpha K}, (R^j_{DR} f_{\alpha *}((\mathcal{E}, \nabla)_{|U_\alpha})^{an}) & & \downarrow \varphi_i \\ \downarrow \varphi''_j & & \\ \oplus_{0\le j\le i} H^{i-j}_{DR}(S^{an}_{\alpha K}, R^j_{DR} f^{an}_{\alpha *}((\mathcal{E}^{an}, \nabla^{an})_{|U^{an}_\alpha}) & \Rightarrow & H^i_{DR}(X^{an}_K, (\mathcal{E}^{an}, \nabla^{an})). \end{array}$$

By (6.4), we know that φ''_j is an isomorphism. By (6.3) and induction on the dimension, we may assume that φ'_{i-j} is an isomorphism. Hence the right vertical arrow φ_i is an isomorphism.

6.6 Proof of Proposition 6.3: As was said before, it suffices to consider the case of $R^1_{DR} f_*$. By quasi-compactness, it is enough to find an open affine neighborhood V of an arbitrary closed point $\xi \in X$, and a coordinatized elementary fibration $f : V \longrightarrow S$ such that $R^1_{DR} f_*((\mathcal{E}, \nabla)_{|V})$ is a coherent $\mathcal{O}_S$-module. We may also assume that X is affine of pure dimension $d \geq 2$ (the cases $d = 0, 1$ are trivial). If we take *any* coordinatized elementary fibration as constructed in (III.1.4), and try to apply the method of (III.4), we encounter two serious obstacles:

i) in general, there is no cyclic vector in the neighborhood of the singular divisors, as was pointed out in (II.7.2.5),
ii) even if one happens to find such a cyclic vector, the leading coefficients of the indicial polynomials will not be units in $\mathcal{O}(S)$ in general.

We shall overcome these difficulties by

1) taking $f : V \longrightarrow S$ *sufficiently general* with respect to the parameters of the Artin construction of good neighborhoods, and
2) applying the method of (III.1.4) to a suitable modification of f.

In the realization of this objective, we shall make full use of our study (II) of irregularity in several variables (specifically, and in order of appearance: (II.5.3.1), (II.7.2.9) and (II.7.1.2)).

6.6.1 Let us consider again in detail the construction of a coordinatized elementary fibration as in (III.1.4, 1.5). It depends on the choice of

i) a projective embedding $\overline{X} \longrightarrow \mathbf{P}^N$ of a projective normal closure of X,
ii) a general linear space $L \subset \mathbf{P}^N$ of codimension $d-1$ passing through ξ,

iii) a general hyperplane $H \subset \mathbf{P}^N$,

and is obtained as a suitable restriction of the linear projection $p : X \subset \mathbf{P}^N \rightharpoonup \mathbf{P}^{d-1}$ with center $L \cap H$. More precisely, let $\epsilon : \tilde{\overline{X}} \to \overline{X}$ denote the blowing up with center $L \cap H \cap \overline{X}$, with exceptional divisor E, and let D denote the divisor $\overline{X} \setminus X$. Then according to *loc. cit.*, there is an affine open neighborhood S of $\zeta := p(\xi)$ in $\mathbf{P}^{d-1}$, a divisor F in $\tilde{\overline{X}}_{|S}$ and a finite morphism $\overline{\pi} : \tilde{\overline{X}}_{|S} \longrightarrow \mathbf{P}^1_S$ such that the induced diagram

$$
(6.6.1.1) \qquad
\begin{array}{ccccccc}
U = \tilde{\overline{X}}_{|S} \setminus Z & \overset{j}{\hookrightarrow} & \tilde{\overline{X}}_{|S} & \hookleftarrow & Z = \left(D \coprod E\right)_{|S} \coprod F \\
\downarrow \pi & & \downarrow \overline{\pi} & & \downarrow \pi' \\
U' & \overset{j'}{\hookrightarrow} & \mathbf{P}^1_S & \hookleftarrow & Z' \\
 & \overset{f'}{\searrow} & \downarrow \mathrm{pr} = \overline{f}' & \overset{g'}{\swarrow} & \\
 & & S & &
\end{array}
$$

pertains to a coordinatized elementary fibration (III.1.3), with $f := p_{|U} = f' \circ \pi$, $\overline{f} := \overline{f}' \circ \overline{\pi}$, and $g := g' \circ \pi'$. Moreover the fiber of $p_{|\tilde{\overline{X}}}$ above ζ identifies with $\overline{X} \cap L$. Note that D is *independent of the choice of* (L, H). Note also that $\overline{\pi}$ is finite étale above $\pi'(D \coprod E)_{|S}$ and that $(\overline{\pi}^{-1}\pi'(D \coprod E)_{|S}) \cap Z = (D \coprod E)_{|S}$ according to (III.1.5.c).

6.6.2 Locally free extensions. There exists a closed subset $T \subset D$ of codimension ≥ 1 such that

i) $D \setminus T = \coprod D_i$, a disjoint union of smooth divisors contained in the smooth part of $\overline{X}$,
ii) the $\mathcal{O}_X$-module $\mathcal{E}$ extends to a locally free module over $\overline{X} \setminus T$.

Since T has codimension ≥ 2 in $\overline{X}$ and since L has codimension $d-1$ in $\mathbf{P}^N$ and is sufficiently general (among those linear spaces passing through ξ), we have $L \cap T = \emptyset$. This implies, after shrinking S if necessary, that $D_{|S} = \coprod D_{i\,|S}$, and that $\mathcal{E}_{|U}$ extends to a locally free $\mathcal{O}_{\tilde{\overline{X}}_{|S}}$-module.

6.6.3 Constancy of Newton polygons. It follows from (II.5.3.1) that we might have chosen the closed subset $T \subset D$ of codimension ≥ 1 in such a way that, moreover, for an affine neighborhood U' of ζ avoiding $\overline{\pi}(T)$ and $U = \pi^{-1}(U')$,

iii) for any locally closed curve C which meets D transversally at some point $Q \in D_i$, there are equalities of Newton polygons:

$$NP_Q((\mathcal{E}, \nabla)_{|C \setminus Q}) = NP_{D_i}((\mathcal{E}, \nabla)_{|U}),$$
$$NP_Q(\underline{End}((\mathcal{E}, \nabla)_{|C \setminus Q})) = NP_{D_i}(\underline{End}(\mathcal{E}, \nabla)_{|U}).$$

We remark, besides, that $\nabla_{|U}$ has no singularity at all along $E_{|S}$ nor along F. We thus have the following property:

(6.6.3.1) for any component Z_i of the divisor Z and any point $Q \in Z_i$, there are equalities of Newton polygons:

$$NP_Q((\mathcal{E},\nabla)_{|f^{-1}(g(Q))}) = NP_{Z_i}((\mathcal{E},\nabla)_{|U}),$$
$$NP_Q((\underline{End}(\mathcal{E},\nabla))_{|f^{-1}(g(Q))}) = NP_{Z_i}(\underline{End}(\mathcal{E},\nabla)_{|U}).$$

We claim that this implies, in turn:

(6.6.3.2) for any component Z'_i of the divisor Z' and any point $Q' \in Z'_i$, there are equalities of Newton polygons:

$$NP_{Q'}(\pi_*(\mathcal{E},\nabla)_{|f'^{-1}(g'(Q'))}) = NP_{Z'_i}(\pi_*(\mathcal{E},\nabla)_{|U'}),$$
$$NP_{Q'}(\underline{End}((\pi_*(\mathcal{E},\nabla))_{|f'^{-1}(g'(Q'))})) = NP_{Z'_i}(\underline{End}(\pi_*(\mathcal{E},\nabla))_{|U'}).$$

This is clear if $Q' \notin \overline{\pi}(D)$, because in this case Z'_i is a regular singularity for $\pi_*(\mathcal{E},\nabla)$. We may thus assume that $\overline{\pi}^{-1}(Q') = \{Q_1,\dots,Q_\delta\}$, with $\delta = deg\,\overline{\pi}$, $Q_1 \in Z_i \subset Z$, $Q_j \notin Z$ if $j > 1$. We set $P = \overline{f}(Q_j) = \overline{f}'(Q')$. In order to justify the claim, we may also replace S by an étale covering, assume that the Z_i (resp. Z'_i) are images of disjoint sections of $\overline{f}$ (resp. $\overline{f}'$), and that $\pi'^{-1}(Z'_i) = \cup_{h=1,\dots,\delta} Z_{ih}$, with $Z_{i1} = Z_i$, $Z_{ih} \cap Z = \emptyset$ if $h > 1$.

For $j > 1$, we then have $NP_{Q_j}((\mathcal{E},\nabla)_{|f^{-1}(P)}) = LQ_\mu$, the *left quadrant with vertex* $(\mu, 0)$ (or else the Newton polygon of the trivial connection of rank μ). We denote provisionally by M the generic fiber of the $\mathcal{O}_{f^{-1}(P)}$-module $\mathcal{E}_{|f^{-1}(P)}$ (viewed as a differential module over $\mathcal{F} = \kappa(f^{-1}(P))$). We introduce the subfield $\mathcal{G} = K_0(x)$ (x being the canonical affine coordinate on $\mathbf{P}^1_S$), endowed with the valuation w induced by Q', and consider the extensions v_j (induced by Q_j) of w to $\mathcal{F}$. We have $({}_{\mathcal{G}}M)\hat{} \cong \oplus_j\, {}_{\hat{\mathcal{G}}}(M^{\wedge(v_j)})$ (I.3.1.3) and $(End({}_{\mathcal{G}}M))\hat{} \cong End(\oplus_j\, {}_{\hat{\mathcal{G}}}(M^{\wedge(v_j)})) \cong \oplus_{i,j} Hom({}_{\hat{\mathcal{G}}}(M^{\wedge(v_i)}),\ {}_{\hat{\mathcal{G}}}(M^{\wedge(v_j)}))$, whence

$$\begin{aligned} NP_{Q'}(\pi_*(\mathcal{E},\nabla)_{|f'^{-1}(P)}) &= NP(({}_{\mathcal{G}}M)\hat{}\,) \\ &= NP({}_{\hat{\mathcal{G}}}(M^{\wedge(v_1)})) + LQ_{(\delta-1)\mu} \\ &= NP_{Q_1}((\mathcal{E},\nabla)_{|f^{-1}(P)}) + LQ_{(\delta-1)\mu} \end{aligned}$$

according to (II.4.3), and

$$\begin{aligned} NP_{Q'}(\underline{End}(\pi_*(\mathcal{E},\nabla))_{|f'^{-1}(g'(Q'))}) &= NP((End({}_{\mathcal{G}}M)\hat{}\,)) \\ = NP_{Q_1}(\underline{End}(\mathcal{E},\nabla)_{|f^{-1}(P)}) &+ 2(\delta-1)\mu NP_{Q_1} \\ ((\mathcal{E},\nabla)_{|f^{-1}(P)}) &+ LQ_{(\delta-1)^2\mu^2}. \end{aligned}$$

By a similar computation,

$$NP_{Z_i'}(\pi_*(\mathcal{E},\nabla)_{|U'}) = NP_{Z_i}(\mathcal{E},\nabla) + LQ_{(\delta-1)\mu},$$
$$NP_{Z_i'}(\underline{End}(\pi_*(\mathcal{E},\nabla))_{|U'}) = NP_{Z_i}(\underline{End}(\mathcal{E},\nabla))$$
$$+ 2(\delta-1)\mu NP_{Z_i}((\mathcal{E},\nabla)) + LQ_{(\delta-1)^2\mu^2},$$

and one obtains (6.6.3.2) (from (6.6.3.1)) by comparison.

6.6.4 A Kummer covering. Let $S_1 \xrightarrow{\tau} S$ be an étale covering such that $U_1 := U' \times_S S_1 \cong \mathbf{P}^1_{S_1} \setminus \coprod_{0,1,\ldots,r} \sigma_i(S_1)$, where σ_i are sections of $\overline{f}'$ given by $x = \theta_i$, $\theta_i \in \mathcal{O}(S_1)$, $\theta_i - \theta_j \in \mathcal{O}(S_1)^\times$ for $0 \neq i \neq j \neq 0$, $\sigma_0(S_1) = \{\infty\} \times S_1$. We shall also denote by τ (by abuse of notation) the induced étale covering $U_1 \longrightarrow U'$. In order to be able to apply our result of (II.7.2) on cyclic vectors, which requires a differential module with Turrittin exponents equal to 1, we shall pull back $\pi_*(\mathcal{E},\nabla)_{|U'}$ to a suitable Kummer étale covering U_2 of U_1.

For $j = 1,\ldots,r$, we denote by $L_j = \mathbf{P}^1_{S_1} \xrightarrow{\lambda_j} \mathbf{P}^1_{S_1}$ the ramified covering of degree $\mu!$ given by $x_j \mapsto x = x_j^{\mu!} + \theta_j$ (totally ramified at $\sigma_j(S_1)$ and at $\sigma_0(S_1)$, unramified elsewhere). We blow up $\Pi_{S_1} L_j = (\mathbf{P}^1)^r_{S_1}$ along the closed subvariety $(\infty,\ldots,\infty)_{S_1}$, and consider the strict transform $\overline{U}_2$ of $(L_1 \times_{\mathbf{P}^1_{S_1}} L_2 \cdots \times_{\mathbf{P}^1_{S_1}} L_r) \subset (\mathbf{P}^1)^r_{S_1}$. We denote by $\overline{\epsilon} : \overline{U}_2 \longrightarrow \mathbf{P}^1_{S_1}$ the canonical finite morphism.

We claim that:

(6.6.4.1) $\overline{U}_2$ is smooth over S_1,

(6.6.4.2) for each $j = 0, 1, \ldots, r$, $\overline{\epsilon}^{-1}(\sigma_j(S_1))$ is a union of smooth divisors Z_{jk}, flat over S_1, with multiplicity $\mu!$,

(6.6.4.3) for each Z_{jk}, $\overline{\epsilon}$ induces a morphism of smooth models $(\overline{U}_2, Z_{ik}) \longrightarrow (\mathbf{P}^1_{S_1}, \sigma_j(S_1))$,

(6.6.4.4) the natural morphism $f_2 : U_2 := \overline{U}_2 \setminus \coprod Z_{ik} \longrightarrow S_1$ is an elementary fibration,

(6.6.4.5) the restriction of $\overline{\epsilon} : U_2 \xrightarrow{\epsilon} U_1$ is a Galois étale covering with group

$$G = Gal(\kappa(S_1)(x, (x-\theta_1)^{\frac{1}{\mu!}}, \ldots, (x-\theta_r)^{\frac{1}{\mu!}})/\kappa(S_1)(x)) \cong (\mathbf{Z}/\mu!\mathbf{Z})^r.$$

Indeed, since the branch loci of the λ_j do not intersect each other at finite distance, (6.6.4.2) and (6.6.4.3) are clear except for $j = 0$, and $\overline{U}_2$ is smooth over S_1 outside $\overline{\epsilon}^{-1}(\sigma_0(S_1)) = \overline{\epsilon}^{-1}(\{\infty\} \times S_1)$. At ∞, we use the coordinates $y_j = 1/x_j$ on L_j and $y = 1/x$ on $\mathbf{P}^1_{S_1}$, so that $y_j^{\mu!} = \frac{y}{1-\theta_j y}$. There is an affine cover $(\mathcal{V}_i)_{i=1,\ldots,r}$ of a neighborhood of $\overline{\epsilon}^{-1}(y=0)$ in $\overline{U}_2$ such that $\{y_i, t_{ij} := y_j/y_i \ (j = 1,\ldots,r,\ j \neq i)\}$ are local coordinates on $\mathcal{V}_i$. We have $y_i^{\mu!} = \frac{y}{1-\theta_i y}$, $t_{ij}^{\mu!} = \frac{1-\theta_i y}{1-\theta_j y}$; in particular, the t_{ij} are unramified at $y = 0$, and properties (6.6.4.1, 2, 3, 4) are now clear.

Finally, (6.6.4.1) implies that $\overline{U}_2$ is normal. It is just the normalization of $\mathbf{P}^1_{S_1}$ in the Kummer extension $\kappa(S_1)(x, (x-\theta_1)^{\frac{1}{\mu!}}, \ldots, (x-\theta_r)^{\frac{1}{\mu!}})/\kappa(S_1)(x)$, which implies (6.6.4.5).

We need the following slight generalization of (III.1.5):

Lemma 6.6.5 *Given an elementary fibration*

(6.6.5.1)
$$\begin{array}{ccccc} X & \hookrightarrow & \overline{X} & \hookleftarrow & Z \\ & \overset{f}{\searrow} & \downarrow \overline{f} & \swarrow & \\ & & S & & \end{array}$$

and a finite set of closed points $\{\xi\} \subset X$*, there is an open neighborhood* U *of* $\{\xi\}$ *(resp.* T *of* $f(\{\xi\})$*), such that* f *induces a coordinatized elementary fibration* $U \longrightarrow T$*.*

Proof: If $\{\xi\}$ has only one element, this is (III.1.5). If $f(\{\xi\})$ has only one element, this is obtained by a trivial modification of the proof of *loc.cit.* In general, let us choose a closed S-embedding $\overline{X} \longrightarrow \mathbf{P}_S^M$. For each $\zeta \in f(\{\xi\})$, we denote by $\{\xi_1^\zeta, \ldots, \xi_r^\zeta\}$ the finite set $f^{-1}\{\zeta\} \cup Z_\zeta$ of closed points of $\overline{X}_\zeta$. There exists an open dense subset $\mathcal{U}^\zeta$ of the Grassmannian of lines in the dual projective space $\check{\mathbf{P}}^M$ such that for any $D_\zeta \in \mathcal{U}^\zeta$:

(a)$_\zeta$ D_ζ is a Lefschetz pencil; in particular, the axis of D_ζ (which has codimension 2 in $\mathbf{P}^M$) does not cut $\overline{X}_\zeta$, so that D_ζ gives rise to a finite morphism $\overline{\pi}_\zeta : \overline{X}_\zeta \longrightarrow \mathbf{P}^1 \simeq D_\zeta$;

(b)$_\zeta$ if $i \neq j$, $\overline{\pi}_\zeta(\xi_i^\zeta) \neq \overline{\pi}_\zeta(\xi_j^\zeta)$;

(c)$_\zeta$ for each i, $\overline{\pi}_\zeta$ is étale at each point of $\overline{\pi}_\zeta^{-1}\overline{\pi}_\zeta(\xi_i^\zeta)$.

There exists a relative line D/S in the dual projective space $\check{\mathbf{P}}_S^M$ such that $D_\zeta \in \mathcal{U}^\zeta$ for each ζ. Up to changing the coordinate in $\mathbf{P}^1 \simeq D_\zeta$, we may assume moreover that

(d)$_\zeta$ the hyperplane H_∞ corresponding to $\infty \in D_\zeta$ cuts $\overline{X}_\zeta$ transversally, and does not meet the points ξ_i^ζ. We then finish the proof as in *loc.cit.* □

6.6.6 Up and down. We now return to the situation of (6.6.4). Since we may shrink S around $\zeta = f(\xi)$ (*i.e.* S_1 around $\tau^{-1}(\{\zeta\})$), we can apply the previous lemma to the elementary fibration $f_2 : U_2 \longrightarrow S_1$ of (6.6.4.4). We find a divisor F_2 in $\overline{U}_2$, which is an étale covering of S_1 and does not contain $\epsilon^{-1}\tau^{-1}\{\pi(\xi)\}$, such that the restriction of f_2 to $V_2 := U_2 \backslash F_2 \xrightarrow{f_2} S_1$ is a *coordinatized* elementary fibration:

(6.6.6.1)
$$\begin{array}{ccccc} V_2 & \overset{j_2}{\hookrightarrow} & \overline{V}_2 = \overline{U}_2 & \hookleftarrow & (\overline{U}_2 \backslash U_2) \coprod F_2 \\ \downarrow \pi_2 & & \downarrow \overline{\pi}_2 & & \downarrow \pi_2' \\ V_2' & \overset{j_2'}{\hookrightarrow} & \mathbf{P}^1_{S_1} & \hookleftarrow & Z_2' \\ & \overset{f_2'}{\searrow} & \downarrow \mathrm{pr} & \overset{g_2'}{\swarrow} & \\ & & S_1 & & \end{array}$$

where, moreover,

(6.6.6.2) $\overline{\pi}_2$ is étale above $\pi'_2(\overline{U}_2 \backslash U_2)$.

We may also replace F_2 by the union of its conjugates under the Galois group G, hence assume that

(6.6.6.3) V_2 is a G-covering of $V_1 := U_1 \backslash \epsilon(F_2)$.

Since we may replace S_1 by any finite étale covering, we may assume that

(6.6.6.4) f'_2 is a rational elementary fibration.

The conditions that $\pi(\xi) \notin \tau\overline{\epsilon}(F_2)$ and that F_2 is disjoint from $\overline{U}_2 \backslash U_2$ imply that $\tilde{F} := \overline{\pi}^{-1}\tau\overline{\epsilon}F_2 \subset \overline{U}$ is a divisor in U which does not contain ξ, and is étale finite over S; in particular, $V := U \backslash \tilde{F} \xrightarrow{f} S$ is an elementary fibration.

The commutative diagram

$$\begin{array}{ccccccc} U & \stackrel{\pi}{\to} & U' & \stackrel{\tau}{\leftarrow} & U_1 = U'_{S_1} & \stackrel{\epsilon}{\leftarrow} & U_2 \\ \downarrow f & & \downarrow f' & & \downarrow f_1 & & \downarrow f_2 \\ S & = & S & \stackrel{\tau}{\leftarrow} & S_1 & = & S_1 \end{array} \tag{6.6.6.5}$$

(where the horizontal maps are étale coverings and the vertical maps are elementary fibrations) induces by restriction a commutative diagram

$$\begin{array}{ccccccc} V = U \backslash \tilde{F} & \stackrel{\pi}{\to} & V' & \stackrel{\tau}{\leftarrow} & V_1 = V'_{S_1} & \stackrel{\epsilon}{\leftarrow} & V_2 = (U_2 \backslash F_2) \\ \downarrow f & & \downarrow f' & & \downarrow f_1 & & \downarrow f_2 \\ S & = & S & \stackrel{\tau}{\leftarrow} & S_1 & = & S_1 \end{array}$$

(with the same properties), which extends to a commutative diagram

$$\begin{array}{ccccccccc} V & \stackrel{\pi}{\to} & V' & \stackrel{\tau}{\leftarrow} & V_1 = V'_{S_1} & \stackrel{\epsilon}{\leftarrow} & V_2 & \stackrel{\pi_2}{\to} & V'_2 \\ \downarrow f & & \downarrow f' & & \downarrow f_1 & & \downarrow f_2 & & \downarrow f'_2 \\ S & = & S & \stackrel{\tau}{\leftarrow} & S_1 & = & S_1 & = & S_1 \end{array} \tag{6.6.6.6}$$

(since there is more danger of inflation than confusion with notation at this point, we use abusively the same letter for various restrictions of the same morphism).

Let us set $(\mathcal{E}_2, \nabla_2) := \epsilon^*\tau^*\pi_*((\mathcal{E}, \nabla)_{|V})$. We have:

$$\begin{aligned} \tau^* R^1_{DR} f_*((\mathcal{E}, \nabla)_{|V}) &\cong R^1_{DR} f_{1*}(\tau^*\pi_*(\mathcal{E}, \nabla)_{|V}) \\ &\cong R^1_{DR} f_{1*}(\epsilon_*(\mathcal{E}_2, \nabla_2))^G, \end{aligned} \tag{6.6.6.7}$$

$$R^1_{DR} f_{1*}(\epsilon_*(\mathcal{E}_2, \nabla_2)) \cong R^1_{DR} f'_{2*}(\pi_{2*}(\mathcal{E}_2, \nabla_2)). \tag{6.6.6.8}$$

6.6.7 Finiteness. We claim that for this choice V of a neighborhood of ξ, $R^1_{DR} f_*((\mathcal{E}, \nabla)_{|V})$ is a coherent $\mathcal{O}_S$-module. In view of (6.6.6.7), (6.6.6.8), it suffices to show that $R^1_{DR} f'_{2*}(\pi_{2*}(\mathcal{E}_2, \nabla_2))$ is a coherent $\mathcal{O}_{S_1}$-module. Let us list a few properties of the latter.

(6.6.7.1) $\pi_{2*}\mathcal{E}_2$ extends to a locally free $\mathcal{O}_{\mathbf{P}^1_{S_1}}$-module.

This follows from the fact that $\mathcal{E}_{|U}$ extends to a locally free $\mathcal{O}_{\tilde{\overline{X}}|S}$-module (6.6.3) and from the fact that all maps in the upper row of (6.6.6.6) extend to finite flat morphisms of relative projective completions.

(6.6.7.2) The Turrittin ramification indices of $\pi_{2*}\nabla_2$ are all 1.

Indeed, by the ramification property (6.6.4.2), the Turrittin indices of $\epsilon^*\tau^*\pi_*((\mathcal{E},\nabla)_{|U})$ at every branch of $\overline{U}_2 \setminus U_2$ are 1, and this connection has no singularity at the branches of F_2; (6.6.7.2) then follows from (6.6.6.2). Let us write $\mathbf{P}^1_{S_1} \setminus V'_2 = \coprod W_i$, where the W_i are images of sections $\sigma_i{}'$ of $f'_2 = \mathrm{pr}$.

(6.6.7.3) For any component W_i and any point $Q'_2 \in W_i$, there are equalities of Newton polygons:

$$NP_{Q'_2}(\pi_{2*}(\mathcal{E}_2,\nabla_2)_{|f_2'^{-1}(g'_2(Q'_2))}) = NP_{W_i}(\pi_{2*}(\mathcal{E}_2,\nabla_2)),$$
$$NP_{Q'_2}(\underline{End}(\pi_{2*}(\mathcal{E}_2,\nabla_2))_{|f_2'^{-1}(g'_2(Q'_2))}) = NP_{W_i}(\underline{End}(\pi_{2*}(\mathcal{E}_2,\nabla_2))).$$

This follows from a calculation completely analogous to the one in (6.6.4). More precisely, we may assume that $Q'_2 \in \overline{\pi}_2(\overline{U}_2 \backslash U_2)$ (other points are regular singular and easier to handle). Then there is a unique point $Q_2 \in \overline{U}_2 \backslash U_2$ such that $\overline{\pi}_2(Q_2) = Q'_2$, and a unique point $Q \in \overline{U} \backslash U$ such that $\overline{\pi}(Q) = \tau\epsilon(Q_2)$. If we set $\delta_2 = \deg \overline{\pi}_2$, and using (II.4.3), we find

$$\begin{aligned} &NP_{Q'_2}(\pi_{2*}(\mathcal{E}_2,\nabla_2)_{|f_2'^{-1}(g'_2(Q'_2))}) \\ &\quad = \varphi_{1,\mu!}NP_Q((\mathcal{E},\nabla)_{|f^{-1}g(P)}) + LQ_{(\delta\delta_2-1)\mu} \\ &\quad = NP_{W_i}(\pi_{2*}(\mathcal{E}_2,\nabla_2)), \end{aligned}$$

$$\begin{aligned} &NP_{Q'}(\underline{End}(\pi_{2*}(\mathcal{E}_2,\nabla_2))_{|f_2'^{-1}(g'_2(Q'_2))}) \\ &\quad = \varphi_{1,\mu!}NP_Q(\underline{End}(\mathcal{E},\nabla)_{|f^{-1}g(Q)}) \\ &\qquad + 2(\delta\delta_2 - 1)\mu NP_Q((\mathcal{E},\nabla)_{|f^{-1}g(Q)}) + LQ_{(\delta\delta_2-1)^2\mu^2} \\ &\quad = NP_{W_i}(\underline{End}(\pi_{2*}(\mathcal{E}_2,\nabla_2))). \end{aligned}$$

These three properties now allow us to apply our result of (II.7.2.9) on the existence of cyclic vectors in neighborhoods of singularities: there exists a smooth K-variety S'_1 and a flat morphism $h : S'_1 \to S_1$ such that

(6.6.7.4) S'_1 is finite over $h(S'_1)$,

(6.6.7.5) $S_1 \backslash h(S'_1)$ has codimension ≥ 2 in S_1,

(6.6.7.6) for any $\zeta' \in S'_1$, there is an affine neighborhood $\mathcal{U}'$ of the finite subset $\{\sigma_i{}'(\zeta')\}$ of $\mathbf{P}^1_{S'_1}$ such that the inverse image of $\pi_{2*}\mathcal{E}_2$ on $\mathcal{U}' \cap (V'_2 \times_{S_1} S'_1)$ is *cyclic* with respect to $(\pi_{2*}\nabla_2)(d/dx)$. Let us fix such a cyclic vector e, and denote by Λ the associated differential polynomial. Property (6.6.7.3) also allows us to apply

(II.7.1.2), and obtain that, up to shrinking $\mathcal{U}'$, the leading coefficient of the indicial polynomial of Λ at $W_i \times_{S_1} S_1'$ is a unit in $\mathcal{O}((W_i \times_{S_1} S_1') \cap \mathcal{U}')$.

We can now conclude, by the method of (III.4) (see especially the Remark III.4.7.iii) that $h^* R^1_{DR} f'_{2*}(\pi_{2*}(\mathcal{E}_2, \nabla_2))$ is coherent over $\mathcal{O}_{S_1'}$. Taking into account the flatness of h and (6.6.7.5), the purity lemma (III.4.9) now shows that $R^1_{DR} f'_{2*}(\pi_{2*}(\mathcal{E}_2, \nabla_2))$ is coherent over $\mathcal{O}_{S_1}$, and we conclude finally that $R^1_{DR} f_*((\mathcal{E}, \nabla)_{|V})$ is a coherent $\mathcal{O}_S$-module.

6.7 Proof of Proposition 6.4: We claim that, with the same $V \xrightarrow{f} S$ as in (6.6.6), the canonical morphism

$$(R^j_{DR} f_*(\mathcal{E}, \nabla)_{|V})^{an} \longrightarrow R^j_{DR} f^{an}_*((\mathcal{E}^{an}, \nabla^{an})_{|V^{an}})$$

is an isomorphism in $\mathbf{MIC}(S^{an})$. This is trivial for $j \neq 0, 1$, and the case $j = 0$ follows from the argument in (5.2.6). Let us now consider the case $j = 1$. In view of (6.6.6.7), (6.6.6.8) and by finite flat descent, it suffices to show that the map

$$(R^1_{DR} f'_{2*}(\pi_{2*}(\mathcal{E}_2, \nabla_2)))^{an} \longrightarrow R^1_{DR} f'^{an}_{2*}(\pi^{an}_{2*}(\mathcal{E}^{an}_2, \nabla^{an}_2))$$

is an isomorphism in $\mathbf{MIC}(S^{an}_1)$. The argument in (5.2.6) also shows that this map is injective. In order to prove the surjectivity, let us consider again the flat morphism h of (6.6.7), and perform the cyclic reduction as above.

We can now conclude, by the method of (5.1), (5.2) (see especially the Remark 5.2.7.ii) that

$$h^{an*}(R^1_{DR} f'_{2*}(\pi_{2*}(\mathcal{E}_2, \nabla_2)))^{an} \longrightarrow h^{an*} R^1_{DR} f'^{an}_{2*}(\pi^{an}_{2*}(\mathcal{E}^{an}_2, \nabla^{an}_2))$$

is surjective - hence in fact an isomorphism of coherent objects in $\mathbf{MIC}(S'^{an}_1)$. Taking into account the flatness of h and (6.6.7.5), the purity lemma (5.3.1) now shows that

$$(R^1_{DR} f'_{2*}(\pi_{2*}(\mathcal{E}_2, \nabla_2)))^{an} \longrightarrow R^1_{DR} f'^{an}_{2*}(\pi^{an}_{2*}(\mathcal{E}^{an}_2, \nabla^{an}_2))$$

is an isomorphism, and we conclude finally that

$$(R^1_{DR} f_*(\mathcal{E}, \nabla)_{|V})^{an} \longrightarrow R^1_{DR} f^{an}_*((\mathcal{E}^{an}, \nabla^{an})_{|V^{an}})$$

is an isomorphism in $\mathbf{MIC}(S^{an})$.

This achieves the proof of (6.4) and of the main theorem (6.1).

6.8 Properties of the GAGA functor

In cohomological degree 0, the statement of (6.1) reduces to the following:

Corollary 6.8.1 *Let $f : X \longrightarrow S$ be a smooth morphism of smooth algebraic K_0-varieties, with $S = A_0(f)$. Then for any coherent $\mathcal{O}_X$-module with integrable connection $(\mathcal{E}, \nabla)$, the natural morphism*

$$(f_* \mathcal{E}^{\nabla_{|\underline{Der}(X/S)}}) \otimes_{\mathcal{O}_S} \mathcal{O}_{S^{an}} \longrightarrow f^{an}_*(\mathcal{E}^{an})^{\nabla^{an}_{|\underline{Der}_{cont}(X^{an}/S^{an})}}$$

is an isomorphism in $\mathbf{MIC}(S^{an})$.

(Simplified) Proof: Indeed, in this special case, one can provide a much shorter argument. With the help of Lemma (III.4.2.1) and Remark (III.3.2.4.ii), we may reduce by dévissage to the case when f is a coordinatized elementary fibration, or even a rational elementary fibration (arguing as in (3.3) or (4.3)). The result then follows by application of (2.1) exactly as in (5.2.6). □

Corollary 6.8.2 *Let X be a smooth algebraic K-variety defined over an algebraically closed subfield $K_0 \subset K$ which does not contain any Liouville number. The functor*

$$\begin{pmatrix}\text{Coherent } \mathcal{O}_X\text{-modules}\\ \text{with integrable connection}\\ \text{defined over } K_0\end{pmatrix} \longrightarrow \begin{pmatrix}\text{Coherent } \mathcal{O}_{X^{an}}\text{-modules}\\ \text{with integrable connection}\end{pmatrix},$$

$$(\mathcal{E}, \nabla) \longmapsto (\mathcal{E}^{an}, \nabla^{an})$$

is fully faithful.

Proof: Apply (6.8.1) to internal $\underline{Hom}$ as in (3.6). □

Remark 6.8.3 This functor is however not essentially surjective, even for $X = \mathbf{A}^1$. Indeed, let h be any non-polynomial entire function. Then there is no entire function g such that $\frac{g'}{g} + h$ is a polynomial. This means that the rank one connection defined by $\nabla(\frac{d}{dx})e = he$ is not algebraizable.

This is in contrast with the complex-analytic situation, where the analogous functor is in fact essentially surjective, but not fully faithful.

Corollary 6.8.4 *Let $(\mathcal{E}, \nabla), (\mathcal{E}', \nabla')$ be coherent $\mathcal{O}_X$-modules with integrable connections over X. Then the natural map*

$$Ext^1_{MIC(X_K)}((\mathcal{E}_K, \nabla_K), (\mathcal{E}'_K, \nabla'_K)) \longrightarrow Ext^1_{MIC(X_K^{an})}((\mathcal{E}_K^{an}, \nabla_K^{an}), (\mathcal{E}_K'^{an}, \nabla_K'^{an}))$$

is a bijection.

Proof: Apply (6.1) with $j = 1$ to internal $\underline{Hom}$ as in (3.7). □

§7 The relative non-archimedean Turrittin theorem

Let $X \longrightarrow S$ be a rational elementary fibration over an affine base S as in (5.1.0) and let $(\mathcal{M}, \nabla)$ be a coherent object of $\mathbf{MIC}(X)$.

Proposition 7.1 *We make the following technical assumption:*

(∗) The horizontal side of the Newton polygon at $P \in Z$ of $End(\mathcal{M}_{|f^{-1}(g(P))})$ does not depend on the point P, except possibly on a subset of codimension 2 in Z.

There exists a finite surjective morphism $S' \longrightarrow S$ and a finite set of elements $a \in \mathcal{O}(S')$ such that, for any $i = 0, \ldots, r$, the $\mathcal{O}(X)[\frac{d}{dx}]$-module $\Gamma(X, \mathcal{M})$ is solvable in

$$\mathcal{O}(S')[[x - \theta_i]][\log(x - \theta_i), (x - \theta_i)^{\alpha}, \exp(a(x - \theta_i)^{-\frac{k}{\mu!}})]_{\alpha \in \Delta,\, k \in \mathbf{Z}_{>0}}.$$

Proof: This follows from the formal decomposition of $\mathcal{M}$ along the divisor Z_i established in (II.6) (for $i = 0$, we tacitly replace θ_i by ∞). □

7.2 Let $\mathcal{S} \subset S^{an}$ be an affinoid subspace, $\mathcal{X} \longrightarrow \mathcal{S}$ be the morphism obtained from f^{an} by base change. The morphism $S'^{an} \longrightarrow S^{an}$ is finite, hence induces a finite morphism of affinoid spaces $\mathcal{S}' \longrightarrow \mathcal{S}$ for any irreducible component $\mathcal{S}'$ of the inverse image of $\mathcal{S}$ in S'^{an} [BGR, 7.1.2, 9.4.4].

We have a Mittag-Leffler decomposition (induced by (III.4.5.2))

$$\mathcal{O}(\mathcal{X}) = \mathcal{O}(\mathbf{A}^{1\,an}_{\mathcal{S}}) \oplus \bigoplus_{i=1}^{r} \frac{1}{x - \theta_i} \mathcal{O}(\mathbf{P}^{1\,an}_{\mathcal{S}} \backslash \sigma_i(\mathcal{S})), \tag{7.2.1}$$

where

$$\mathcal{O}(\mathbf{A}^{1\,an}_{\mathcal{S}}) = \left\{ \sum_{n \geq 0} a_{n,\infty} x^n \in \mathcal{O}(\mathcal{S})[[x]] \,,\ \lim_{n \to \infty} ||a_{n,\infty}||^{1/n} = 0 \right\},$$

$$\mathcal{O}(\mathbf{P}^{1\,an}_{\mathcal{S}} \setminus \sigma_i(\mathcal{S})) = \left\{ \sum_{n \geq 0} \frac{a_{n,i}}{(x - \theta_i)^n} \in \mathcal{O}(\mathcal{S}) \left[\left[\frac{1}{x - \theta_i} \right]\right],\quad \lim_{n \to \infty} ||a_{n,i}||^{1/n} = 0 \right\}$$

for some (any) Banach norm $||\ \ ||$ on the Tate K-algebra $\mathcal{O}(\mathcal{S})$, and similarly for $\mathcal{S}'$ instead of $\mathcal{S}$. For any $\epsilon > 0$, we introduce the affinoid tube in $S'^{an} \times (\mathbf{A}^1)^{an}$

$$T_{\mathcal{S}',i,\epsilon} = \{(s, x),\ s \in \mathcal{S}',\ |x - \theta_i(s)| \leq \epsilon\},$$

and

$$T^*_{\mathcal{S}',i,\epsilon} = \{(s, x),\ s \in \mathcal{S}',\ 0 < |x - \theta_i(s)| \leq \epsilon\}.$$

We define

$$C'_{\mathcal{S}',i} = \bigcup_{\epsilon > 0} \mathcal{O}(T^*_{\mathcal{S}',i,\epsilon}),$$

$$A_{\mathcal{S}',i} = \bigcup_{\epsilon > 0} \mathcal{O}(T_{\mathcal{S}',i,\epsilon})[\log(x - \theta_i), (x - \theta_i)^{\alpha}, \exp(a(x - \theta_i)^{-\frac{k}{\mu!}})]_{\alpha \in \Delta,\, k \in \mathbf{Z}_{>0}}.$$

As in (5.2.6), we consider the Banach field K', completion of the fraction field of $\mathcal{O}(\mathcal{S}')$ with respect to the spectral norm, and let $\mathcal{M}^{(K')}_{\theta_i}$ be the field of meromorphic functions at θ_i on the analytic curve over K', obtained by extension of $\mathcal{X}$ to K'. Then, as in (5.2.4), we show that

$$A'_{\mathcal{S}',i} := C'_{\mathcal{S}',i} \cap A_{\mathcal{S}',i} \subset \mathcal{M}^{(K')}_{\theta_i}.$$

This in turn implies

$$A'_{\mathcal{S}',i} = \bigcup_{\epsilon>0} \mathcal{O}(T_{\mathcal{S}',i,\epsilon})\left[\frac{1}{x-\theta_i}\right]$$
$$= \left\{ \sum_{n>>-\infty} a_{n,i}(x-\theta_i)^n \in \mathcal{O}(\mathcal{S}')((x-\theta_i)),\ \underline{\lim}_{n\to\infty}||a_{n,i}||^{-1/n} > 0 \right\}.$$

We adopt similar definitions for $i = 0$, upon replacing θ_i by ∞, and set

$$A_{\mathcal{S}'} = \prod_{i=0}^{r} A_{\mathcal{S}',i} \quad C'_{\mathcal{S}'} = \prod_{i=0}^{r} C'_{\mathcal{S}',i} \quad \partial = \frac{d}{dx}.$$

Let us remark that

- $A_{\mathcal{S}',i}$ is free over $A'_{\mathcal{S}',i}$, and the $(x-\theta_i)$-adic completion of $A'_{\mathcal{S}',i}$ is $\mathcal{O}(S')((x-\theta_i))$; it follows that $A_{\mathcal{S}'}$ is faithfully flat over $A^{\partial}_{\mathcal{S}'} = C^{\partial}_{\mathcal{S}'} = \mathcal{O}(\mathcal{S}')^{r+1}$;
- $\mathcal{O}(\mathcal{X})$ embeds diagonally in an obvious way into $C'_{\mathcal{S}'}$, and $A_{\mathcal{S}'} \cap \mathcal{O}(\mathcal{X}) = \mathcal{O}(\mathcal{S})[x, \frac{1}{\prod(x-\theta_i)}] \cong \mathcal{O}(X) \otimes_{\mathcal{O}(S)} \mathcal{O}(\mathcal{S})$, as follows from the Mittag-Leffler decomposition (5.1.6.1).

The following result is a generalization of the *p-adic Turrittin theorem* [B1] *in the relative case*:

Theorem 7.3 *Under assumption* $(*)$, *the* $\mathcal{O}(X)[\frac{d}{dx}]$*-module* $\Gamma(X,\mathcal{M})$ *is solvable in* $A_{\mathcal{S}'}$.

Proof: We have to show that $\Gamma(X,\mathcal{M})$ is solvable in $A_{\mathcal{S}',i}$, for every $i = 0,\ldots,r$. From (7.1), the problem becomes to show that all formal series $\in \mathcal{O}(S')[[x-\theta_i]]$ involved in the solutions converge, in the sense that they define elements of $\mathcal{O}(T_{\mathcal{S}',i,\epsilon})$ for ϵ small enough, i.e. are of the form

$$\sum_{n\geq 0} a_{n,i}(x-\theta_i)^n,\ \liminf_{n\to\infty} ||a_{n,i}||^{-1/n} > 0.$$

As in [B1] (or (II.6.ii)), we remark that these series are formal solutions of $\Gamma(X, \underline{End}\ \mathcal{M})$, and that the Turrittin exponents of $\underline{End}\ \mathcal{M}$ belong to $\Delta \subset K_0$. Then the same argument of indicial polynomials as in *loc. cit.* (due to Clark [Cl]) allows us to conclude from the fact that K_0 does not contain Liouville numbers. □

Remark 7.4 It is tempting to try to apply (2.4) with $C = A_{S'}C'_{S'}$ instead of the differential ring C being considered in (5.2.6) in order to obtain the surjectivity of φ_1 without using (5.1.1). However, a problem arises with the stability of the differential ring $A_{S',i}$ under integration: we have to consider, as in (5.2.4.1), formal expressions of the type $y = u^{-1}z^{-\alpha}\ e^{-a\,z^{-k}} \int u\, z^{\alpha}\ e^{a\,z^{-k}} dz$, with $z = (x - \theta_i)^{\frac{1}{\mu!}}$, $u \in A'_{S',i}$ or $C'_{S',i}$, $\alpha \in \Delta$, and remark that y satisfies a homogenous differential equation of second order, whose indicial polynomial $\phi_{z=0}(T)$ equals $ak(T - k - 1)$ if $ka \neq 0$. This would lead us to assume that any non-zero element of the additive monoid generated by the elements $a \in \mathcal{O}(S')$ occuring in (7.1) are units.

Appendix E: Riemann's "existence theorem" in higher dimension, an elementary approach

Theorem E.1 *Let X be a smooth complex algebraic variety. Then the natural functor*

$$\{\text{algebraic étale coverings of } X\}$$
$$\xrightarrow{(*)} \{\text{topological unramified finite coverings of } X(\mathbf{C})\}$$

is an equivalence of categories.

The problem of essential surjectivity of this functor, sometimes called Riemann's existence problem, was settled by Grauert and Remmert [GR] and subsequently by Grothendieck using Hironaka's resolution of singularities.

In this appendix, we present an "elementary proof" by reduction to the one-dimensional case.

E.2 Recall that a topological unramified covering of $X(\mathbf{C})$ admits a canonical structure of an analytic variety $\mathcal{Y}$, endowed with a holomorphic étale morphism $\pi : \mathcal{Y} \longrightarrow X^{an}$. In the finite case, we have equivalences of categories

$$\{\text{finite étale coverings } \pi : \mathcal{Y} \longrightarrow X^{an}\} \longrightarrow$$

$$\{\text{coherent } \mathcal{O}_{X^{an}}\text{-modules with integrable connection} (\mathcal{E}, \nabla), \text{ endowed with a horizontal (commutative) } \mathcal{O}_{X^{an}}\text{-algebra structure)} \times : \mathcal{E}^{\otimes 2} \longrightarrow \mathcal{E}\}\}$$

$$\longrightarrow \{\text{étale } \mathcal{O}_{X^{an}}\text{-algebras } (\mathcal{E}, \times), \text{ coherent as } \mathcal{O}_{X^{an}}\text{-modules}\}$$

given by

$$\pi \longmapsto (R^0_{DR}\pi_*(\mathcal{O}_{\mathcal{Y}}, d), \times) = (\pi_*\mathcal{O}_{\mathcal{Y}}, \nabla, \times) \longmapsto (\pi_*\mathcal{O}_{\mathcal{Y}}, \times),$$

where $\times$ is deduced from the $\pi^{-1}\mathcal{O}_{X^{an}}$-algebra structure of $\mathcal{O}_{\mathcal{Y}}$. These equivalences also take place in the algebraic categories:

$$\{\text{finite étale coverings } \pi : Y \longrightarrow X\} \longrightarrow$$

$$\{\text{coherent} \mathcal{O}_X\text{-modules with integrable } \textit{regular} \text{ connection} (\mathcal{E}, \nabla), \text{ endowed with a horizontal (commutative) } \mathcal{O}_X\text{-algebra structure } \times : \mathcal{E}^{\otimes 2} \longrightarrow \mathcal{E}\}$$

$$\longrightarrow \{\text{finite étale } \mathcal{O}_X\text{-algebras } \mathcal{E}\}.$$

They permit us to deduce that the functor $(*)$ is fully faithful, from the comparison Corollary 3.6. (But this fact is of course much more elementary, *cf.* [SGA 4, Example XI, Proposition 4.3. ii]).

E.3 We will need an "extension lemma" for étale coverings, which, thanks to the dictionary E.2, will be obtained as a special case of a general extension lemma for regular connections.

Extension Lemma E.4 *Let X be a smooth complex algebraic variety, and let $\mathcal{L}$ be a local system of finite dimensional complex vector spaces on $X(\mathbf{C})$. Assume that there is a dense open subset $U \subset X$ and a regular connection $(\mathcal{E}_U, \nabla_U)$ on U such that $(\mathcal{E}_U^{an})^{\nabla_U^{an}} \cong \mathcal{L}_{|U(\mathbf{C})}$. Then there is a regular connection $(\mathcal{E}, \nabla)$ on X such that $(\mathcal{E}^{an})^{\nabla^{an}} \cong \mathcal{L}$ and $(\mathcal{E}, \nabla)_{|U} \cong (\mathcal{E}_U, \nabla_U)$.*

Proof: Let Z be the smooth, purely 1-codimensional part of $X \setminus U$. We denote by $j : U' := U \amalg Z \hookrightarrow X$ the canonical immersion, and by $(\tilde{\mathcal{E}}_U, \tilde{\nabla}_U)$ the τ-extension of $(\mathcal{E}_U, \nabla_U)$ to U' (I.4.7). Because $(\mathcal{E}_U^{an})^{\nabla_U^{an}}$ extends to a local system on $U'(\mathbf{C})$ (even on $X(\mathbf{C})$), the local monodromy around Z is trivial. With the help of [De, II, 3.11], and taking into account the fact that $\tau(0) = 0$, we deduce that $\mathrm{Res}_Z \tilde{\nabla}_U = 0$, hence $\tilde{\nabla}_U$ is a genuine connection on U' (without pole at Z). We claim that $(\mathcal{E}, \nabla) := j_*(\tilde{\mathcal{E}}_U, \tilde{\nabla}_U)$ satisfies the requirements. Indeed, since $X \setminus U'$ has codimension ≤ 2, $\mathcal{E}$ is coherent, according to [EGA IV, 5.11.4], and the natural map $j^*(\mathcal{E}, \nabla) \longrightarrow (\tilde{\mathcal{E}}_U, \tilde{\nabla}_U)$ is an isomorphism (which restricts to an isomorphism $(\mathcal{E}, \nabla)_{|U} \longrightarrow (\mathcal{E}_U, \nabla_U)$ over U). On the other hand, $(\mathcal{E}^{an})^{\nabla^{an}}_{|U'(\mathbf{C})} \cong (\tilde{\mathcal{E}}_U^{an})^{\tilde{\nabla}_U^{an}} \cong \mathcal{L}_{|U'(\mathbf{C})}$. Since $(X \setminus U')(\mathbf{C})$ has topological codimension ≤ 4 in $X(\mathbf{C})$, $\pi_1((X \setminus U')(\mathbf{C}), x) \longrightarrow \pi_1(X(\mathbf{C}), x)$ is an isomorphism, and we conclude that $(\mathcal{E}^{an})^{\nabla^{an}} \cong \mathcal{L}$. □

Extension Lemma E.5 *Assume that there is an étale dominant morphism $X' \xrightarrow{\varphi} X$ finite over its image, and an étale covering $Y' \longrightarrow X'$, such that Y'^{an} and $\mathcal{Y} \times_{X^{an}} X'^{an}$ are isomorphic coverings of X'^{an}. Then there exists an étale covering $Y \longrightarrow X$ such that Y^{an} and $\mathcal{Y}$ are isomorphic coverings of X^{an}.*

Proof: Let us first assume that φ is surjective, hence an étale covering. Let Y''/Y' be an étale covering such that the induced covering Y''/X is Galois with group G. Then by the usual Galois theory of topological coverings, there is a subgroup $H \subset G$ such that $\mathcal{Y} \cong Y''^{an}/H$. It then suffices to take $Y := Y''^{an}/H$.

This first step allows us to assume that φ is an open dominant immersion $X' \hookrightarrow X$. Let $(R^0_{DR}\pi_*(\mathcal{O}_\mathcal{Y}, d), \times)$ be the triple associated to π as in D.2. We know from the assumption that the restriction to X' of this triple is algebraizable; in particular, $R^0_{DR}\pi_*(\mathcal{O}_\mathcal{Y}, d)$ comes from a regular connection on X'. In virtue of the extension Lemma E.4, $R^0_{DR}\pi_*(\mathcal{O}_\mathcal{Y}, d)$ itself comes from a regular connection $(\mathcal{E}, \nabla)$ on X. By the comparison Corollary 3.6, the horizontal multiplication $\times : \mathcal{E}^{an\otimes 2} \longrightarrow \mathcal{E}^{an}$ comes from a horizontal $\mathcal{O}_X$-linear map $\times : \mathcal{E}^{\otimes 2} \longrightarrow \mathcal{E}$, which endows $\mathcal{E}$ with the structure of a finite $\mathcal{O}_X$-algebra. It then suffices to take $Y := \mathrm{Spec}\mathcal{E}$. □

E.6 Let us prove the essential surjectivity of $(*)$ by induction on $\dim X$, assuming the (elementary) case $\dim X = 1$. According to E.5, we may replace X by any open dense subset. Hence we may assume that X is the total space of an elementary fibration

$$\begin{array}{ccccc} X & \overset{j}{\hookrightarrow} & \overline{X} & \hookleftarrow & Z \\ & \overset{f}{\searrow} & \downarrow \overline{f} & \swarrow & \\ & & S. & & \end{array}$$

Let K be an algebraically closed subfield of $\mathbf{C}$ of finite transcendence degree over $\mathbf{Q}$, such that this elementary fibration comes from an elementary fibration

$$\begin{array}{ccccc} X_K & \overset{j_K}{\hookrightarrow} & \overline{X}_K & \hookleftarrow & Z_K \\ & \overset{f_K}{\searrow} & \downarrow \overline{f}_K & \swarrow & \\ & & S_K & & \end{array}$$

defined over K. Let $s \in S(\mathbf{C})$ induce a geometric *generic* point of S_K, and let us denote by K' the algebraic closure of the image of $\kappa(S_K)$ in $\mathbf{C}$ (via the embedding given by s). We notice that the smooth projective curve $\overline{X}_s$ comes from $\overline{X}_K \times_{S_K} \operatorname{Spec} K'$ by extension of scalars $K' \hookrightarrow \mathbf{C}$.

E.7 In virtue of Riemann's existence theorem in dimension 1, we know that the finite étale covering of X_s^{an} induced by $\mathcal{Y}$ is algebraizable, and in fact comes from a finite covering $\overline{T} \longrightarrow \overline{X}_s$ unramified outside Z_s (inducing a finite étale covering $T \longrightarrow X_s$). By a standard argument of (Weil) descent, such a covering is already defined over K', i.e. comes from a finite covering $\overline{T}_{K'} \longrightarrow \overline{X}_K \times_{S_K} \operatorname{Spec} K'$, unramified outside $Z_K \times_{S_K} \operatorname{Spec} K'$ (thus inducing an étale covering $T_{K'} \longrightarrow X_K \times_{S_K} \operatorname{Spec} K'$).

Moreover, by "spreading-out", one can find an étale morphism $S'_K \longrightarrow S_K$, finite over its image, an embedding $\kappa(S'_K) \hookrightarrow K'$, a finite étale morphism $T'_K \longrightarrow X_K \times_{S_K} S'_K$ and an isomorphism of coverings $T_{K'} \overset{\sim}{\longrightarrow} T'_K \times_{S'_K} \operatorname{Spec} K'$. Let $s' \in S'_K(\mathbf{C})$ be the point above s defined by the embedding $\kappa(S'_K) \hookrightarrow K' \hookrightarrow \mathbf{C}$. Then the fiber at s' of the morphism $T'_K \longrightarrow S'_K$ satisfies

$$T'^{an}_{Ks'} \cong T^{an} \cong \mathcal{Y}_s \cong (\mathcal{Y} \times_{S^{an}} S'^{an})_{s'} .$$

E.8 According to E.5, we can again replace X by an étale dominant neighborhood finite over its image in X, for instance $(X_K \times_{S_K} S'_K) \otimes_K \mathbf{C}$ or a Galois covering of this space lying above $T'_K \otimes_K \mathbf{C}$. This allows us to assume that $\mathcal{Y}_s$ is a trivial covering of X_s. Because any elementary fibration is locally topologically trivial, we have a homotopy long exact sequence, a fragment of which being

$$\pi_1(X_s^{an}, x) \longrightarrow \pi_1(X^{an}, x) \longrightarrow \pi_1(S^{an}, s) \longrightarrow 1$$

for any $x \in X(\mathbf{C})$ lying above s. We see that $\mathcal{Y}$ comes from an analytic étale covering of S^{an}, which is algebraizable by the induction hypothesis. Therefore $\mathcal{Y}$ is algebraizable. □

It would be interesting to give a similar proof of the p-adic avatar, namely Lütkebohmert's theorem [Lü2].

References

[ABa] Y. André and F. Baldassarri, "Geometric theory of G-functions", in *Arithmetic Geometry*, F. Catanese Ed., Cambridge University Press 1997.

[A.C.] N. Bourbaki, *Algèbre Commutative*, Masson, Paris 1985.

[Ad] A. Adolphson, "An index theorem for p-adic differential operators", *Trans. Amer. Math. Soc.*, **216**(1976), 279–293.

[B1] F. Baldassarri, "Differential modules and singular points of p-adic differential equations", *Advances in Math.*, **44**(1982), 155–179.

[B2] F. Baldassarri, "Comparaison entre la cohomologie algébrique et la cohomologie p-adique rigide à coefficients dans un module différentiel I", *Inv. Math.*, **87**(1987), 83–99.

[B3] F. Baldassarri, "Comparaison entre la cohomologie algébrique et la cohomologie p-adique rigide à coefficients dans un module différentiel II", *Math. Ann.*, **280**(1988), 417–439.

[Bk] V. Berkovich, *Spectral Theory and Analytic Geometry over a Non-Archimedean Field*, Math. Surveys and Monographs, **33**, AMS 1990.

[Bn] J. Bernstein, *Lectures on $\mathcal{D}$-modules*, mimeographed notes.

[Be1] P. Berthelot "Cohomologie rigide et cohomologie rigide à support propre", to appear.

[Be2] P. Berthelot "$\mathcal{D}$-modules arithmétiques II. Descente par Frobenius", *Mémoires Soc. Math. de France*, **81**(2000).

[BO] P. Berthelot and A. Ogus, *Notes on Crystalline Cohomology*, Math. Notes, **21**, Princeton Univ. Press 1978.

[BS] E. Bombieri and S. Sperber, "On the p-adic analyticity of solutions of linear differential equations", *Illinois J. of Math.*, **26**(1982), 10–18.

[Bo] A. Borel, *Algebraic D-modules*, Perspectives in Mathematics, **2**, Academic Press 1987.

[BGR] S. Bosch, U. Güntzer and R. Remmert, *Non-Archimedean Analysis*, Grundlehren der Math. Wissenschaften **261**, Springer-Verlag 1984.

[BLR] S. Bosch, W. Lütkebohmert and M. Raynaud, *Néron Models*, Ergebnisse der Math. u. ihrer Grenzgebiete, Ser. 3, N. 21, Springer-Verlag 1990.

[CE] H. Cartan, and S. Eilenberg, *Homological Algebra*, Princeton Mathematical Series **19**, Princeton University Press 1956.

[C] B. Chiarellotto, "Sur le théorème de comparaison entre cohomologies de De Rham algébrique et p-adique rigide", *Ann. Inst. Fourier*, **38**(1988), 1–15.

[CD] G. Christol and B. Dwork, "Modules différentiels sur des couronnes", *Annales Inst. Fourier, Grenoble*, **44**,3(1994), 663–701.

[Cl] D. Clark, "A note on the p-adic convergence of solutions of linear differential equations", *Proc. Am. Math. Soc.*, **17**(1966), 262–269.

[De] P. Deligne, *Equations Différentielles à Points Singuliers Réguliers*, Lecture Notes in Math. **163**, Springer-Verlag, 1970.

[Dw] B. Dwork, *Lectures on p-adic Differential Equations*, Grundleheren der math. Wiss., **253**, Springer-Verlag, 1982.

[DGS] B. Dwork, G. Gerotto and F. Sullivan, *An Introduction to* G*-functions*, Annals of Mathematical Studies, **133**, Princeton University Press, Princeton N.J., 1994.

[EGA I] A. Grothendieck and J. Dieudonné, *Éléments de Géométrie Algébrique I*, Grundleheren der math. Wiss., **166**, Springer-Verlag 1971.

[EGA II] A. Grothendieck and J. Dieudonné, *Éléments de Géométrie Algébrique II. Étude Globale Élementaire de Quelques Classes de Morphismes*, Publications Mathématiques IHES **8**, 1961.

[EGA III] A. Grothendieck and J. Dieudonné, *Éléments de Géométrie Algébrique III. Étude Cohomologique des Faisceaux Cohérents*, Publications Mathématiques IHES **11**, 1961, *ibid.* **17**, 1963.

[EGA IV] A. Grothendieck and J. Dieudonné, *Éléments de Géométrie Algébrique IV. Étude Locale des Schémas et des Morphismes de Schémas*, Publications Mathématiques IHES **20**, 1964, *ibid.* **24**, 1965, *ibid.* **28**, 1966, *ibid.* **32**, 1967.

[F] L. Fuchs, "Zur Theorie der linearen Differentialgleichungen mit veränderlichen Coefficienten", *J. für reine u. angew. Math.*, **66**(1866), 121–160 and **68**(1868), 354–385.

[GL] R. Gérard and A.H.M. Levelt, "Sur les connexions à singularités régulières dans le cas deplusieurs variables", *Funkcial. Ekvac.*, **19**(1976), 149–173.

[GR] H. Grauert and R. Remmert, "Komplexe Räume", *Math. Ann.*, **136**(1958), 245–318.

[G1] A. Grothendieck, "On the De Rham cohomology of algebraic varieties", *Publications Mathématiques IHES*, **29**(1966), 93–103.

[G2] A. Grothendieck, "Crystals and the De Rham cohomology of schemes", in *Dix Exposées sur la Cohomologie des Schémas*, North Holland (1968).

[H1] R. Hartshorne, *Residues and Duality*, Lecture Notes in Math. **20**, Springer-Verlag 1966.

[H2] R. Hartshorne, "On the De Rham cohomology of algebraic varieties", *Publications Mathématiques IHES*, **45**(1975), 5–99.

[H3] R. Hartshorne, *"Algebraic Geometry"*, Springer Graduate Texts in Mathematics **52**, Springer Verlag 1977.

[H4] R. Hartshorne, "Cohomology with compact supports for coherent sheaves on an algebraic variety", *Math. Ann.*, **195**(1972), 199–207.

[HK] O. Hyodo and K. Kato, "Semi-stable reduction and crystalline cohomology with logarithmic poles", in *Périodes p-adiques*, Astérisque, **223**(1994).

[HL] M. Herrera and D. Lieberman, "Duality and the De Rham cohomology of infinitesimal neighborhoods", *Inv. Math.*, **13**(1971), 97–124.

[I1] L. Illusie, " Géométrie logarithmique", mimeographed notes of a course held at IHP, Paris, Spring 1997.

[I2] L. Illusie, "An overview of the work of K. Fujiwara, K. Kato and C. Nakayama on logarithmic étale cohomology", mimeographed notes of a course held at IHP, Paris, Spring 1997.

[Kas] M. Kashiwara, "Algebraic study of systems of partial differential equations", (master's thesis, Tokyo Univ. 1970), English translation in Mémoire **63** de la S.M.F., t. 123, fasc. 4.

[KK] K. Kato, "Logarithmic structures of Fontaine-Illusie", in *Algebraic Analysis, Geometry and Number Theory*, The Johns Hopkins Univ. Press (1989), 191–224.

[Ka1] N. Katz, "Nilpotent connections and the monodromy theorem. Applications of a result of Turrittin", *Publ. Math. IHES*, **39**(1970), 175–232.

[Ka2] N. Katz, "The regularity theorem in algebraic geometry", *Actes du Congrès Intern. Math.*, 1970, T.1, 437–443.

[Ka3] N. Katz, "A simple algorithm for cyclic vectors", *Amer. J. of Math.*, **109**(1987), 65–70.

[Ka4] N. Katz, *Exponential Sums and Differential Equations*, Annals of Math. Studies, **124**, Princeton University Press (1990).

[KaO] N. Katz, and T. Oda, "On the differentiation of De Rham cohomology classes with respect to parameters", *J. Math. Kyoto Univ.*, **8**,2(1968), 199–213.

[Ki1] R. Kiehl, "Theorem A und Theorem B in der nichtarchimedischen Funktionentheorie", *Inv. Math.*, **2**(1967), 256–273.

[Ki2] R. Kiehl, "Die De Rham Kohomologie algebraischer Mannigfaltigkeiten über einem bewerteten Körper", *Publ. Math. IHES*, **33**(1967), 5–20.

[Kl] S. Kleiman, "The transversality of a general translate", *Compos. Math.*, **28**,3(1974), 287–297.

[L] Y. Laurent, "Polygone de Newton et b-fonction pour les modules microdifférentiels", *Ann. Scient. Norm. Sup.* (4)**20**(1987), 391–441.

[LMe] Y. Laurent and Z. Mebkhout, "Pentes algébriques et pentes analytiques d'un $\mathcal{D}$-module", *Ann. Scient. Norm. Sup.* (4) **32**(1999), 39–69.

[Le] A. Levelt, "Jordan decomposition for a class of singular differential operators", *Ark. Math.*, 13–1(1975), 1–27.

[Li] H. Lindel, "On projective modules over polynomial rings over regular rings", in *Algebraic K-Theory, Proc. Oberwolfach 1980*, R.K. Dennis, Ed., Lecture Notes in Math. **966**, Springer-Verlag 1980.

[Lü1] W. Lütkebohmert, "Der Satz von Remmert-Stein in der nichtarchimedischen Funktionentheorie", *Math. Z.*, **139**(1974), 69–84.

[Lü2] W. Lütkebohmert, "Riemann's existence problem for a p-adic field", *Invent. Math.*, **111**(1993), 309–330.

[Mal] B. Malgrange, *Equations Différentielles à Coefficients Polynomiaux*, Progress in Math., **96**, Birkhäuser (1991).

[Man] Y. Manin, "Moduli fuchsiani", *Annali Scuola Normale Sup.*, Pisa III **19**(1965), 113–126.

[Me1] Z. Mebkhout, "Sur le théorème de semicontinuité des équations différentielles", *Astérisque*, **130**(1985), 365–417.

[Me2] Z. Mebkhout,"Le théorème de positivité de l'irrégularité pour les $\mathcal{D}_X$-modules", in *The Grothendieck Festschrift* Vol. III, 83–132, Progress in mathematics, Birkhäuser, 1990.

[Me3] Z. Mebkhout,"Le théorème de comparaison entre cohomologies de De Rham d'une variété algébrique complexe et le théorème d'existence de Riemann", *Publ. Math. IHES*, **69**(1989), 47–89.

[Me4] Z. Mebkhout, "Le polygone de Newton d'un $\mathcal{D}_X$-module", in *Conférence de la Rabida*, III, Progress in mathematics **134**, Birkhäuser, 1996., 237–358.

[Me5] Z. Mebkhout, *Le Formalisme des Six Opérations de Grothendieck pour les $\mathcal{D}_X$-modules Cohérents*, Travaux en Cours **35**, Hermann, 1989.

[N] C. Nakayama, "Logarithmic étale cohomology", *Math. Ann.*, **308**(1997), 365–404.

[Og] A. Ogus, "F-crystals, Griffiths transversality, and the Hodge decomposition", *Astérisque*, **221**(1994), 1–183.

[P] H. Poincaré, "Sur les integrales irrégulières des équations linéaires", *Acta Math.*, **8**(1886), 295–344.

[R] P. Robba, "Lemme de Hensel pour des opérateurs différentiels", *Ens. Math.*, **26**, fasc. 3–4(1980), 279–311.

[S] C. Sabbah, "Equations différentielles à points singuliers irréguliers et phénomène de Stokes en dimension 2", Preprint Ecole Polytechnique, April 1997.

[Sa] M. Saito, "Induced D-modules and differential complexes", *Bull. Soc. Math. France*, **117**(1989), 361–387.

[Se] J.P. Serre, "Prolongement de faisceaux analytiques cohérents", *Ann. Inst. Fourier Grenoble*, **16**,1(1966), 363–374.

[SGA 1] A. Grothendieck, *Revêtements Étales et Groupe Fondamental*, Lecture Notes in Math. **224**, Springer-Verlag 1971.

[SGA 4] M. Artin, A. Grothendieck and J.-L. Verdier, *Théorie des Topos et Cohomologie Étale des Schémas. Tome 3*, Lecture Notes in Math. **305**, Springer-Verlag 1973.

[SGA 7,II] P. Deligne and N. Katz, *Groupes de Monodromie en Géométrie Algébrique*, Lecture Notes in Math. **340**, Springer-Verlag 1973.

[T] T. Tsuji, "Saturated morphisms of logarithmic schemes", preprint, June 30, 1997 .

[T.S.] N. Bourbaki, *Théories Spectrales*, Hermann, Paris, 1967.

[vR] A.C.M. van Rooij, *Non-Archimedean Functional Analysis*, Monographs and Textbooks in Pure and Applied Mathematics, **51**, Marcel Dekker, 1978.

Index

A
adapted (basis of derivations), 11
adapted (basis of formal derivations), 12
adapted (étale coordinates), 3
adapted (formal étale coordinates), 12
almost every (curve on a variety), 83
Artin set, 109

B
base change theorem, 141

C
characteristic abelian sheaf (of a log scheme), 42
characteristic monoid (of a log scheme), 42
closed immersion (of models), 3
coherent (module with connection), 110
complex of Dworks algebraic dual theory, 149
connection, 15
coordinatized (elementary fibration), 106
coordinatized (formal model), 3
coordinatized (K-model), 3
coordinatized (tubular neighborhood), 6
cyclic vector, 111
cyclic with respect to f (module with connection), 110

D
De Rham cohomology with compact supports (algebraic notion), 157
De Rham complex with logarithmic singularities, 13
differential module (algebraic notion), 15
differential module (formal notion), 15
direct image with compact supports (algebraic notion), 151
divisorial (valuation), 2
divisorially valued (function field), 2

E
elementary fibration, 105
exponents (Fuchs) (of a differential module, general case), 97
exponents (Fuchs) (of a differential operator), 32
exponents (of a regular differential module; formal one variable notion), 33
exponents (of a regular differential module; global notion), 35
exponents (Turrittin) (of a differential operator), 97
exponents along a divisor, 34
exponents at v (several variables), 34

F
finiteness theorem, 141
formal functions (field of), 2
formal model (smooth, affine), 3
free (module with connection), 110
fuchsian, 29
fuchsian part (of a differential module), 97
function field, 2

G
Gauss map, 80
Grothendieck's linearization of differential operators, 154

H
height (of a tower of elementary fibrations), 107
Hodge-De Rham spectral sequence, 116
horizontal (map), 16

I
indicial polynomial, 32
integrable (connection), 15
integral curves, 8
irregularity (of a differential operator in one variable), 73

J
jets (on a relative log scheme), 46

K
Katz' lemma of the cyclic vector, 98
K-model (smooth, affine), 2
Kummer étale (morphism of models), 4

L
left perfect, 148
local for the étale topology (property), 117
locally closed immersion (of models), 3
log dominant (morphism of models), 4
lower semicontinuous, 75

M
model, 2
monodromy theorem, 141
morphism (of log schemes), 42
morphism (of models), 3

N

Newton polygon (of a differential module), 78
Newton polygon (of a differential module) at v, 79
Newton polygon (of a differential operator in one variable), 73
Newton polygon (principal), 79
Newton polygons (secondary), 84
non-Liouville (number), 175

P

p-adic Turrittin theorem, 204
Poincaré-Katz rank (at a divisor), 67
Poincaré-Katz rank (at a valuation), 58
pro-object (Artin-Rees), 151
purely of Poincaré-Katz rank σ, 72

Q

quasi-coherent (module with connection), 110

R

ramification index (of a morphism of models), 4
rational (elementary fibration), 106
reflexive, 23
regular (connection) at (or along) Z, 19
regular (differential module; formal one variable notion), 16
regular (differential module; formal several variables notion), 19
regular (differential module; global notion), 29
regular at v (differential module; one variable notion), 17
regular at v (differential module; several variables notion), 19
regularity theorem, 141
relative derivations (for a relative log scheme), 47
relative differential operators (for a relative log scheme), 47
relative ramification index, 51

S

semicontinuity theorem, 84
simple (module with connection), 110
simple zero (of a derivation) at a valuation, 11
simple zero (of a vector field) along a divisor, 8
slope filtration, 70
spectral norm, 53
spectral valuation, 55
stable under finite direct image (property), 117
strongly equivalent (norms), 51

T

transposition isomorphism, 142
transversal (derivation) to a valuation, 11
transversal (vector field) to a divisor, 8
tubular neighborhood, 6
Turrittin index, 85
Turrittin-Levelt-Hukuhara decomposition, 70

U

unramified (morphism of models), 4

V

v-adic generic radius of convergence, 63
valuative (Banach) norm, 55
vertical (morphism of log schemes), 4

Y

Young's theorem (equicharacteristic analogue), 64

Z

Zariski spectral sequence, 114
τ-extension, 23